S0-CAH-205

Vibronic Processes in Inorganic Chemistry

NATO ASI Series

Advanced Science Institutes Series

A Series presenting the results of activities sponsored by the NATO Science Committee, which aims at the dissemination of advanced scientific and technological knowledge, with a view to strengthening links between scientific communities.

The Series is published by an international board of publishers in conjunction with the NATO Scientific Affairs Division

A Life Sciences **B Physics**	Plenum Publishing Corporation London and New York
C Mathematical and Physical Sciences **D Behavioural and Social Sciences** **E Applied Sciences**	Kluwer Academic Publishers Dordrecht, Boston and London
F Computer and Systems Sciences **G Ecological Sciences** **H Cell Biology**	Springer-Verlag Berlin, Heidelberg, New York, London, Paris and Tokyo

Series C: Mathematical and Physical Sciences - Vol. 288

Vibronic Processes in Inorganic Chemistry

edited by

Colin D. Flint
Birkbeck College, University of London,
London, U.K.

Kluwer Academic Publishers

Dordrecht / Boston / London

Published in cooperation with NATO Scientific Affairs Division

Proceedings of the NATO Advanced Study Institute on
Vibronic Processes in Inorganic Chemistry
Riva del Sole, Tuscany, Italy
September 7–18, 1988

03790617
CHEMISTRY

Library of Congress Cataloging in Publication Data

NATO Advanced Study Institute on Vibronic Processes in Inorganic
Chemistry (1988 : Riva del Sole, Italy)
Vibronic processes in inorganic chemistry : proceedings of the
NATO Advanced Study Institute, September 7-18, 1988 / edited by
Colin D. Flint.
p. cm. -- (NATO ASI series. Series C, Mathematical and
physical sciences ; no. 288)
ISBN 0-7923-0436-5
1. Chemistry, Inorganic--Congresses. 2. Jahn-Teller effect-
-Congresses. I. Flint, Colin D., 1943- . II. Title.
III. Series.
QD461.N36 1988
541--dc20 89-36644

ISBN 0-7923-0436-5

Published by Kluwer Academic Publishers,
P.O. Box 17, 3300 AA Dordrecht, The Netherlands.

Kluwer Academic Publishers incorporates the publishing programmes of
D. Reidel, Martinus Nijhoff, Dr W. Junk and MTP Press.

Sold and distributed in the U.S.A. and Canada
by Kluwer Academic Publishers,
101 Philip Drive, Norwell, MA 02061, U.S.A.

In all other countries, sold and distributed
by Kluwer Academic Publishers Group,
P.O. Box 322, 3300 AH Dordrecht, The Netherlands.

Printed on acid free paper

All Rights Reserved
© 1989 by Kluwer Academic Publishers
No part of the material protected by this copyright notice may be reproduced or utilized in any form or by any means, electronic or mechanical, including photocopying, recording or by any information storage and retrieval system, without written permission from the copyright owner.

Printed in The Netherlands

QD461
N36
1988
CHEM

TABLE OF CONTENTS

PREFACE

This volume reports the main lectures and seminars given at the NATO Advanced Study Institute on Vibronic Processes in Inorganic Chemistry held at Riva del Sole, Tuscany, Italy between 7th and 18th September 1988. In addition to the about 40 hours of lectures represented by this volume, a further fifteen lectures on current research topics were given by the other participants.

Many factors contributed to the decision to hold this ASI but the final trigger was given at a meeeting in Padova when Marco Bettinelli, Lorenzo Disipio and Gianluigi Ingletto asked me to recommend a text where the diverse conceptual, spectroscopic and structural consequences of the impossibility of treating the motions of the electrons and nuclei independantly in inorganic compounds were presented. There seemed to be no suitable comprehensive text where the relationship between the relatively simple theoretical ideas and the huge range of their application in inorganic chemistry and physics was developed. The Institute and this text are a contribution to filling this gap.

Seventy-nine participants from fifteen countries attended the Institute. Topics raised in the lectures and from the participants own research frequently led to discussions which went on long into the night. I have never been to meeting where I lost so many pens and pencils in such a diversity of places! The essence of the NATO ASI Programme is the opportunity for less experienced scientists from many countries to spend days with those responsible for the current status of the theory and its experimental manifestations. This opportunity was thoroughly utilized. Hotel Riva del Sole proved to be an excellent venue and this added significantly to the success of the meeting.

I should like to thank the members of the organising Committee for all their hard work; in particular Carl Ballhausen for keeping us all on our academic toes at every scientific session and Marco Bettinelli for the local organisation and transportation. I also wish to thank all of the speakers for the care they took in presentation of the material and the preparation of the text of this volume. Above all I thank every participant for contributing to such a stimulating and enjoyable Institute.

The financial support and the encouragement of the NATO Advanced Study Institute Programme is greatfully acknowledged. Financial support from the University of Padova and the CNR, Rome contributed greatly to the success of the Institute.

Colin D. Flint
London, 1989.

OVERTURE
INTRODUCING VIBRONIC COUPLINGS

C.J. BALLHAUSEN
University of Copenhagen
Institute for Physical Chemistry
Universitetsparken 5
DK–2100 Copenhagen Ø
Denmark

The Perturbation Hamiltonian.

In a polyatomic molecule the wavefunctions are in the Born–Oppenheimer approximation described as a product of rotational, vibrational and electronic wavefunctions. These three types of motions are indeed separable to zeroth order because $E_{rot} < E_{vib} < E_{elect}$. The theme of this school is now to elucidate the higher order couplings which occur between the vibrational and electronic motions; for short vibronic interactions, a term coined by Mulliken[1)] in 1937.

Two types of phenomena are intimately related to the vibronic couplings, namely the intensities of "forbidden" electronic transitions and molecular stability. Traditionally both happenings are handled by means of perturbation theory. The change in the electronic energy with the movements of the nuclei are nowadays often handled by direct computation of the electronic energy as a function of the structual parameters of the molecule. However, perturbation theory is still unsurpassed when it comes to bring out the general features of the vibronic couplings. This point is also clearly brought home when we consider the role symmetry considerations play in the theory.

To zeroth order the vibronic eigenfunctions are therefore in the socalled crude Born–Oppenheimer approximation written as a product

$$\Psi_{iv} = \psi_i(q,o)\chi_v(Q) \tag{1}$$

where ψ_i is the electronic wavefunction in the electronic coordinates q at the nuclear equilibrium configuration point $Q = o$ and $\chi_v(Q)$ is the nuclear vibrational wavefunction. The electronic function ψ_i may be classified according to the point group of the molecule pertinent to $Q = o$, as may the 3N–6 vibrational symmetry coordinates Q.

Vibronic mixings are determined by the mixings of $\Psi_{iv''}$ and $\Psi_{jv'}$ under a perturbation operator, $\mathcal{H}^{(1)}$. Consider an electron in an orbital centered on M,

C. D. Flint (ed.), Vibronic Processes in Inorganic Chemistry, 1–5.
© 1989 by Kluwer Academic Publishers.

(Figure 1) with an other center L_1 at a distance $\mathbf{a}_1$ from M. We now displace M by $\mathbf{s}_m$ and L_1 by $\mathbf{s}_1$. Then from Figure 1

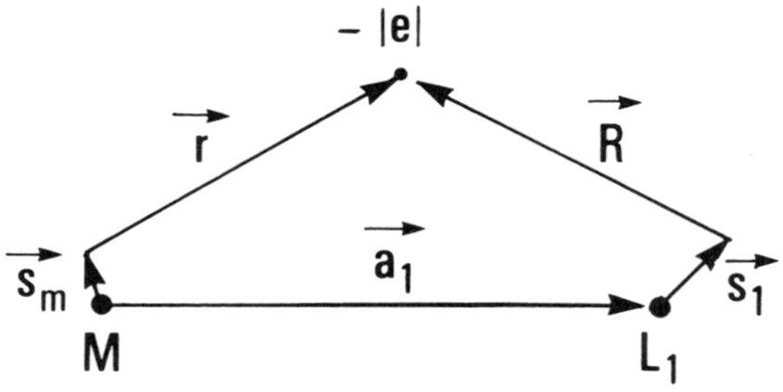

Figure 1. Displacement coordinates $\mathbf{s}_m$ and $\mathbf{s}_1$.

$$\mathbf{a}_1 + \mathbf{s}_1 + \mathbf{R} - \mathbf{r} - \mathbf{s}_m = 0 \tag{2}$$

or

$$\frac{1}{|\mathbf{R}|} = \frac{1}{|(\mathbf{r}-\mathbf{a}_1) + (\mathbf{s}_m-\mathbf{s}_1)|} \tag{3}$$

$$= [(\mathbf{r}-\mathbf{a}_1)^2 + 2(\mathbf{r}-\mathbf{a}_1)\cdot(\mathbf{s}_m-\mathbf{s}_1) + (\mathbf{s}_m-\mathbf{s}_1)^2]^{-\frac{1}{2}}$$

By expansion

$$\frac{1}{|\mathbf{R}|} = \frac{1}{|\mathbf{r}-\mathbf{a}_1|}\left[1-\frac{1}{2}\,\frac{2(\mathbf{r}-\mathbf{a}_1)\cdot(\mathbf{s}_m-\mathbf{s}_1) + (\mathbf{s}_m-\mathbf{s}_1)^2}{|\mathbf{r}-\mathbf{a}_1|^2} + \cdots\right]$$

Summing over $L_1 \cdots L_N$ we find to first order in $(\mathbf{s}_m-\mathbf{s}_n)$, $n = 1 \cdots N$

$$\mathcal{H}^{(1)} = \sum_{n=1}^{N} \nabla_{\mathbf{a}_n} \frac{(\mathbf{s}_n-\mathbf{s}_m)}{|\mathbf{r}-\mathbf{a}_n|} \tag{4}$$

or to the t'th order in $(\mathbf{s}_n-\mathbf{s}_m)$

$$\mathcal{H}^{(1)} = \sum_{t=1}^{\infty} \sum_{n=1}^{N} \frac{(\mathbf{s}_n-\mathbf{s}_m)^t}{t!} \nabla^t_{\mathbf{a}_n} \frac{1}{|\mathbf{r}-\mathbf{a}_n|} \tag{5}$$

The matrix elements of $\mathcal{H}^{(1)}$ inside an electronic manifold are therefore to first order in $(\mathbf{s}_n-\mathbf{s}_m)$ given by

$$\langle \psi_i(q,o) | \mathcal{H}^{(1)} | \psi_j(q,o) \rangle = c_{ij} \sum_{k=1}^{3N} Q_k \tag{6}$$

where the Q_k's are the symmetry determined displacement coordinates of the N nuclei including the three rotations and the three translations.

The Vibronic Intensities.

Consider a "forbidden" electronic dipole transiton $\psi_o \rightarrow \psi_i$. Using perturbation theory we have with $W_o \ll W_i < W_j$

$$\langle \psi_i \chi_{v''} | er | \psi_o \chi_{v'} \rangle = \sum_j \frac{\langle \psi_i \chi_{v''} | \mathcal{H}^{(1)} | \psi_j \chi_{v'''} \rangle}{W_j - W_i} \langle \psi_j \chi_{v'''} | er | \psi_o \chi_{v'} \rangle$$

$$= \sum_j \frac{c_{ij}}{W_j - W_i} \langle \psi_j | er | \psi_o \rangle \langle \chi_{v'''}(Q) | \sum_k Q_k | \chi_{v'}(Q) \rangle \tag{7}$$

Provided $\psi_o \rightarrow \psi_j$ is allowed as an electronic dipole transition, the $\psi_o \rightarrow \psi_i$ transition can therefore "steal" intensity herefrom, via a "vibronic" mechanism, by exciting a proper vibration in the molecule.

The Jahn–Teller Couplings.

At the equilibrium point $Q = o$, all linear terms in Q_k in the energy expressions has by definition been removed. However, we observe from eq (6) that provided $\psi_i(q,o)$ and $\psi_j(q,o)$ are members of a degenerate set of electronic wavefunctions linear terms in Q_k are introduced in the energy expression. In this situation Jahn and Teller[19] showed in 1937 that one cannot simultaneously eliminate the linear terms in the vibrational coordinates for all of the electronic components. A Jahn–Teller coupling occurs between the electronic and nuclear motions centered around $Q = o$ which confirmation gave rize to the electronic degeneracy.

In the calculations of the Jahn–Teller couplings a distincion is made between the static– and the dynamic Jahn–Teller couplings. The static case elucidates the potential surfaces, looking apart from the nuclear kinetic term in the Hamiltonian. In the dynamic coupling calculations this term is included, thereby giving us the quantized energy levels of the coupled system.

Chemists and Physicists often talk about the "Jahn–Teller effect". Such a usage is to be avoided since there is no "effect". We cannot observe the molecule or ion without the system having coupled its electronic and vibrational motions. All we can observe is how the coupling is reflected, for instance in the spectroscopic behaviour of the system. In situations were $W_i^o \neq W_j^o$ linear coupling terms also occur. Usually this coupling is quenched by the state energy difference. If, however, of importance such cases are labelled "Pseudo" Jahn–Teller couplings.

Linear molecules with two–fold degenerate electronic states, as e.g. Π states, do not experience first order Jahn–Teller couplings. However, as shown by Renner[20] in 1934, three years before Jahn and Teller's paper, second order terms in the vibrational "bending modes" couples the two components of the Π state together. Such a coupling may – or may not – preserve the linear shape of the molecule. These couplings are usually called Renner couplings. Second and higher order coupling terms may of cource also occur in non–linear Jahn–Teller coupling cases. The only electronic degeneracies which are completely immune towards vibronic couplings are the so–called Kramers doublets.

Provided additional off–diagonal perturbations are active in a Jahn–Teller coupled system, as for instance a spin–orbit coupling, the vibrational overlap factor will quench the manifestations of the Jahn–Teller coupling. The generality of this phenomenum was first recognized in 1965 by Ham[17], and it is named after him. The first "mathematical" proof of the Jahn–Teller theorem was given by Ruch and Schönhofer[21] in 1965.

The vibronic levels in a Jahn–Teller coupled system are traditionally calculated numerically on big computers as functions of the coupling strength. It was Brian Judd[18] who in 1979 first obtained analytical solutions for a certain number of parameter values, and great efforts have since gone into obtaining analytical solutions for some simple model systems.

Some Historical important Calculations.

The order of magnitude to be expected for vibronic intensity in absorption bands of complexes of the rare earths was first estimated by Van Vleck[2] in 1937. However, the experimental proof of such an intensity–giving mechanism was first demonstrated by Sponer, Nordheim, Sklar and Teller[3] in 1939 by explaining the detailed absorption band of the "forbidden" ${}^1A_{1g} \rightarrow {}^1B_{2u}$ transition in benzene. A semi–empirical calculation for the intensity of this transition was done by Craig[4] in 1950, the first non–empirical by Liehr[5] in 1955.

The first non–empirical calculation of the intensities of the parity forbidden "ligand field bands" of Ti^{+3} and Cu^{+2} complexes was performed by Liehr and Ballhausen[6] in 1956. This paper also contained the coth formula for the variation of the vibronic intensity with temperature. In 1958 Koide and Pryce[7] calculated the oscillator strenghts of ${}^2A_{1g} \rightarrow {}^4A_{1g}$ and ${}^4E_g$ in manganous salts, thereby including the effects of spin–orbit coupling. In 1976 Mason and Seal[8] considered the contributions to the intensities using a ligand field polarization model.

The static Jahn–Teller couplings in O_h symmetry were calculated by Van Vleck in 1939[9] for E, T_1 and T_2 states. The results were used[10] to estimate the magnetic susceptibilities of Vanadium–, Titanium– and Chrome Alum. It is for the latter system Van Vleck remarks "It is a great merit of the Jahn–Teller effect that it disappears when not needed".

The first order static Jahn–Teller couplings in octahedral complexes were also looked into by Öpik and Pryce[11] in 1957, and Liehr and Ballhausen[12] in 1958 for the first time included the second order static couplings in $E \otimes \epsilon, a_1$. The dynamic $E \otimes \epsilon$ coupling was investigated by Moffitt and Liehr[13] in 1957 and by Moffitt and Thorson[14] and Longuet–Higgins, Öpik, Pryce and Sack[15] in 1958. Also in 1958 Pople and Longuet–Higgins[16] calculated the Renner couplings in the II state of the NH_2 radical.

The themes which I have touched upon will form the leading motives throughout the coming days. They occupy a central place in molecular spectroscopy. Furthermore, collective static Jahn–Teller couplings play an important role when it comes to the understanding of the crystal structures of transitionmetal complexes. We have a big programme in front of us!

References.

1) R.S. Mulliken: J. Phys. Chem. 41 (1937) 159.
2) J.H. Van Vleck: J. Phys. Chem. 41 (1937) 67.
3) H. Sponer, G. Nordheim, A.L. Sklar and E. Teller: J. Chem. Phys. 7 (1939) 207.
4) D.P. Craig: J. Chem. Soc. 1950, 59.
5) A.D. Liehr: Thesis 1955 and Zeit. f. Naturforsch. 13a (1958) 311.
6) A.D. Liehr and C.J. Ballhausen: Phys. Rev. 106 (1957) 1161.
7) S. Soide and M.H.L. Pryce: Phil Mag. 3 (1958) 607.
8) S.F. Mason and R.H. Seal: Mol. Phys. 31 (1976) 758.
9) J.H. Van Vleck: J. Chem. Phys. 7 (1939) 72.
10) J.H. Van Vleck: J. Chem. Phys. 7 (1939) 61.
11) U. Öpik and M.H.L. Pryce: Proc. Roy. Soc. A238 (1957) 425.
12) A.D. Liehr and C.J. Ballhausen: Ann. Physics [N.Y.] 3 (1938) 304.
13) W. Moffitt and A.D. Liehr: Phys. Rev. 106 (1957) 1195.
14) W. Moffitt and W. Thorson: Calcul des Fonctions D'Onde Moleculaire, Edition du Centre National de la Recherche Scientifique 82 (1958) 141.
15) H.C. Longuet–Higgins, U. Öpik, M.H.L. Pryce and R.A. Sack: Proc. Roy. Soc. A244 (1958) 1.
16) J.A. Pople and H.C. Longuet–Higgins: Mol. Phys. 1 (1958) 372.
17) F.S. Ham: Phys. Rev. 138 (1965) A1727.
18) B. Judd: J. Phys. Chem. Solid. 12 (1979) 1685.
19) H.A. Jahn and E. Teller: Proc. Roy. Soc. A161 (1937) 220.
20) E. Renner: Zeit. f. Physik 92 (1934) 172.
21) E. Ruch and A. Schönhofer: Theoret. Chim. Acta. 3 (1965) 291.

VIBRONIC COUPLING, BASES

GAD FISCHER
Department of Chemistry
The Australian National University
GPO Box 4, Canberra, ACT 2601
Australia

ABSTRACT. A survey of approaches to the determination of molecular vibronic energies and wavefunctions is presented. The various adiabatic approximations as well as the diabatic and the Generator Coordinate Method are considered. Herzberg-Teller coupling is dealt with in the context of the crude adiabatic approximation.

1. Introduction

Frequently our description of natural phenomena is hindered by our inability to effect the necessary separation of variables. The subject of vibronic coupling is a very well-known example of this and concerns the interaction between the electronic and nuclear motions. Vibronic coupling is introduced as a correction to the assumption that the vibrational motion takes place on the one electronic potential energy surface.

We are all familiar with the determination of the energy levels of the hydrogen atom. The Schrödinger equation for the H atom, a two-particle problem, can be solved exactly and analytical solutions can be derived. It is customary to divide the original two-particle problem into two single-particle problems. This is achieved by a coordinate transformation which allows separation of the Hamiltonian into two parts, one relating to the motion of the centre-of-mass of the system, and the other to the energy of the electron in the field of the proton. The solution of the latter problem is generally carried out by transforming to polar coordinates, the coordinates most suitable for a problem displaying spherical symmetry. Three separate equations in the three polar coordinates r, θ, and $\emptyset$ are obtained. Strictly, not even for the simplest molecule, the hydrogen molecule ion H_2^+, are analytical solutions possible and approximate methods must be employed.

From early spectroscopic studies of molecules, it emerged that the molecular energy levels could be very successfully described in terms of a specific hierarchy. Rotational levels belong to particular vibrational levels which in turn are associated with specific electronic states. Thus, already the early experimental observations suggested that the electronic and nuclear coordinates could, in fact, be treated separately. In their celebrated paper, Born and Oppenheimer (BO) attempted to provide theoretical justification for these observations. They based their argument on the smallness of the ratio of the electronic to nuclear masses and showed that the term values could be split into parts of different orders of magnitude corresponding to electronic, vibrational and rotational energies. However, their approach did not really explain why electronic and

C. D. Flint (ed.), Vibronic Processes in Inorganic Chemistry, 7–19.
© 1989 by Kluwer Academic Publishers.

nuclear motions could be separated, and over the intervening years, some unease has been expressed about the physical interpretation of their treatment.

Nevertheless, in common with Born and Oppenheimer, the approach most frequently discussed in textbooks has it that because of their extreme lightness compared to the nuclei, the electrons move so much more rapidly that so far as the nuclei are concerned, the system of electrons behaves like a gas. It is the energy of this electron gas which then provides the potential energy for the nuclear motion. Born and Huang (BH) have given a mathematical formulation for this approach and have shown under what conditions it is valid. The BH approach will be discussed in more detail in section 3.1.

For molecules, in parallel with the treatment of the energy levels of the H atom, the centre-of-mass (C of M), motion, (translational motion), may first be removed. A consequence of the constraint imposed on the internal system by separation of the C of M motion is the appearance of mass polarization effects. Depending on the choice of coordinates, the mass polarization terms may derive from cross-terms in the electronic coordinates $\nabla_{q\alpha} \cdot \nabla_{q\beta}$ and cross-terms in electronic and nuclear coordinates $\nabla_q \cdot \nabla_Q$. However, for most atoms and molecules, the contributions arising from the removal of the translational motion are small and less than the inaccuracies introduced by the use of approximate wavefunctions. The C of M is essentially determined by the nuclei and they do not move to any large extent. For the H_2 molecule (similarly H_2^+ a different situation exists. The calculations are so accurate, at least as accurate as the experimentally determined energy levels, that not only must allowance be made for the translational motion but also relativistic and radiative (Lamb shift) corrections must be taken into account.

In this discussion our interest is in vibronic coupling in representative molecules for which any corrections due to the inclusion of the translational motion are considerably smaller than the accuracy of the calculations. So we can satisfactorily ignore the translational motion. We are somewhat less justified in removing the rotational motion.

For molecules in states with electronic angular momentum, the coupling of the rotational and electronic angular momenta may need to be considered. This is particularly pertinent to the linear molecules. However, only for a small number of non-linear polyatomic molecules is there sufficient symmetry to permit the existence of electronic angular momentum. Furthermore, the off-diagonal coupling matrix elements depend inversely on the reduced masses. Similarly, we ignore the interaction between the rotational and vibrational motions, the Coriolis coupling. It should be noted that the Coriolis energy is small compared to the vibrational energy, and in general, small compared to corrections resulting from vibronic coupling.

2. The Schrödinger Equation

The molecular stationary states are found from the time-independent Schrödinger equation

$$H(q,Q)\Psi(q,Q) = \varepsilon\Psi(q,Q), \tag{1}$$

where the vibronic wavefunctions are represented by $\Psi(q,Q)$, the eigenergies by ε, and H is the molecular Hamiltonian given by

$$H(q,Q) = T(q) + T(Q) + U(q,Q) \tag{2}$$

The electronic and nuclear kinetic energies, T(q) and T(Q) respectively, are given by expressions of the type

$$T(q) = -\frac{\hbar^2}{2m}\sum_{\alpha} \nabla^2_{q_\alpha} \tag{3}$$

where the sets $\{q_\alpha\}$ and $\{Q_k\}$ are electronic and nuclear internal coordinates respectively, and generally mass-weighted nuclear coordinates are used. In Eqn (2), U(q,Q) includes the potential energy of interaction between the electrons, between the electrons and the nuclei, and between the nuclei. Any other terms in the Hamiltonian which are functions of the electronic coordinates, such as spin-orbit coupling, can be included in U(q,Q) if desired.

Eqn (1) for the exact eigenstates cannot be solved analytically because as discussed earlier, the Hamiltonian, Eqn (2) does not allow a separation of the variables q and Q. The general approach is to expand the full molecular wavefunction in some basis set of *electronic* functions $\psi_n(q,Q)$,

$$\Psi_i(q,Q) = \sum_n \psi_n(q,Q)\chi_{ni}(Q) \tag{4}$$

where the coefficients $\chi_{ni}(Q)$ are functions of the nuclear coordinates only. Substituting the expansion, Eqn (4) for the molecular wavefunction in the Schrödinger equation (1) we obtain

$$[H_e(q,Q) + T(Q)]\sum_n \psi_n(q,Q)\chi_{ni}(Q) = \varepsilon_i \sum_n \psi_n(q,Q)\chi_{ni}(Q) \tag{5}$$

where H_e is the electronic Hamiltonian and comprises the terms in the Hamiltonian dependent on the electronic coordinates, namely

$$H_e(q,Q) = T(q) + U(q,Q) \tag{6}$$

Now recalling that

$$T(Q)\psi_n\chi_{ni} = (T(Q)\psi_n)\chi_{ni} + \psi_n T(Q)\chi_{ni} - \hbar^2\sum_k(\partial\psi_n/\partial Q_k)(\partial\chi_{ni}/\partial Q_k) \tag{7}$$

Eqn (5) may be rewritten

$$\sum_n[\psi_n(q,Q)T(Q)+(H_e+T(Q))\psi_n(q,Q)-\hbar^2\sum_k(\partial\psi_n(q,Q)/\partial Q_k)(\partial/\partial Q_k)]\chi_{ni}$$
$$= \varepsilon_i \sum_n \psi_n(q,Q)\chi_{ni}(Q) \tag{8}$$

Multiplying on the left by $\psi_n^*(q,Q)$ and integrating over the electronic coordinates leads to the following coupled differential equations for the expansion coefficients (the nuclear wavefunctions),

$$[T(Q)+\langle n(q,Q)|T(Q)+H_e(q,Q)|n(q,Q)\rangle - \varepsilon_i]\chi_{ni}Q$$
$$+ \sum_{m\neq n} [\langle n(q,Q)|T(Q)|m(q,Q)\rangle -- \sum_k \hbar^2\langle n(q,Q)|\partial/\partial Q_k|m(q,Q)\rangle\partial/\partial Q_k$$
$$+ \langle n(q,Q)|H_e|m(q,Q)\rangle]\chi_{mi}(Q) = 0 \tag{9}$$

It should be noted that in writing Eqn (9), the diagonal element $<n|\partial/\partial Q|n>$ has been omitted since for any bound state the expectation value of linear momentum vanishes. Furthermore, it must be emphasized that no approximations have been made in deriving Eqn (9) for the coefficients χ_{ni}. It is essentially this feature which makes the practical application of these equations useless. To overcome such a serious shortcoming, a number of approximate approaches have been employed. They are considered in the following.

3. Bases and Approximations

We begin with the Born and Huang approach.

3.1. ADIABATIC, BORN-HUANG (ABH)

It is the procedure that mathematically is the most lucid and in the past decades has been adopted in most discussions and treatments of vibronic coupling.

The eigenfunctions of the molecular Hamiltonian, Eqn (2), are expanded in terms of the complete set of electronic wavefunctions $\psi_n(q,Q)$, but, where now the $\psi_n(q,Q)$ are eigenfunctions of the electronic Hamiltonian and satisfy the Schrödinger equation

$$H_e(q,Q)\psi_n(q,Q) = E_n(Q)\psi_n(q,Q) \tag{10}$$

$E_n(Q)$ is the n-th eigenvalue and depends on Q as a parameter. That is, for every nuclear configuration specified by Q, an eigenvalue equation analogous to Eqn (10) can be written down. Upon substitution for the molecular wavefunctions in Eqn (1), multiplication on the left by $\psi_n^*(q,Q)$ and integration over the electronic coordinates the following coupled differential equations are obtained for the expansion coefficients:

$$(T(Q) + E_n(Q) + T_{nn} - \varepsilon_i)\chi_{mi}(Q) + \sum_{m\neq n} (T_{nm} - \sum_k \hbar^2 < n(q,Q)|\partial/\partial Q_k|m(q,Q)>\partial/\partial Q_k)\chi_{mi}(Q) = 0 \tag{11}$$

where $T_{nm} = <n(q,Q)|T(Q)m(q,Q)>$.

When the coupling terms off-diagonal in the electronic state indices in Eqn (11) are neglected, that is when

$$<n(q,Q)|T(Q)m(q,Q)> = 0, \qquad n\neq m \tag{12}$$

and $$<n(q,Q)|\partial/\partial Q_k|m(q,Q) = 0, \tag{13}$$

the expansion coefficients vanish except for m = n. Under these assumptions, the molecular wavefunction may be written as a simple poduct,

$$\Psi^A_{ni}(q,Q) = \psi_n(q,Q)\chi^A_{ni}(Q) \tag{14}$$

and the coefficients $\chi^A_{ni}(Q)$ are expressed as the eigenfunctions of the following Schrödinger equation

$$(T(Q) + E_n(Q) + T_{nn} - \varepsilon^A_{ni})\chi^A_{ni}(Q) = 0 \qquad (15)$$

In Eqn (15) the subindex n indicates that the molecular wavefunction is associated with electronic state n and the superscript A refers to the approximations made, ABH, Eqns (12) and (13), in arriving at Eqn (14). By invoking the assumptions, Eqns (12) and (13) a separation of the electronic and nuclear coordinates has been effected. When the electronic wavefunctions $\psi_m(q,Q)$ and $\psi_n(q,Q)$ vary slowly with nuclear displacements the assumptions are satisfactory because then the integrals, Eqns (12) and (13), are small compared with the separations of electronic states.

We see that in the ABH approach to the nuclear motion in polyatomic molecules a series of coupled equations, (11), are obtained. By introducing the approximations, Eqns (12) and (13), the differential equation (11) can be uncoupled and the molecular wavefunction can be written as a simple product, Eqn (14). This mathematical formulation is in accord with the physical description of nuclear motion on one potential energy surface.The specific assumptions that need to be made in order to achieve nuclear motion separability are linked to the choice of basis states, and the most appropriate choice of basis states is in turn governed by the characteristics of the problem. Thus, in the approach discussed above (ABH), a basis of electronic states that are eigenstates of the electronic Hamiltonian is used. This basis is convenient provided the weak coupling limit applies, that is provided the electronic states are well separated relative to a typical vibration frequency.

3.2. ADIABATIC, BORN-OPPENHEIMER (ABO)

The same basis of electronic states is used as in the BH approach but an additional approximation is made. It is assumed that further to the assumptions concerning the integrals Eqns (12) and (13) of the BH adiabatic approximation the diagonal elements of the nuclear kinetic energy are neglected, that is

$$T_{nn} = 0 \qquad (16)$$

Note, that it is these elements that are responsible for the differences between isotopes.

A basis that is, and has been widely used and also belongs to the weak coupling limit is the crude adiabatic (CA) or clamped nuclei basis. The CA wavefunctions are approximate adiabatic (ABH or ABO) wavefunctions. The CA basis is the set of electronic wavefunctions determined for some chosen nuclear configuration, generally the ground state equilibrium structure. Herein lies the attraction of this basis since molecular symmetry can be employed to determine those matrix elements that are vanishing.

3.3. CRUDE ADIABATIC

The molecular wavefunction is expanded in terms of the set $\{\psi_n(q,Q_0)\}$

$$\Psi_i(q,Q) = \Sigma\, \psi_n(q,Q_0)\chi_{ni}(Q) \qquad (17)$$

where the electronic wavefunctions satisfy a 'clamped nuclei' Schrödinger equation analogous to Eqn (10) namely,

$$H_e(q,Q_o)\psi_n(q,Q_o) = E_n(Q_o)\psi_n(q,Q_o) \tag{18}$$

Eqn (18) is sometimes called the static equation and the adiabatic version, Eqn (10), the dynamic equation.
The two Hamiltonians are related as follows,

$$\begin{aligned} H_e(q,Q) &= T(q) + U(q,Q) \\ &= T(q) + U(q,Q_o) + \Delta U(q,Q) \\ &= H_e(q,Q_o) + \Delta U(q,Q) \end{aligned} \tag{19}$$

It is important to recognize that either type of wavefunction forms a complete set and therefore either set is equally valid as an expansion basis.

In a manner paralleling the derivation of the coupled equations (11) but instead employing the CA basis $\{\psi(q,Q_o)\}$ the following coupled differential equations are obtained for the expansion coefficients

$$(T(Q) + E_n(Q_o) + \Delta U_{nn} - \varepsilon_i)\chi_{ni}(Q) + \sum_{m \neq n} \Delta U_{nm}\chi_{mi}(Q) = 0 \tag{20}$$

Also here, neglect of the off-diagonal coupling, that is setting

$$\Delta U_{nm}(\equiv \langle n(q,Q_o)|\Delta U(q,Q)|m(q,Q_o)\rangle) = 0 \tag{21}$$

allows the molecular wavefunction to be written as a simple product

$$\Psi^C_{ni}(q,Q) = \psi_n(q,Q_o)\chi^C_{ni}(Q) \tag{22}$$

The one remaining expansion coefficient is given by

$$(T(Q) + E_n(Q_o) + \Delta U_{nn} - \varepsilon^C_{ni})\chi^C_{ni}(Q) = 0 \tag{23}$$

The superscript C refers to the CA approximation, Eqn (21), made in arriving at the vibrational Schrödinger equation, Eqn (23), and in allowing the molecular wavefunction to be expressed as the product, Eqn (22).

Some confusion exists in the literature as to what is meant by the various adiabatic approximations. We have given the most commonly used definitions and the ones most strongly recommended. A summary is presented in Table 1.

3.4. HERZBERG-TELLER (HT) COUPLING

The CA wavefunctions are essentially approximate adiabatic (BH or BO). By defining the electronic wavefunctions at some specific nuclear configuration Q_o all the Q dependence has been removed. The correction to be discussed here concerns the reintroduction of the Q dependence in the electronic wavefunction.

The CA wavefunctions are eigenfunctions of the Q independent, electronic Hamiltonian $H_e(q,Q_o)$, Eqn (18). The potential energy term $\Delta U(q,Q)$, Eqn (19), can be treated as a perturbation which can couple different CA wavefunctions. Thus, as was shown by Herzberg and Teller, the electronic wavefunction for any nuclear configuration Q, can be expressed in terms of electronic wavefunctions at Q_o, according to Eqn (24),

$$\psi_f(q,Q) = \sum_n a_{fn}(Q)\psi_n(q,Q_o) \tag{24}$$

TABLE 1. Three most frequently used adiabatic approximations

	CA	BO	BH
Vibronic wavefunctions	$\Psi^{C}_{ni}(q,Q) = \psi_n(q,Q_o)\chi^{C}_{ni}(Q)$	$\Psi^{BO}_{ni}(q,Q) = \psi_n(q,Q)\chi^{BO}_{ni}(Q)$	$\Psi^{A}_{ni}(q,Q) = \psi_n(q,Q)\chi^{A}_{ni}(Q)$
Electronic Schrödinger equation	$[T(q)+U(q,Q_o)]\psi(q,Q_o) = E_n(Q)\psi(q,Q_o)$	$[T(q)+U(q,Q)]\psi(q,Q) = E_n(Q)\psi(q,Q)$	$[T(q)+U(q,Q)]\psi(q,Q) = E_n(Q)\psi(q,Q)$
Vibrational Schrödinger equation	$[T(Q)+E_n(Q_o) + \langle n\|\Delta U(q,Q)\|n\rangle]\chi^{C}_{ni}(Q) = \varepsilon^{C}_{ni}\chi^{C}_{ni}(Q)$	$[T(Q)+E_n(Q)]\chi^{BO}_{ni}(Q) = \varepsilon^{BO}_{ni}\chi^{BO}_{ni}(Q)$	$[T(Q)+E_n(Q) + \langle n(q,Q)\|T(Q)\|n(q,Q)\rangle]\chi^{A}_{ni}(Q) = \varepsilon^{A}_{ni}\chi^{A}_{ni}(Q)$

In this expression the coefficients $a_{fn}(Q)$ now contain the nuclear coordinate dependence. In order to determine the coefficients, the usual practice is to expand $\Delta U(q,Q)$ (or $H_e(q,Q)$) as a Taylor series about Q_o,

$$\Delta U(q,Q) = U(q,Q) - U(q,Q_o)$$

$$= \sum_r \left(\frac{\partial U(q,Q)}{\partial Q_r}\right)_{Q_o} Q_r + \tfrac{1}{2} \sum_r \sum_s \left(\frac{\partial^2 U(q,Q)}{\partial Q_r \partial Q_s}\right)_{Q_o} Q_r Q_s + \cdots . \tag{25}$$

Employing perturbation theory the approximate adiabatic wavefunction can be written

$$\psi_f(q,Q) = \psi_f(q,Q_o) + \sum_{n\neq f} a_{fn}(Q)\psi_n(q,Q_o) \tag{26}$$

where the Q-dependent coefficients, correct to first order of perturbation theory, are given by

$$a_{fn}(Q) = \frac{\langle n|\Delta U(q,Q)|f\rangle}{E_f(Q_o) - E_n(Q_o)} \tag{27}$$

Substitution of Eqn (25) for $\Delta U(q,Q)$ in Eqn (27) leads to the following expression for the coupling constants

$$a_{fn}(Q) = \sum_r \frac{\langle n| \left(\frac{\partial U(qQ)}{\partial Q_r}\right)_{Q_o} |f\rangle}{E_f(Q_o) - E_n(Q_o)} Q_r + \frac{1}{2} \sum_r \sum_s \frac{\langle n| \left(\frac{\partial^2 U(qQ)}{\partial Q_r \partial Q_s}\right)_{Q_o} |f\rangle}{E_f(Q_o) - E_n(Q_o)} Q_r Q_s + \dots \quad (28)$$

The corrected wavefunction (26) is frequently called the HT or HT corrected *adiabatic wavefunction*. The inclusion of the Q dependence in Eqn (26) is also known as the breakdown of the *Condon approximation*. This is treated more explicitly in the consideration of transition moments in the chapter on Vibronic Intensities.

The correction of the CA wavefunctions expressed by Eqn (26), is one type of correction which comes under the general heading of *vibronic coupling*. More precisely, it is called HT vibronic coupling. However, it should be clearly recognized that the terminology is somewhat misleading since vibronic coupling is ascribed to the coupling between the electronic and vibrational wavefunctions, that is to say coupling between the electronic and vibrational motions. It is this coupling which prevents expression of the vibronic wavefunction as a simple product. This is not the case in HT vibronic coupling since the correction is to the electronic wavefunction only,

$$\Psi_{fi}(q,Q) = [\psi_f(q,Q_o) + \sum_{n \neq f} a_{fn}\psi_n(q,Q_o)]\chi_{fi}(Q) \quad (29)$$

In other words, the corrected vibronic wavefunction is still adiabatic Eqn (26).

Second order HT coupling, as distinct from second order perturbation theory, concerns inclusion of the quadratic term in the potential energy expansion, Eqn (25), in the coupling. It is of importance when the first-order term is small or zero. Then the coefficients a_{fn}, are determined by the quadratic terms in Eqn (28).

3.5. THE DIABATIC BASIS

In the intermediate and strong coupling domains, that is when neglect of the off-diagonal elements in the adiabatic bases, Eqns (12), (13) and (21), is not justified, poor vibrational energies and wavefunctions are obtained. Perturbation theory is unsatisfactory to handle the coupling.

It has been proposed that replacement of the adiabatic basis (ABH, ABO and CA) by a basis set of product functions (diabatic, DBO) which exactly, or approximately diagonalize the nuclear kinetic energy terms in the molecular Hamiltonian allows for a more satisfactory treatment of the strong coupling domain. This is for the reason that in the vicinity of narrowly avoided crossings of potential energy curves it is unrealistic to confine the vibrational motion to the one potential energy surface, since the nuclear momentum neglected in the adiabatic approximation can cause strong coupling of the vibrational motions belonging to different potential energy surfaces. The DBO representation which has this interaction included in its zeroth order is superior to the adiabatic representation near avoided crossings and leads to freely crossing potential energy curves. But it is not as good elsewhere due to its more approximate treatment of the potential energy terms.

A serious drawback of the DBO basis is its lack of uniqueness. No procedure is available for selecting the best DBO product function for a given molecular state. This contrasts the unique adiabatic product function which can be obtained from the variation principle. The only available explicit method for generating a DBO function involves the CA functions.

One can write the two expansion schemes as follows,

$$\Psi = \sum_i \emptyset_i^A \chi_i^A = \sum_i (\sum_j c_{ij} \emptyset_j^C) \chi_i^A$$

and
$$\Psi = \sum_i \emptyset_i^D \chi_i^D = \sum_i \emptyset_i^C \chi_i^D \qquad (30)$$

In the DBO basis the following coupled Schrödinger equations, analogous to Eqn (20) are obtained. They determine the coefficients χ_{mi}^D

$$(T(Q) + E_n(Q_0) + U_{nn}(Q) - \varepsilon_i)\chi_{ni}^D(Q)$$

$$+ \sum_{m \neq n} U_{nm}(Q)\chi_{mi}^D(Q) = 0 \qquad (31)$$

The classical treatment of the colliding system (diatomic) of Na and Cl atoms provides an illuminating example of the use of DBO states.

For a sodium and a chlorine atom moving in each other's vicinity, their electronic state may be described at large or moderate interatomic separations R, as either covalent (Na + Cl) or ionic ($Na^+ + Cl^-$). The corresponding wavefunctions are ψ_{cov} and ψ_{ion} respectively. At very large separations the covalent state is lower in energy, as depicted in Fig 1., where the expectation values

$$E_{cov} = \langle \psi_{cov} | H_e | \psi_{cov} \rangle$$

and
$$E_{ion} = \langle \psi_{ion} | H_e | \psi_{ion} \rangle \qquad (32)$$

are plotted as a function of R. Note that the two curves of Fig. 1 cross at R_c despite the fact that both states are of the same symmetry. They do this in violation of the non-crossing rule, but may do so since they are not eigenfunctions of the electronic Hamiltonian H_e. In this two-state basis, the H_e matrix has the form

$$\begin{bmatrix} E_{cov} & E_{cov,ion} \\ E_{ion,cov} & E_{ion} \end{bmatrix}$$

which is not diagonal. The off-diagonal elements ($E_{cov,ion}$) though small, are not negligible and are nevertheless much larger than the off-diagonal elements of the nuclear kinetic energy, Eqns (12 and (13).

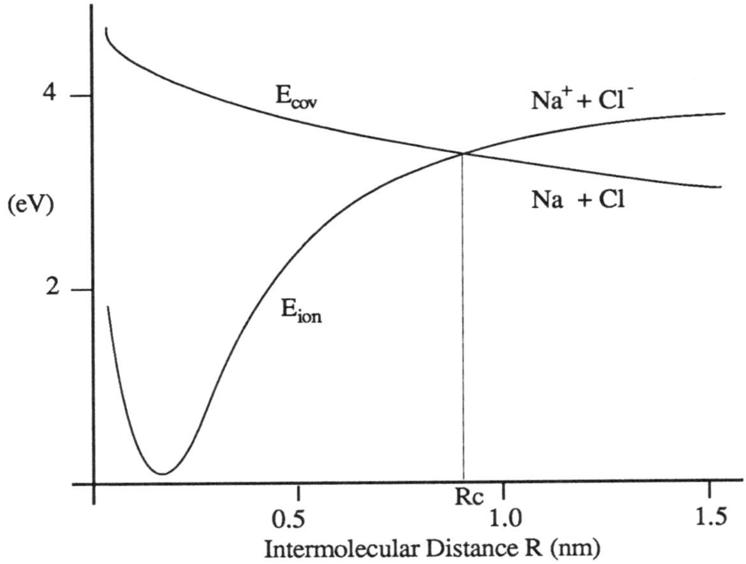

Figure 1. Electronic energies of the ionic and covalent states of NaCl as a function of internuclear separation. Rc marks the crossing point of the two curves.

Zener expanded the full molecular wavefunction in terms of the diabatic functions ψ_{cov} and ψ_{ion} as follows,

$$\Psi(r,R) = \psi_{cov}\chi_{cov}(R) + \psi_{ion}\chi_{ion}(R) \tag{33}$$

Since, as mentioned earlier, the BO approximation applies the following coupled equations may be written for the coefficients $\chi_{cov}(R)$ and $\chi_{ion}(R)$

$$(T(R) + E_{cov}(R)-\varepsilon)\chi_{cov}(R) + E_{cov,ion}\chi_{ion}(R) = 0$$

$$(T(R) + E_{ion}(R)-\varepsilon)\chi_{ion}(R) + E_{ion,cov}\chi_{cov}(R) = 0 \tag{34}$$

The role that each element of the electronic energy matrix plays in the above equations is clear. The diagonal elements provide the potential energy for elastic motion in the particular state concerned and the off-diagonal elements provide the coupling between the two states. And it is this coupling which allows for transitions between the states.

Using Eqn (34), Zener calculated the transition probability and found that it was greatest in the vicinity of R_c, the point of intersection of the two potential energy curves. Furthermore, as is to be expected, the probability was small for all but the lowest interatomic velocities.

3.6. GENERATOR COORDINATE METHOD

Doubts about the validity of the physical picture underlying the separation of variables, in particular the semiclassical notions of potential energy surfaces and molecular shapes, have prompted different approaches to the solutions of the Schrödinger equation.

In the Generator Coordinate Method (GCM), a reformulation of the theory is proposed in which the nuclei are treated as quantum particles from the outset. The theory also disposes of the notion of molecular structure. The idea of separating the electronic and nuclear motions is abandoned from the beginning. However, the simplifying features of the BO and BH adiabatic procedures, such as the separation of the total energy into electronic, vibrational and rotational contributions, are not lost.

In the limited time available, it is not our intention to discuss the GCM in any depth but rather to acknowledge that it provides a powerful approach to the theory of nuclear motion in molecules. It is particularly useful in the region where the BO approximation fails. A brief description follows.

We consider a set of functions $\emptyset(x|\alpha)$ depending on the dynamical variables x of a many-body system and a real parameter α. α lies in the interval [a,b] and corresponds to a molecular structure. The $\emptyset(x|\alpha)$ are intrinsic state functions and the most general trial functions $\Psi(x)$ are formed by taking linear combinations of these intrinsic state functions with weight factors $f(\alpha)$. Thus

$$\Psi(x) = f_0\emptyset(x|\alpha_0) + f_1\emptyset(x|\alpha_1) + \ldots$$

$$= \sum_i f_i\emptyset(x|\alpha_i) \tag{35}$$

In the limit where all members of the set $\{\emptyset(\alpha)\}$ are included in the trial function we can write

$$\Psi(x) = {}_a\!\int^b f(\alpha)\emptyset(x|\alpha)d\alpha \tag{36}$$

where the integration is over the parameter domain [a,b]. The function $f(\alpha)$ in Eqn (36) can be viewed as a continuously *labelled* set of superposition coefficients. The energy found by minimizing with respect to $f(\alpha)$ (variational principle) is a function of $f(\alpha)$,

$$E[f(\alpha)] = \frac{\iint f^*(\alpha)\, H_{\alpha\beta}\, f(\beta)d\alpha d\beta}{\iint f^*(\alpha)\, \Delta_{\alpha\beta}\, f(\beta)d\alpha d\beta} \tag{37}$$

where

$$H_{\alpha\beta} = \int\emptyset^*(\alpha)H\emptyset(\beta)dx \tag{38}$$

$$\Delta_{\alpha\beta} = \int\emptyset^*(\alpha)\emptyset(\beta)dx \tag{39}$$

and H is the many-body Hamiltonian. The variationally optimal weight factors satisfy the integral equation

$$\int[H_{\alpha\beta} - E\Delta_{\alpha\beta}]f(\beta)d\beta = 0 \tag{40}$$

The term intrinsic states is related to the need in the variational method for choosing a set of basis functions that are of intrinsic relevance. The real parameter α may be considered as an extra coordinate which serves to generate the trial function through the integration, Eqn (36). Hence it is called the generator coordinate. In principle any parameter can be chosen as a GC, but in practice the choice is restricted to parameters such that the integral Eqn (36) describes the desired features. Finally, the matrices $H_{\alpha\beta}$ and $\Delta_{\alpha\beta}$ have their parallel in the Hamiltonian and overlap matrices respectively.

As an illustrative example of a simple application of the GC method, Lathouwers and van Leuven considered the hydrogen atom problem. For the description of the s-states they used scaled Gaussians as their intrinsic states

$$\phi(r|\alpha) = e^{-\alpha r^2}. \tag{41}$$

The GC is the Gaussian scaling factor α. For the unormalized ground state they found

$$\begin{aligned}\Psi_o(r) &= e^{-r}\\ &= \int F_0(\alpha)e^{-\alpha r^2}d\alpha\end{aligned} \tag{42}$$

with a weight function

$$F_0(\alpha) = \frac{1}{2\sqrt{\pi}}\alpha^{-3/2}\, e^{-1/4\alpha} \tag{43}$$

that gives the exact ground state of the hydrogen atom.
The important feature of the GCM wavefunctions is that they contain no dependence on α. The molecular structures appear as integration variables only and thus the introduction of a definite molecular structure is avoided.
The GCM contains as a limiting case, the adiabatic approximation. The adiabatic functions, Eqn (14), can be rewritten as follows

$$\begin{aligned}\Psi^A(q,Q) &= \int f_A(\alpha)\psi(q|\alpha)\delta(Q-\alpha)d\alpha\\ &= \psi(q|Q)f_A(Q)\end{aligned} \tag{44}$$

Eqn (41) shows that the adiabatic wavefunctions are in fact generator coordinate functions in which the nuclear positions are the GCs, the adiabatic vibrational functions are the weight functions and the intrinsic states are the so-called 'clamped nuclei' states

$$\chi_\delta(q,Q|\alpha) = \psi(q|\alpha)\delta(Q-\alpha) \tag{45}$$

Thus the adiabatic approximation is a fixed nucleus version of the GCM.

BIBLIOGRAPHY

Born, M. and Huang, K. (1954) Dynamical Theory of Crystal Lattices, Clarendon Press, Oxford, England.
Ballhausen, C.J. and Hansen, A.E. (1972) "Electronic Spectra', Ann. Revs. Phys. Chem. 23, 15-38.
Essén, H. (1977) 'The Physics of the Born-Oppenheimer Approximation', Int. J. Quantum Chem. 12, 721-735.
Fischer, G. (1984) Vibronic Coupling, Academic Press, London.
Lathouwers, L. and van Leuven, P. (1982) 'Generator Coordinate Theory of Nuclear Motion in Molecules', Advs. in Chem. Phys. 49, 115-189.
Longnet-Higgins, H.C. (1961) 'Theory of Molecular Energy Levels', Advs. in Spectrosc. 2, 429-472.
O'Malley, T.F. (1971) 'Diabatic States of Molecules', Advs. in Atomic and Molecular Physics, 7, 223-249.
Özkan, I. and Goodman, L. (1979) 'Coupling of Electronic and Vibrational Motions in Molecules', Chem. Revs. 79, 275-284.

VIBRATIONS OF CRYSTAL LATTICES

V. SCHETTINO
Dipartimento di Chimica
Universita' di Firenze
Via G. Capponi 9, 50121 Firenze
ITALY

ABSTRACT. The theory of lattice dynamics and the construction of the dymamical matrix is reviewed with particular emphasis on molecular crystals. The results of infrared, Raman and neutron spectroscopy of crystals are illustrated and compared with calculations based on the use of model potentials. Vibrational relaxation and dephasing in solids are discussed.

1. Introduction

Many physical properties of crystals are connected with the small oscillations of the constituent atoms about their equilibrium positions. Among these are included the thermodynamic properties, elasticity, thermal and electrical conductivity, dielectric and optical properties and others. Of major interest to chemists is that vibrational, like other solid state, properties are intimately connected to the strenght and type of intermolecular interactions. Vibrational properties depend on second and higher order derivatives of the intermolecular potential with respect to appropriate coordinates and give thus information on the shape of the potential energy surfaces. Vibrations of crystal lattices are studied by a variety of tehcniques and mainly by infrared, Raman and neutron spectroscopy. Progresses in the experimental techniques and in the theory of lattice dynamics, where improved models of the intermolecular potential have become available, have facilitated the establisment of lattice dynamics as an independent field of research. The purpose of this paper is to review the treatment of the dynamics of crystal lattices, of the experimental results available and their interpretation and of some recent developments in this field. Emphasis will be on the vibrations of crystal lattices made of molecular units. By this we mean neutral molecules, molecular ions or even sections of polymeric chains.

C. D. Flint (ed.), Vibronic Processes in Inorganic Chemistry, 21–51.
© 1989 by Kluwer Academic Publishers.

2. Theory of lattice dynamics

2.1. VIBRATIONAL COORDINATES

We consider an infinite crystal with Z molecules per unit cell and N atoms per molecule. The existence in this type of crystals of well defined molecular units implies that the forces within the units are much stronger than the intermolecular interactions and that these latter can be considered as a perturbation. It is convenient to adopt a system of displacement coordinates that from the beginning takes this peculiarity into account. The original treatment of lattice dynamics [1,2] formulated in terms of cartesian displacements coordinates is not suited for this purpose.

For the "molecular" type of crystals we are dealing with it is possible in most cases to distinguish between vibrations internal to the units, that are little affected by the intermolecular interaction, and external vibrations corresponding to translations and librations of the molecular units as rigid bodies. For the internal vibrations the intermolecular forces produce small frequency shifts, splitting into multiplets and dispersion of the crystal frequencies. On the contrary external vibrations are entirely governed by the intermolecular interactions and occur at lower frequencies. The most convenient set of vibrational coordinates should allow to the maximum extent the separation of internal from external (translational and rotational) motions. These coordinates are well known in the theory of molecular vibrations [3,4]. In essence we may proceed as follows [5-7].

We adopt a molecule fixed and a crystal fixed coordinate system and define, for each molecule, the following quantities :

u_i^o equilibrium position vector of atom i in the molecular frame
u_i cartesian displacement vector of atom i in the molecular frame
U_i cartesian displacement vector of atom i in the crystal frame
U center of mass displacement vector in the crystal frame
$\boldsymbol{\Gamma}$ instantaneous direction cosines between molecule and crystal frame
$\boldsymbol{\Lambda}$ equilibrium direction cosines between molecular and crystal frame

The cartesian displacement vectors in the two systems are related by

$$U_i = U + \boldsymbol{\Gamma} u_i + (\boldsymbol{\Gamma}-\boldsymbol{\Lambda}) u_i^o \qquad (1)$$

It can be shown [5] the $(\boldsymbol{\Gamma}-\boldsymbol{\Lambda})$ can be expressed in terms of three independent rotational angles about the axes of the molecular frame. It is therefore seen that, for each molecule of N atoms, the 3N cartesian displacements are equivalent to a) three translations of the center of mass; b) three rotations about molecular axes; c) 3N cartesian displacements in the molecular frame. There are thus six redundancies in this new set of coordinates.As it is usually done in the theory of molecular vibrations the subset c) can be substituted by 3N-6 internal coordinates (variations of bond lenghts and angles) and six null coordinates defining the redundancy conditions. This set of coordinates allows to first order the complete separation of internal and external motions [5]. To second order there is a coupling between rotational and internal coordinates but this can be neglected for most purposes.

2.2. THE DYNAMICAL EQUATIONS.

Having chosen the set of molecular coordinates it is possible to write the equations of motions for the vibrations of the crystal.

The kinetic energy after appropriate normalization of the coordinates assumes the form [5]

$$2T = \Sigma_{\text{all molecules}} \Sigma(\dot{t}_i^{\,2} + \dot{r}_i^{\,2} + g_{ll'} \dot{S}_l \dot{S}_{l'}) \quad (2)$$

where t_i, r_i and S_l are translational, rotational and internal coordinates, respectively, and g are elements of the inverse kinetic energy matrix well known in the theory of molecular vibrations [3,4]. The kinetic energy is not diagonal in this set of coordinates, except for the external part. It is, however, possible to have the kinetic energy in a diagonal form by a linear trasformation into the normal coordinates q of the isolated molecules. In this new coordinate system the kinetic energy has the form

$$2T = \Sigma_{ls\alpha} \dot{q}_{ls\alpha}^{\,2} \quad (3)$$

where l, α and s are indices for the unit cell, site amd coordinate, respectively. The coefficient of the linear trasformation between internal and normal coordinates is generally known from the solution of the dynamical problem for the isolated molecule [3,4].

The potential energy can be expanded in a power series of the vibrational coordinates. The expansion is performed about the equilibrium position and therefore the linear terms in the expansion are zero. In the harmonic approximation the expansion is limited to the quadratic terms. The effect of higher order terms in the expansion will be considered later. In the harmonic limit the potential energy has thus the following form in the normal coordinates

$$2V = \Sigma_{l\alpha s} \Sigma_{l'\sigma' s'} [F_{ss'}(l\alpha, l'\alpha') + \delta_{ss'} \delta_{ll'} \delta_{\alpha\alpha'} \omega_s^2] q_{l\alpha s} q_{l'\alpha' s'} \quad (4)$$

where the ω are the frequencies of the isolated molecule.

The force constants defined in (4) are not completely independent quantities. Besides being, by definition, symmetric in the exchange of indices, they must fulfill a series of conditions deriving from the periodicity of the crystal lattice and from the invariance of the crystal potential under infinitesimal translations or rotations [2,5].

To take advantage of the molecular nature of the crystal we may express the crystal potential as the sum of a molecular V_M and an intermolecular V_I part

$$V = V_M + V_I \quad (5)$$

V_M is expressed as the sum of the internal potentials of all the molecules in the crystal. Since the internal potential is not generally known as an analytic function it can be expanded in terms of the internal coordinates and the force constants, considered as adjustable parameters, can be obtained from the best fit between calculated and observed internal frequencies of the molecule.

The intermolecular potential can be conveniently expressed as the sum of pairwise interactions between the molecules a and b

$$V_I = \Sigma_{a,b} V_I(a,b) \qquad (6)$$

The intermolecular potential can be treated in the same way as the internal part defining a set of force constants to use as adjustable parameters, subjected to the symmetry and invariance conditions discussed above. This, however, is inconvenient because the number of independent force constants is exceedingly high and because this would leave the vibrational properties of the crystal rather uncorrelated with other properties. It is therefore customary to derive the intermolecular force constants from analytical forms of the intermolecular potential. As it will be discussed later and as suggested by the theory of intermolecular forces [8-10] the intermolecular potential is expressed as a sum of an electrostatic (charge-charge or multipolar interactions) and an atom-atom (Lennard-Jones or Buckingham) part.

Once the potential and kinetic energy are available as quadratic forms it is possible to write the equations for the vibrational motion of the crystal as

$$q_{l\alpha s} + \Sigma_{l'\alpha's'}[\, F_{ss'}(l\alpha,l'\alpha') + \delta_{ss'}\delta_{\alpha\alpha'}\delta_{ll'}\ \omega_s^2\ ] = 0 \qquad (7)$$

So far we have assumed that the crystal is infinite. We have therefore an infinite set of coupled equations whose solution will lead to a secular equation of infinite dimensions. To make the problem tractable we exploit the periodicity of the lattice and look for solutions of the type

$$q_{l\alpha s} = A_{\alpha s}\exp(i\omega t)\exp(i\mathbf{k}\cdot R_l) \qquad (8)$$

that are oscillatory solutions of amplitude A with periodicity in time , with frequency $\nu=2\pi/\omega$, and in space with wavelenght $\lambda = 2\pi/k$. Introducing (8) in (7) we obtain the following eigenvalue equation for the normalized amplitudes **e**

$$\Sigma_{s'\alpha'} D_{ss'}(\alpha\alpha'|k)\ e(s'\alpha'|pk) = \omega_{pk}^2\ \ e(s\alpha|pk) \qquad (9)$$

It is therefore seen that the infinite dimensional problem has separated into problems of dimension 3ZN, one for each value of the wave vector **k**. The elements of the dynamical matrix D are

$$D_{ss'}(\alpha\alpha'|k)=\Sigma_l\ [F_{ss'}\ (l\alpha,l'\alpha')+\delta_{ss'}\ \delta_{\alpha\alpha'}\ \delta_{ll'}\ \omega_s^{\ 2}\ ]\ \exp(ik\ R_l\ ) \qquad (10)$$

The eigenvectors **e** diagonalize the dynamical matrix and define the trasformation from the local normal coordinates to the crystal normal coordinates Q

$$q_{l\alpha s} = L^{-1/2}\, \Sigma_{pk}\, \mathbf{e}(s\alpha|pk)\, Q_{pk}\, \exp(-i\mathbf{k}\cdot\mathbf{R}_l) \qquad (11)$$

We still are faced with the problem of solving an infinite number of equations of the type (9). To circumvent this difficulty we introduce the Born-Van Karmann cyclic boundary conditions [1]. We divide the infinite crystal into blocks with L_1, L_2 and L_3 unit cells in the three directions and impose that equivalent atoms separated by L_1, L_2 or L_3 basic translations in each direction carry the same movement.The cyclic boundary conditions reduce the dynamical problem to that of a finite crystal without introducing surface effects. If t_1, t_2 and t_3 are the three basic translations of the crystal the boundary conditions have the form

$$\exp(i\mathbf{k}\cdot t_1 L_1) = \exp(i\mathbf{k}\cdot t_2 L_2) = \exp(i\mathbf{k}\cdot t_3 L_3) = 1 \qquad (12)$$

and impose restrictions on the possible values of the wavevector k that can now take $L = L_1L_2L_3$ independent values.

We can look at the procedure described above in the following way. The allowed values of the wavevector **k** are defined as vectors of the reciprocal space confined within the reciprocal unit cell or, more conveniently, within the first Brillouin zone. According to the space group theory the allowed values of k label the irreducible representations of the translational group of the crystal. The procedure followed is therefore essentially equivalent to a transformation to a site symmetrized set of coordinates

$$q_{s\alpha}(\mathbf{k}) = L^{-1/2}\, \Sigma_l\, q_{l\alpha s}\, \exp(i\mathbf{k}\cdot\mathbf{R}_l) \qquad (13)$$

and to a subsequent formulation of the dynamical problem in the symmetry coordinates. The factorization of the dynamical problem into separate blocks, one for each value of k, is therefore a consequence of the translational symmetry of the crystal since in the potential energy there cannot be coupling terms between coordinates corresponding to different representations of the translational group.

2.3. DISPERSION CURVES AND DENSITY OF STATES.

From the solution of the dynamical equations we obtain the crystal frequencies as a function of the wave vector **k**. These relations are called dispersion relations and can be represented graphically as dispersion curves plotting ω versus **k**. In this way the crystal frequencies are represented into 3ZN branches of the dispersion curves with L points in each branch. Because of the hermiticity and symmetry of the dynamical matrix the crystal frequencies will be real with the additional relation $\omega_{pk} = \omega_{p-k}$ and it will therefore be sufficient to represent the dispersion curves for positive values of k.

A typical set of the dispersion curves is shown in Fig.1[11]. In all crystals the frequencies of three of the branches of the dispersion curves are zero at k = 0. These are called acoustic branches and are connected with the sound propagation in the crystal. In the long wavelenght limit (k=0) they correspond to crystal vibrations where all atoms of the unit cell move in the same direction as rigid bodies. As the wavevector increases this character can be partially lost as mixing occurs with other kinds of crystal vibrations. All other branches of the dispersion curves have non zero limiting frequencies at k = 0 and correspond to the so called optical modes of vibration. In crystals with molecular units it is in most cases possible to make

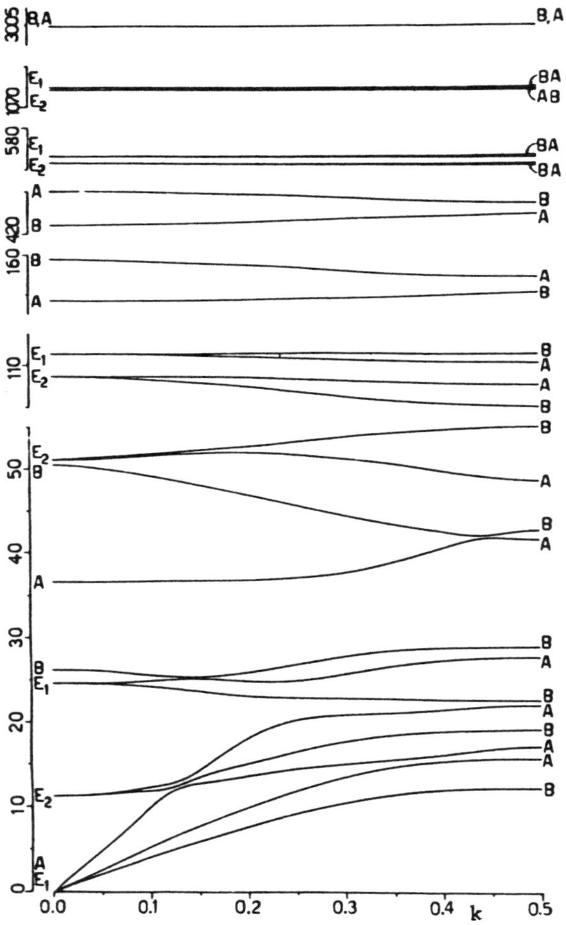

Fig.1 - Dispersion curves of crystalline iodoform.

further distinction within the optical branches, a distinction that shows the advantage of the molecular coordinates introduced above. In fact 6Z - 3 branches can occur at much lower frequencies (0-150 cm-1) than the others. These correspond to crystal motions where all the molecules translate or rotate as rigid units about the equilibrium positions. The remaining Z(3N - 6) branches correspond to internal deformations of the molecular units without overall translation or rotation. This separation is not rigorous for complicated molecular systems with low frequency internal modes. In this case internal and and external vibrations are mixed into more general crystal motions. However, whenever the separation of internal and external degrees of freedom is possible, the so called rigid body approximation can be used to handle the dynamics of the crystal. According to this approximation the dynamical equations are factorized into a 6Zx6Z block corresponding to the external (translational and rotational) degrees of freedom and a Z(3N-6)xZ(3N-6) block corresponding to the internal modes. For these latter, when an internal mode is weakly coupled to the other modes, it is also possible to study in detail the perturbation of the crystal field on this mode separately. For instance, this can be done using the vibrational exciton model [5,12,13].

In several cases it is convenient to classify the crystal modes as longitudinal or transverse. These correspond to modes where the atomic displacements occur in direction parallel or perpendicular, respectively, to the propagation direction of the mode (direction of the **k** vector). This distinction is particularly important for dipolar modes, i.e. for mode that develop an electric dipole moment and are therefore infrared active. The long range dipolar interaction produces an additional force constant for the longitudinal mode. Therefore in the long wavelenght limit longitudinal (LO) modes occur at higher frequency than the transverse (TO) modes, the difference being a measure of the strenght of the transition dipole moment. While in cubic crystals the longitudinal or transverse character of a polar mode is preserved in every direction the same does not occur for crystals of lower symmetry. In this case, in certain crystal directions modes of mixed TO-LO character are possible. As a consequence the frequency of a polar mode in uniaxial or biaxial crystals can be dependent on the direction of propagation of the mode. This is referred to as angular dispersion and an example is shown in Fig. 2 for the $NaNO_2$ crystal [14].

If the dynamical equations (9) are solved for a uniform grid of **k** values within the Brilluoin zone it is possible to obtain a distribution of vibrational frequencies of the crystal. This is referred to as the phonon density of states $n(\omega)$ and is defined in a way that $n(\omega)d\omega$ represents the number of crystal frequency in the range between ω and $\omega+d\omega$. The density of states is defined as

$$n(\omega) = \Sigma_{pk}\ \delta(\omega - \omega_{pk}) \tag{14}$$

To avoid spurious features the density of states must be calculated for a sufficiently large number of k values and various extrapolation procedures are available. The phonon density of states is an important quantity in connection with the thermodynamic and

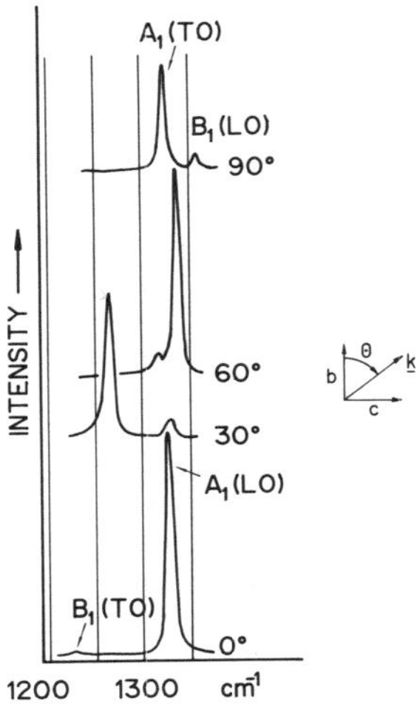

Fig. 2. Angular dispersion of optic modes in crystalline $NaNO_2$

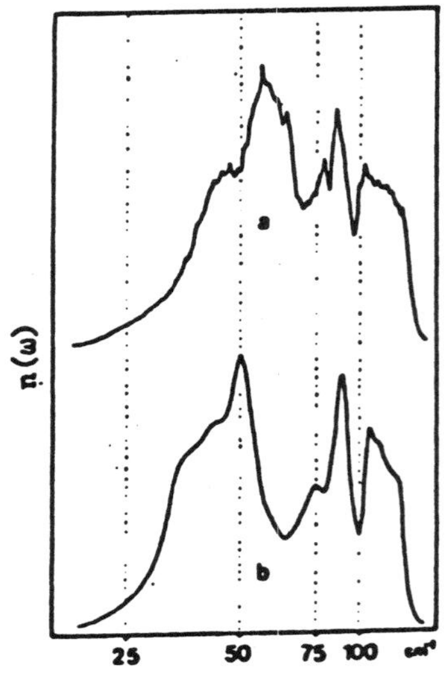

Fig. 3. Observed (a) and calculated (b) phonon density of states of deuteronaphthalene.

anharmonic properties of the crystal and the inelastic neutron scattering.As an example the experimental and calculated phonon density of states for the naphthalene crystal is shown in Fig. 3 [15,16].

3. Quantization and the concept of phonon.

In section 1.2 the dynamical problem was treated classically but the results obtained are easily translated in quantum language. First we express the classical Hamiltonian in terms of the crystal normal coordinates. Using the relation (11) and expressions (3) and (4) for the kinetic and potential energy it is easy to show that the hamiltonian assumes the form

$$H = \Sigma_{pk} (\dot{Q}_{pk}^2 + \omega_{pk}^2 \; Q_{pk}^2) \qquad (15)$$

or in terms of the conjugate momenta P

$$H = \Sigma_{pk} (P_{pk}^2 + \omega_{pk}^2 \; Q_{pk}^2) \qquad (16)$$

It is seen that the crystal behaves as a set of independent harmonic oscillators. Substituting the classical quantities with the corresponding operators the quantum mechanical hamiltonian operator is obtained. The Schrodinger equation for the vibrating crystal can then be separated into independent harmonic oscillator wave equations , one for each crystal normal coordinate. Using the harmonic oscillator result the total crystal energy is given by

$$E = \Sigma_{pk} (n_{pk} + 1/2) \hbar\omega_{pk} \qquad (17)$$

and the crystal wavefunction is the product of harmonic oscillator wavefunctions.

In this representation the vibrational state of the crystal is specified by a state vector $| \; n_{pk} \; >$ that through 3ZNL quantum numbers gives the degree of excitation of each oscillator.

For many purposes it is convenient to use creation $a^\dagger$ and annihilation a operators defined by the relations

$$\begin{aligned} a^\dagger_{pk} &= (1/2\hbar\omega_{pk})^{1/2} (\omega_{pk} \; Q_{p-k} - iP_{pk}) \\ a_{pk} &= (1/2\hbar\omega_{pk})^{1/2} (\omega_{pk} \; Q_{pk} + i \, P_{p-k}) \end{aligned} \qquad (18)$$

In the new operators the hamiltonian assumes the form

$$H = \Sigma_{pk} \hbar\omega_{pk} \; (a^\dagger_{pk} a_{pk} + 1/2) \qquad (19)$$

The operators a and $a^\dagger$ have the following effect on the state vector

$$\begin{aligned} a_{pk} \; | \; n_{pk} \; > &= n_{pk}^{1/2} \; | \; (n_{pk} -1) \; > \\ a^\dagger_{pk} \; | \; n_{pk} \; > &= (n_{pk} - 1)^{1/2} \; | \; (n_{pk} +1) \; > \end{aligned} \qquad (20)$$

The operator $n = a^{\dagger} a$ (called number operator) has the property

$$a^{\dagger}_{pk}\, a_{pk} \mid n_{pk} \rangle = n_{pk} \mid n_{pk} \rangle \qquad (21)$$

In the previous description n_{pk} was used to define the state of excitation of the oscillator pk. Alternatively we may think of n_{pk} as an operator counting the number of particles with energy $\hbar\omega_{pk}$ in the vibrational state pk. These quasi particles are called phonons and obey the Bose-Einstein statistics. The vibrational state of the crystal is therefore specified by giving the number of phonons of each type present in the crystal. A phonon is characterized by its energy and wavevector . This description is useful in many cases and in particular in the study of anharmonic interactions and in the treatment of the interaction of lattice vibrations with photons.

4. Symmetry of lattice vibrations

The number of vibrational degrees of freedom of a crystal is very high and therefore the symmetry classification of the crystal vibrations is extremely useful. The symmetry of a crystal is described by the space group. The application of the space groups theory to lattice dynamics is rather complex and we can only review a few basic concepts and the interested reader is referred to several detailed treatments of the subject available [17-21].

We may distinguish the whole of the symmetry operations of an infinite crystal into two parts. The first comprises the translational symmetry operations that bring the crystal into self coincidence. The translational group is the direct product of three cyclic groups whose generators are the unit cell translations t_1, t_2 and t_3. Using the cyclic boundary conditions the translation group reduces to a finite group and has only one dimensional irreducible representations. The irreducible representations of the translation group are labelled by the **k** vectors defined in the first Brillouin zone and the character associated with the translational operation $\mathbf{r} = n_1 t_1 + n_2 t_2 + n_3 t_3$ in the $\mathbf{k}^{th}$ representation is $\exp(-i\mathbf{k}\cdot\mathbf{r})$. We have already exploited the translational symmetry to factorize the dynamical matrix into blocks, one for each allowed value of **k**.

In most cases the crystal is also invariant under rotational symmetry operations. These latter can be composite operations including a rotation and a fractional translation. If T denotes the translational group and $R_1, \ldots\ldots, R_h$ a set of representative rotational operations the full symmetry of the crystal is described by the space group defined as

$$G = TR_1 + TR_2 + \ldots\ldots\ldots + TR_h \qquad (22)$$

The space group is thus decomposed into the invariant subgroup T and a number of residual cosets. Multiplication of a rotation R_i by the translation group implies that from a representative origin cell the rotational operation is translated into equivalent positions of all the unit cells of the crystal. All the elements of each coset can be defined as a single operation thus defining a new group called, in this particular case, the unit cell group. In the unit cell group memory is lost of the individual cells as if after any symmetry operation each point were brought back into the equivalent point of the origin cell. The unit cell group is isomorphic with one of the 32 point groups and its irreducible representations are known.

Crystal states must be classified in the irreducible representations of the full space group for whose construction we refer to the general references given above. We only recall here that for each **k** vector we may define :

the point group of **k**, $G_o(\mathbf{k})$, including all the rotational elements β_i that leave **k** invariant;

the space group of **k**, $G(\mathbf{k}) = T\beta_1 + \ldots\ldots\ldots + T\beta_m$, which is a subgroup of G;

the star of **k**, as the set of **k** vectors obtained from **k** acting on it by the rotational elements α_j not included in $G_o(\mathbf{k})$;

Then the space group can be expressed as

$$G = G(\mathbf{k})\alpha_1 + \ldots\ldots\ldots\ldots + G(\mathbf{k})\alpha_l \quad (23)$$

The irreducible representations of the space group are easily obtained if those of G(**k**) are known. If the space group is not symmorphic (i.e. the rotations do not include fractional translations) or if **k** is not on the surface of the Brillouin zone the irreducible representations of G(**k**) are simply related to those of the point group $G_o(\mathbf{k})$. If the space group is symmorphic and **k** in on the surface of the Brillouin zone the procedure is more complicated and cannot be described here.

The crystal vibrations classify in the irreducible representations of G(**k**). These latter may exhibit higher degeneracy (fourfold or sixfold) that the molecular groups. In addition the time inversion symmetry must be taken into account. This can have an effect when in the space group there are rotational operations that change **k** into -**k** . The time inversion symmetry increases the actual symmetry of the crystal states and may produce additional degeneracies.

The classification of crystal vibrations in the irreducible representations of G(**k**) at **k**=0 is of particular importance for optical spectroscopy. At **k**=0 the representations of G(**k**) coincide with those of the unit cell group and the symmetry classification is easily obtained using the methods of point group theory [3]. For a molecular crystal the procedure can be further simplified if the symmetry classification of the modes (internal and external) of the isolated molecule in the molecular point group is known [22]. In fact we may define a site group describing the local symmetry at the molecular site in the crystal. This describes the symmetry of the crystal field viewed from that particular site. The site group is a subgroup of the molecular group and therefore a correlation between the irreducible

Table 1. Correlation diagram for the naphthalene crystal.

Molecular group D_{2h}			Site group C_i	Unit cell group C_{2h}			
i	ii	a	a	a	iii	iv	v
	9	a_g		A_g	24	R_x,R_y,R_z	aa,bb,cc,ac
R_x	3	b_{1g}	g				
R_y	4	b_{2g}		B_g	24	R_x,R_y,R_z	ab,bc
R_z	8	b_{3g}					
	4	a_u					
T_y	8	b_{1u}	u	A_u	24	T,T	b
T_x	8	b_{2u}					
T_z	4	b_{3u}		B_u	24	T	a,c

a - irreducible representations
i - classification of rotations and translations of the free molecule
ii - number of internal vibrations of the free molecule
iii - number of internal vibrations in the unit cell
iv - number and type of the external modes of the unit cell
v - components of polarizability tensor and dipole moment.

representations of the molecular and site group can be easily found and is reported in many textbooks [3,4]. If there are more than one molecule per unit cell, the unit cell group will contain other symmetry operation than those of the site group and these interchange non equivalent molecules. These symmetry operations define the so called interchange group and the unit cell group is the direct product of the site and interchange groups. Therefore the site group is necessarily a subgroup of the unit cell group and a correlation exists between their irreducible representations. The ratio of the orders of the unit cell and site group is equal to the number of molecules per unit cell.

As an example the correlation diagram for the naphthalene crystal is reported in Table 1. Naphthalene crystallizes in the space group C_{2h}^5 with two molecules in the unit cell on sites with C_i symmetry . Table 1 shows, for instance, that a molecular internal vibration of symmetry a_g because of the reduced symmetry in the crystal falls in the site group in the g species together with other vibrations of g type. Therefore a_g, b_{1g}, b_{2g} and b_{3g} modes are not anymore independent modes in the crystal and can mix to a degree that is determined by the strenght of the intermolecular potential. The correlation diagram also shows that each internal mode splits in the unit cell into two components (Davydov components) that classify in the irreducible representation A_g and B_g of the unit cell group, for the g-type modes, or in the A_u and Bu species, for the u-type modes. The splitting into two components is bound to the double occupancy of the unit cell. The selection rules in optical experiments for the various Davydov components are determined by the unit cell group with the usual procedures of point group theory. In much the same way from the correlation diagram the symmetry classification of the external (translation and rotation) modes can be obtained. Care should be taken to eliminate the overall translations of the crystal corresponding to the acoustic modes of zero frequency at **k**=0.

5. Infrared absorption, Raman scattering and inelastic scattering of neutrons by lattice vibrations.

The vibrations of crystal lattices are more conveniently probed by absorption of infrared radiation, by Raman scattering of visible light and by inelastic scattering of thermal neutrons. These techniques are complementary in the sense that they can give different information and the results obtained by one technique can be useful for planning and interpreting experiments made with the other techniques. The basics of the techniques and the relations among them are best illustrated by comparing the relevant selection rules .

In an absorption experiment an infrared photon of energy $\hbar\omega$ and momentum $\hbar q$ is absorbed by the crystal and results in the creation of a phonon pk. The absorption is possible because the energy of infrared photons is in the same range as the energy of lattice vibrations. The photon wavevector in the infrared region (typically 10^3 cm^{-1}) is very small compared to that of phonons that equals the reciprocal lattice spacing (10^8 cm^{-1}). This means that the infrared wavelenght is very long compared to the lattice spacing. In the absorption experiment energy and momentum must be conserved. The first rule implies

$$\hbar\omega = \hbar\omega_{pk} \tag{24}$$

and the second

$$\hbar q = \hbar k + \hbar G \tag{25}$$

where G is a reciprocal lattice vector. The quantity $\hbar k$, called pseudo-momentum or crystal momentum, must be conserved in the experiment in the sense specified by equation (25). Since the dynamical equations are periodic in the reciprocal lattice it is seen that the only vibrations of the crystal that can be probed are the k = 0 modes. Therefore, by infrared absorption we can obtain information on the very limited number of the crystal vibrations that, as seen in the previous paragraph, classify in the irreducible representation of the unit cell group. The meaning of the momentum selection rule is the following. A crystal vibrational mode can absorb energy from the electromagnetic radiation if there is a modulation of the dipole moment of the crystal by the atomic movement, i.e. if the first derivative of the crystal dipole moment with respect to the normal coordinate of interest is different from zero. This will occur only if the atomic movement of the various unit cells are all in phase. This corresponds to the condition k = 0.

It is clear that not all the k = 0 modes will necessarily develop a transition dipole moment. In fact the k=0 selection rules derives from the translational symmetry of the crystal. If the crystal has

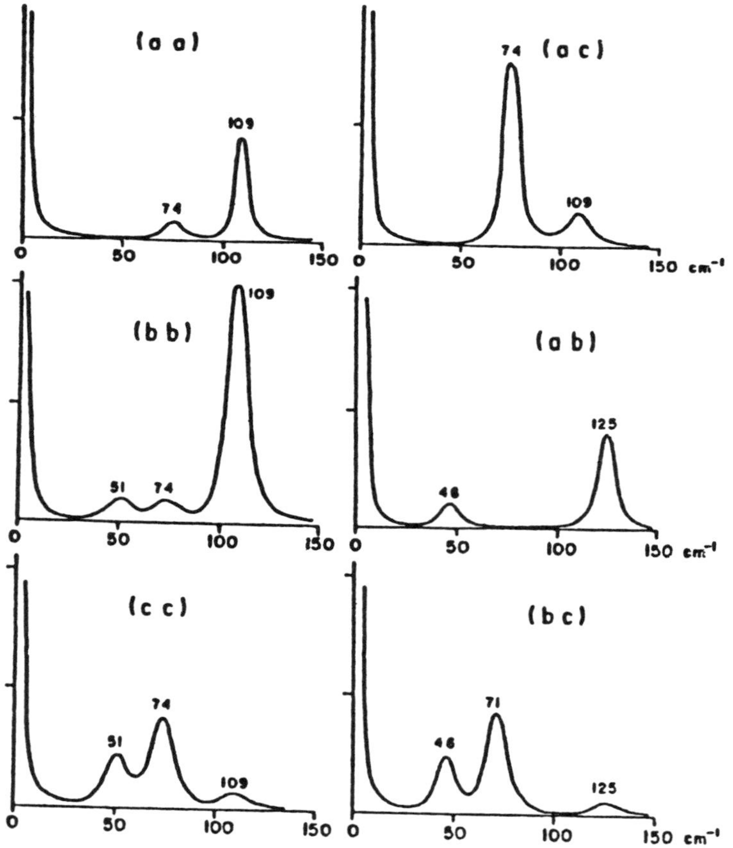

Fig. 4. The polarized Raman spectrum of crystalline naphthalene.

additional rotational symmetry, as defined by the unit cell group, additional selection rules may arise. Once the symmetry classification of the normal vibrations is available these selection rules are easily obtained. For instance, considering the case of the naphthalene crystal Table 1 shows that an internal vibration of symmetry b_{1u} of the free molecule splits in the crystal into two **k** = 0 mode of symmetry A_u and B_u. These are both active in the infrared spectrum and the transition dipole moments developped are parallel and perpendicular to the b crystal axis, respectively. Table 1 shows also that among the external modes of the naphthalene crystal three are infrared active and correspond to translational motions of the rigid molecules. Three external bands are therefore observed in the far infrared spectrum of the naphthalene crystal and, as it can be seen from Table 1, two of them will be polarized parallel to the b axis and the third perpendicular.

This description shows that in order to obtain the most complete information infrared experiments should be carried on single crystals in polarized light. This is not always possible since in most cases the obtainment of very thin samples, as required, is very difficult.

The interpretation of the intensities in the infrared spectra of crystals poses some problems [5]. The simplest model is the so called oriented gas model [22] were the transition dipole moment of the unit cell is interpreted as the sum of the transition dipoles of the

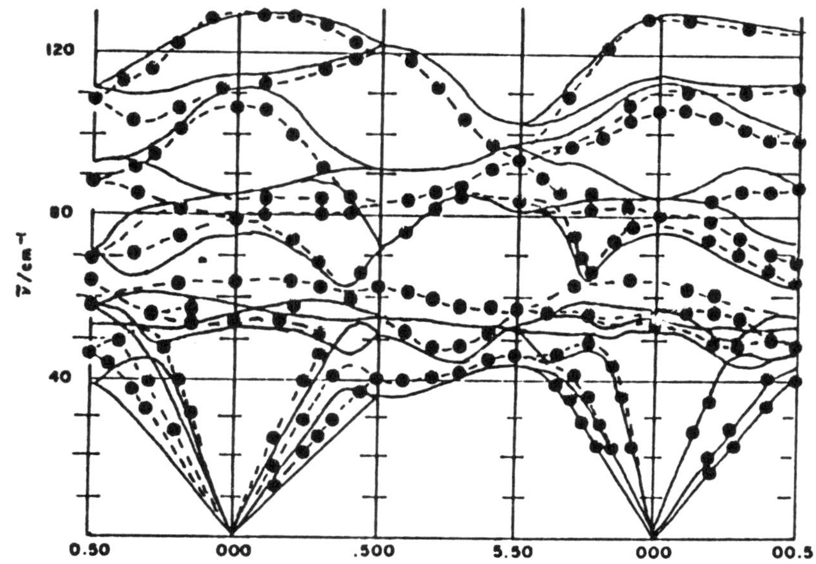

Fig. 5. Observed and calculated dispersion curves of deuteronaphthalene

constituent molecules with neglect of intermolecular interaction. As such the oriented gas model has been found very useful for the interpretation of the crystal spectra but many deviations from the model have been observed [5].

In the Raman experiment [23] a visible photon of energy $\hbar\omega_o$ is scattered inelastically by the crystal and is observed at a shifted frequency ω_s. The Raman shift is determined by the creation or annihilation of a crystal vibration of frequency ω_{pk}. The energy conservation requires

$$\hbar\omega_s = \hbar\omega_o \pm \hbar\omega_{pk} \tag{26}$$

with the + and - signs referring to the Raman antistokes and stokes scattering, respectively. Also for visible photons the wavevector **q** is very small compared to the reciprocal lattice spacing. Therefore the momentum conservation

$$\hbar q_s - \hbar q_o = \hbar \mathbf{k} + \hbar G \tag{27}$$

again requires that **k** = 0. Therefore, also in the Raman experiment only the crystal vibrations at **k** = 0 can be probed. However, the Raman selection rules can be different from the infrared and different informations are obtained in the two experiments. In fact for the Raman activity of a mode it is required that the crystal polarizability be modulated by the atomic movements. Since the polarizability is a tensorial quantity the directional properties of the Raman scattering are more varied than those of infrared absorption.

Considering again the naphthalene crystal,Table 1 shows that among the external vibrations those of rotational character are active in the Raman spectrum. This should be constrasted with the infrared case described above. Therefore six Raman external modes are predicted by symmetry and the components of the polarizability tensor associated with each of them are shown in Table 1 . The Raman spectrum of the naphthalene external modes in the various polarizations [24] is shown in Fig. 4 as an example.

In the neutron scattering spectroscopy a beam of thermal neutrons is scattered inelastically by the crystal [25]. Thermal neutron have velocity v of the order of 2 km/s and therefore their momentum k_o = mv/h is of the order of the reciprocal lattice spacing. The energy of the thermal neutrons $\hbar\omega_o = 1/2\, mv^2 = \hbar^2/2m k_o^2$ is of the same magnitude as the energy of crystal vibrations. Thermal neutrons therefore match both the energy and the momentum of the phonons. In the experiment where the neutron is scattered by a phonon ω_{pk} the energy conservation implies that the energy transferred to the crystal be

$$\hbar\omega = \hbar^2/2m\,(k_o^2 - k_1^2) = \pm\hbar\omega_{pk} \tag{28}$$

where k_i is the wavevector of the scattered neutron. The momentum conservation requires

$$\hbar k_t = \hbar(k_o - k_1) = \hbar \mathbf{k} + \hbar G \tag{29}$$

In the neutron scattering experiment the momentum trasfer can be large. Measuring the change in wavelength, energy or velocity of the scattered neutron and the directional properties of the inelastic scattered intensity offers a means of measuring the frequency of phonons as a function of the wavevector. As compared to optical experiments the neutron scattering gives information on the whole dispersion curves.

Actually the neutron scattering cross section can be separated into two parts, the first depending on the relative position of the atomic nuclei. This part is that relevant to the coherent inelastic scattering that gives information on the k dependence of the phonon frequencies. The coherent inelastic cross section depends on the eigenvectors of the crystal vibrations. Therefore in neutron scattering spectroscopy information on the crystal modes obtained by optical methods or by model calculations can be very useful. As an example the phonon dispersion curves measured for the perdeuterated naphthalene [15] are shown in Fig. 5. The second part of the scattering cross section refers to the incoherent scattering. It can be shown that the incoherent scattering intensity depends in a complicated way on the phonon density of states.

6. Intermolecular potentials and lattice dynamics calculations

Considerable progresses have been made in recent years in lattice dynamics calculations [5]. Calculations of crystal frequencies and intensities have been initially instrumental to interpretation of experimental data but later it has been found that they were useful to obtain information on intermolecular potentials.

In order to perform lattice dynamics calculations a suitable form of intermolecular potential must be available. From the theory of intermolecular forces [8-10] the intermolecular potential is expressed as the sum of two contributions :

a) the electrostatic interaction between the molecular charge distributions;

b) the short range dispersion and repulsive interactions.

The electrostatic interaction can be expressed assigning an effective charge to each atom and considering the sum of the charge-charge interactions. Since the actual charge distribution in the molecule is far more complicated than that represented by point charges on the atomic positions, in many cases it has been found necessary to use additional point charges located within or outside the chemical bonds. An alternative form of the electrostatic interaction is obtained by the multipolar expansion with multipoles located generally on the molecular center of mass. This is convenient since the multipolar interaction has a form suitable for lattice dynamics calculations [5]. However, for practical reasons only the first terms of the expansion can be taken into consideration and in actual cases only the dipole-dipole, dipole-quadrupole, quadrupole-quadrupole and

Table 2. Observed and calculated lattice frequencies of NH_3 (cm^{-1})

Mode	Exp.	Calculated			
		a	b	c	d
ω_1	---	95	84	117	143
ω_2	310	196	194	203	306
ω_3	107	122	122	119	122
ω_4	298	192	195	278	299
ω_5	140	132	135	125	130
ω_6	183	186	183	187	181
ω_7	260	128	129	130	273
ω_8	358	203	206	327	360
ω_9	533	265	316	483	524
Lattice energy (kcal/mole)		-3.19	-4.40	-6.63	-8.13

a : atom-atom potential
b : a + dipole-dipole
c : b + dipole-quadrupole and quadrupole quadrupole
d : c + dipole-octupole

dipole-octupole terms have been considered. The electrostatic interaction is long range and special care should be taken in evaluating the summation of the interactions over the crystal. The actual procedure is to consider interactions within a sphere centered about an origin molecule. This, however, may not be satisfactory for charge-charge and dipole-dipole interactions. Appropriate summation procedures are available to fully consider the long range character of the electrostatic interactions [1,2].

When the molecular charge distributions do not overlap significantly, the short range part of the interaction potential can be expressed as the sum of atom-atom interactions between atoms of different molecules. There are convenient analytical forms for the atom-atom potential and those that have most commonly be used in lattice dynamics are the Lennard-Jones or Buckingham potentials. The parameters of the potential are obtained empirically from independent

Table 3. Lattice frequencies (cm^{-1}) of ferroelectric $NaNO_2$.

Mode		Exp.	Calc[a]	Calc[b]
B_2	LO	254	283	256
	TO	227	185	211
	LO	201	166	189
	TO	149	126	145
B_1	LO	236	275	244
	TO	184	147	162
	LO	165	124	119
	TO	154	106	119
A_1	LO	270	300	269
	TO	190	179	188
A_2		119	151	133

a - Rigid ion model
b - Shell model

sources or used as adjustable parameters to fit together the dynamical and the static (structure, sublimation energy) properties of the crystal. The advantage of the atom-atom potential is that it can be transferred between similar systems with a reasonable degree of confidence. Because of their short range character the atom-atom interactions are simply summed over a sphere of 6-10 A radius. In more sofisticated calculations it has been found that the atom-atom interaction is too isotropic in many cases. This is partly compensated by the electrostatic interaction that has a more directional character. However, in many cases more sofisticated forms of the atom-atom interaction have been proposed [26] to account for the anisotropy of experimental parameters.

It is convenient to consider the use of intermolecular potentials in lattice dynamics with reference to some typical applications. As a first example we consider the results obtained for the ammonia crystal reported in Table 2 [27]. The intermolecular potential includes atom-atom interactions and dipole-dipole, dipole-quadrupole. dipole-octupole and quadrupole-quadrupole terms. From the table it is seen that short range terms are insufficient to reproduce satsfactorily the sublimation energy and the lattice frequencies and that the electrostatic interaction plays a relevant role both in the statics and the dynamics of the crystal. One important remark is that the various terms of the electrostatic potential can act specifically on certain vibrational modes.

The importance of the electrostatic terms will obviously be of primary importance in the dymanics of ionic crystals. In these cases it has also been found necessary to consider that the atomic charges or molecular multipoles are polarizable. As an example of the importance of polarization interactions in ionic crystals some results obtained for the $NaNO_2$ crystal [14] are shown in Table 3.

The deficiencies of the simplest forms of the atom-atom potential can be illustrated with reference to the CS_2 crystal [28]. Isotropic atom atom potentials of the Buckingham type were found unable to reproduce the crystal structure of this system. As a consequence imaginary vibrational frequencies were calculated. To reproduce the experimental data satisfactorily it was found necessary to express the electrostatic interaction by means of distributed multipoles (quadrupoles centered on the S atoms) and use atom atom interactions with more pronounced directional properties.

7. Measurement of the lifetimes and linewidths of vibrational states in crystals.

The relaxation of excess vibrational energy in crystals is of considerable importance in connection with both the chemical reactivity in solids and the understanding of the elementary processes responsible for the energy transfer in these materials. In the previous paragraphs the vibrating crystal has been considered in the harmonic approximation as a system of non interacting phonons. In this approximation when, by an appropriate excitation technique, some excess energy is put in a vibrational state of the crystal there could not be a mechanism for transfer of the excess energy in other vibrational states and the energy would remain in the originally populated state for an infinite time. By the uncertainty principle this infinite lifetime would correspond to zero linewidths of the vibrational transitions. Actually it is found that the spectral linewidth are not zero and consequently the lifetimes of the phonon states are finite. This implies that there are in the crystal hamiltonian terms that couple the phonon states. These terms are the anharmonic terms of the intermolecular potential.

Information on vibrational relaxation in crystals can be obtained by various experimental techniques in the frequency or in the time domain [29-31]. The techniques used so far have been spectroscopic techniques and therefore the information available concern the optically active $k = 0$ crystal vibrations. Frequency domain experiments measure the band profile, and in particular the full width at half maximum (FWHM) γ, while time domain experiments give the time decay curve of some optical signal. The two types of experiments are in principle equivalent since the Fourier transform of the time decay curve gives the band profile in the frequency domain. In some cases the relation can be simple. An exponential relaxation with time constant T_2 corresponds to a lorentzian profile with linewidth

$$\gamma = 1/(\pi c T_2) \tag{30}$$

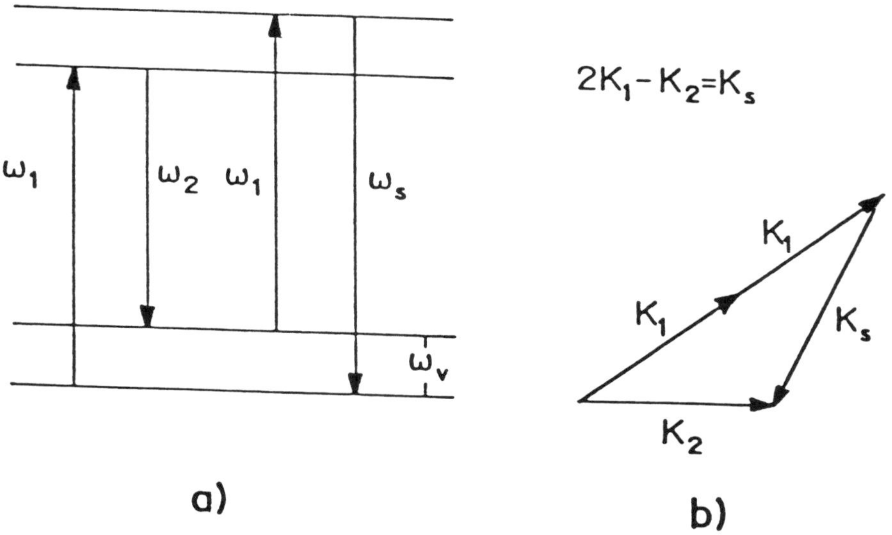

Fig. 6. Energy level diagram (a) and phase matching beams alignment (b) in a CARS experiment.

Phonon lifetimes or linewidths are very sensitive to temperature. Vibrational relaxation times in crystals at low temperature have been found to range tipically from 1 ps to 1 ns, with corresponding linewidths from 5 to 5 10^{-3} cm^{-1}. The measurement of phonon linewidths may thus require instruments with high resolution. Modern Fourier transform infrared spectrometers can attain a resolution of about 10^{-3} cm^{-1} and are thus particularly suited for the measurements of vibrational linewidths in crystals. However, infrared spectroscopy has not been used much for this purpose. The reason is likely due to experimental difficulties in preparing single crystals of suitable thickness for this kind of measurement. Infrared linewidths can more easily be measured of polycrystalline thin films, but various experiments have shown that in such cases unwanted inhomogeneous contributions to the line broadening are present.

For these reasons measurements of linewidths of phonon states in crystals have been mainly measured by Raman techniques. A spontaneous Raman instrument can reach a resolution of about .5 cm^{-1}. Therefore spontaneous Raman measurements are suitable at high temperature. The low temperature Raman linewidths have been satisfactorily measured using a Raman spectrometer operated with a single mode laser line and coupled with a Fabry-Perot interferometer [32]. The tandem Raman spectrometer is capable of a resolution better than 10^{-2} cm^{-1}.

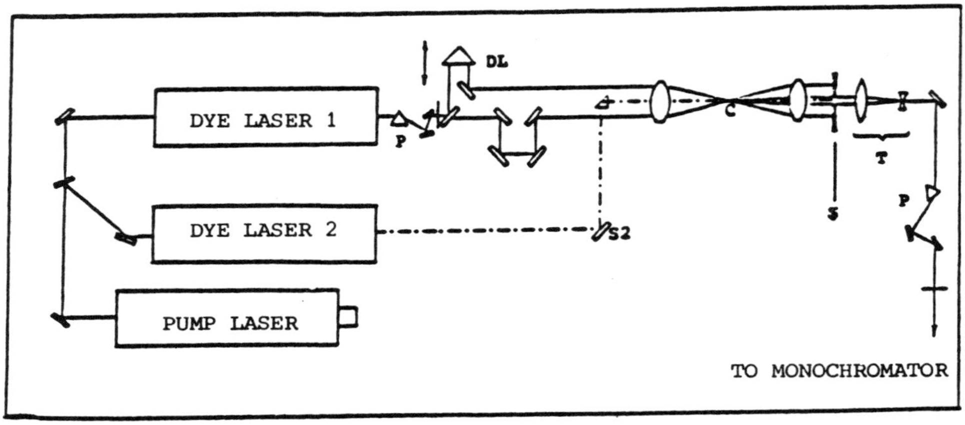

Fig. 7. Schematic experimental set up for CARS experiments.

Linewidths of phonon states can also be measured by non linear Raman techniques and in particular by cw CARS (Coherent Antistokes Raman Spectroscopy) [29,30]. The technique can be illustrated with reference to Fig. 6a. Two laser beams of frequency ω_1 and ω_2 are focussed on the crystal of interest. If the beams have a sufficient intensity they produce a coherent excitation of the crystal in the vibrational state ω_v. A third beam of frequency ω_1 produce an anstistokes signal from the vibrationally excited state at frequency $2\omega_1 - \omega_2$. The intensity of the antistokes signal is greatly enhanced when the resonance condition $\omega_1 - \omega_2 = \omega_v$ is fulfilled. The important fact is that the CARS experiment is coherent and an appropriate phase matching of the interacting beams is required as shown, for instance, in Fig. 6b. Under phase matching conditions the CARS signal is collected entirely in a single direction. Therefore in the CARS technique the sensitivity is several orders of magnitude higher than in spontaneous Raman spectroscopy. The CARS signal arises from the interaction of various beams in the sample and is due to the non linear cubic susceptibility of the material. The intensity of the CARS signal is proportional to the square of the third order susceptibility. The latter contains a resonant complex component and a non resonant contribution with possible interference between the two. Therefore, the CARS signal has a complex shape that must be analyzed in order to obtain useful spectroscopic information. Varying the frequency ω_1 while ω_2 is fixed the spectrum in the region of interest is scanned. The resolution of the CARS experiment is determined by the width of the ω_1 laser beam. If, as usually, this is the output of a dye laser pumped by a nanosecond source the resolution of the experiment can be of the order of .05 cm^{-1}.

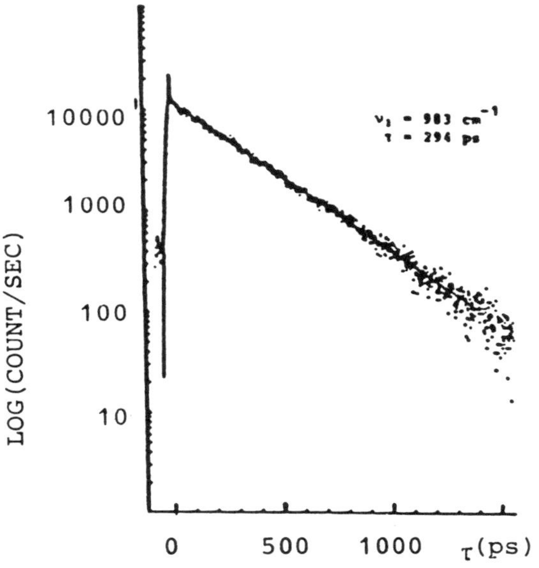

Fig.8. Vibrational decay curve for the ν_1 mode of K_2SO_4.

The lifetimes of phonon states can be measured directly with time resolved CARS techniques [29,30]. The relevant diagram for the experiment is again as in Fig. 6, but in the present case the laser beams have a duration short as compared to the phonon lifetimes, for instance are picosecond laser beams. The ω_1 and ω_2 beams producing the coherent excitation of the sample at $\omega_v = \omega_1 - \omega_2$ are focused on the crystal sample simultaneously at same initial time t = 0. The probe beam producing the coherent antistokes signal is shot on the crystal after a delay time τ. If the excited state ω_v relaxes with some kind of mechanism the CARS signal will be a function of the delay time. The CARS signal as a function of the delay time gives therefore a decay curve of the vibrational state population or coherence. An experimental set up is shown in Fig. 7 and an example of decay curve for a phonon in the K_2SO_4 crystal [33] is shown in Fig. 8. The sensitivity of the technique can be appreciated from the fact that the signal can be measured easily over various decades.

8. Anharmonicity and mechanisms for vibrational relaxation in crystals.

As already noted the observation that phonon states have non zero linewidths is a consequence of the fact that there are terms in the hamiltonian that couple the various harmonic normal coordinates.

Therefore the vibrating crystal cannot be considered as a system of non interacting phonons with infinite lifetimes. We consider therefore higher order terms in the expansion of the intermolecular potential. Considering only the lowest order terms the crystal hamiltonian can be written as [5]

$$H = H_0 + H_3 + H_4 \tag{31}$$

where H_0 is the quadratic harmonic part already discussed,

$$H_3 = \Sigma V_3(pk,p'k',p''k'') Q_{pk} Q_{p'k'} Q_{p''k''} \tag{32}$$

and

$$H_4 = \Sigma V_4(pk,p'k',p''k'',p'''k''') Q_{pk} Q_{p'k'} Q_{p''k''} Q_{p'''k'''} \tag{33}$$

H_3 and H_4 can be taken as sufficiently small as compared to H_0 to be considered as a perturbation of the harmonic hamiltonian. They can therefore be treated either with the usual perturbation techniques or by Green function methods. The effect of the anharmonic terms of the potential is to change the frequencies of the lattice vibrations. Up to fourth order the anharmonic phonon frequency can be expressed as

$$\omega_1 = \omega_1^{(0)} + \omega_1^{(3)} + \omega_1^{(4)} \tag{34}$$

where $\omega_1^{(3)}$ and $\omega_1^{(4)}$ are the cubic and quartic corrections, respectively. The anharmonic corrections are in general complex quantities. The real part of the correction gives the contribution to the phonon frequency shift and the imaginary part is the contribution to the halfwidth. The cubic contribution to the linewidth derives from processes where the phonon decomposes into two lower frequency phonons according the diagram of Fig. 9a. In addition we must consider processes were a phonon inelastically interacts with another phonon to produce a higher frequency phonon according to the diagram of Fig. 9b. These fission or fusion processes thus produce a change in the population of the initial state. The contribution to the linewidth for

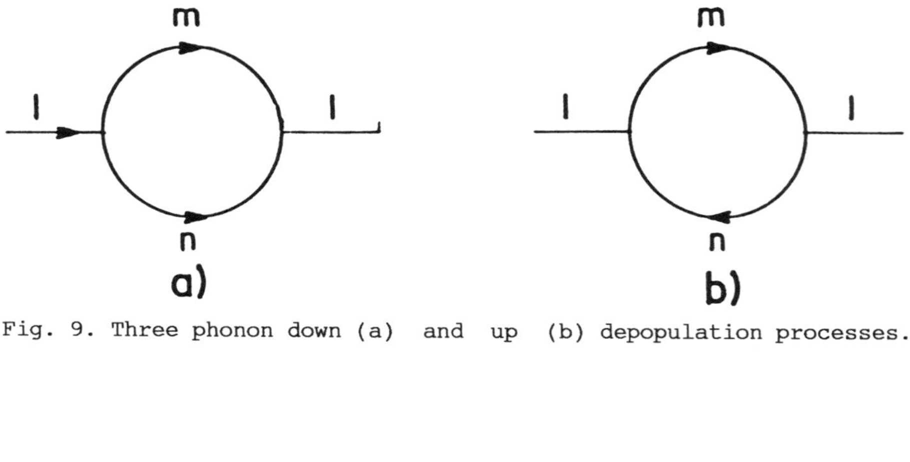

Fig. 9. Three phonon down (a) and up (b) depopulation processes.

down conversion processes is of the type

$$\gamma_l^d = \Sigma_{lmn} |V_3(lmn)|^2 (n_m + n_n + 1)\delta(\omega_l - \omega_m - \omega_n) \quad (35)$$

and for up conversion processes is

$$\gamma_l^u = \Sigma_{lmn} |V_3(lmn)|^2 (n_m - n_n)\delta(\omega_l + \omega_m - \omega_n) \quad (36)$$

where n is the phonon occupation number. The δ's in (35) and (36) indicate that energy and momentum must be conserved in the scattering processes. This implies limitations on the possible relaxation routes. The relaxation rates are therefore governed on one side by the strength of the anharmonic coefficients and on the other on the number of available routes (density of states). Since the occupation numbers are zero at T = 0 K it can be seen that only the down conversion processes contribute to the zero temperature linewidths. In the high temperature limit the cubic contribution to the linewidth will be linear with the temperature.

Quartic terms can also contribute to the depopulation of the phonon states and therefore to the broadening of the lines. These are processes where, for instance, a phonon dissociates into three lower frequency phonons. Since these processes involve the interaction of a larger number of phonons they are less probable, at least at moderate temperature.

In addition quartic terms of the hamiltonian contribute to the broadening of the lines through different processes that do not result in a change of the population of the phonon states but randomize the phase of the state. The simplest dephasing process due to the quartic terms is represented in the diagram of Fig. 10 where the frequency of the phonon l is modulated by energy exchange with all the phonons of the thermal bath. The pure dephasing contribution to the linewidth due to this diagram is of the type

$$\gamma_l^{(4)} = \Sigma_{mn} |V_4(llmn)|^2 n_n(n_m + 1)\delta(\omega_m - \omega_n) \quad (37)$$

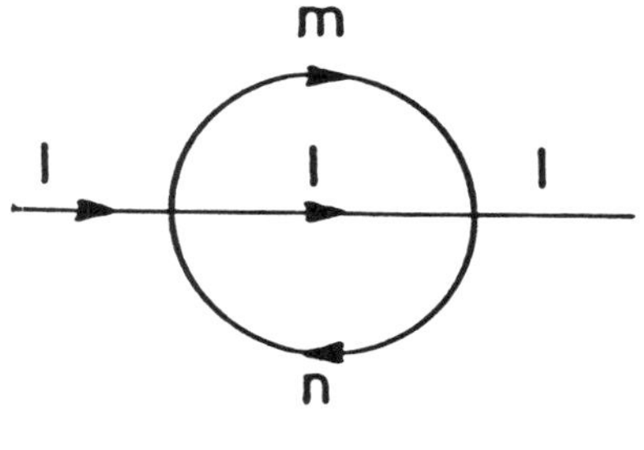

Fig. 10. Quartic pure dephasing processes of phonon states.

Although of higher order than the cubic processes this pure dephasing contribution to the linewidth can be substantial since the energy and momentum conservation are automatically fulfilled by all the phonons of the thermal bath if m=n. The pure dephasing linewidth is zero at T = 0 and at high temparature is quadratic in T.

There are other contributions to the linewidth arising from high order terms of the hamiltonian and corresponding to more complicated diagram but these cannot be analyzed here. The line broadening processes considered so far contribute to the homogeneous linewidth of the phonon states. In addition there can be inhomogeneous contributions due the presence of defects or impurities in the crystal. The latter are particularly important since real crystals are contaminated at least by isotopic impurities at natural abundance. Example of the inhomogeneous broadening will be discussed in the following section.

9. Experimental results and calculations of phonon linewidths.

One important topic that has been discussed in connection with experimental results is the dependence of the phonon linewidth on the excess vibrational energy. This problem can be illustrated with reference to the l-alanine crystal phonon linewidths [34] reported in Table 4. From calculations and experiments in various crystals it has

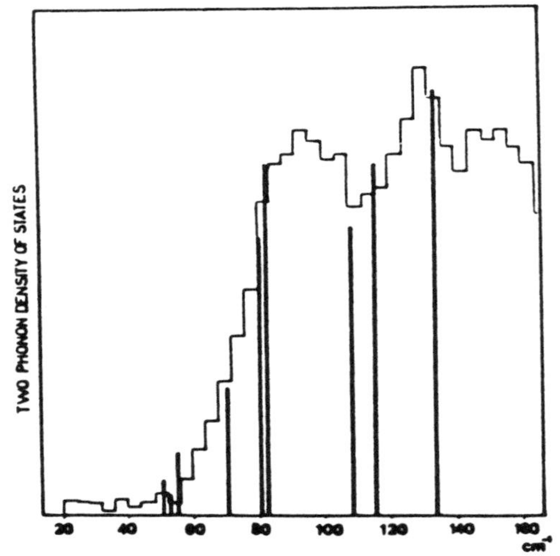

Fig. 11. Two-phonon density of states of crystalline naphthalene. The height of vertical bars is proportional to the phonon linewidth.

Table 4. Phonon vibrational relaxation times in the l-alanine crystal at 10 K.

Frequency (cm^{-1})	Decay time (ps)
40	>4000
50	1800
75	550
104	180
105	120
108	180
121	10
137	<10

Table 5. Vibrational relaxation times of internal modes in the benzene crystal at 4 K.

mode	symmetry	frequency (cm^{-1})	decay time (ps) pure crystal	natural crystal
ω_6	e_{2g}	606	2650	95
ω_{10}	e_{1g}	854	884	379
ω_1	a_g	991	62	39
ω_9	e_{2g}	1174	51	27
ω_8	e_{2g}	1584	23	16

been found that for the low frequency lattice phonons, corresponding to translation and librational motions, the phonon linewidths follow closely the behaviour of the two phonon density of states. This can be seen for the naphthalene crystal in Fig. 11 where the phonon linewidths at T = 0 K are compared with the two phonon density of states [35]. This means that the linewidth is determined essentially by the number of available decay channels that conserve the energy and momentum and not on the value of the coupling coefficients. The latter therefore can be considered approximately constant and independent from the type of phonon under consideration.

In some cases the phonon linewidths for lattice phonons have been measured as a function of temperature. As an example the results for the CO_2 crystal [36] are shown in Fig. 12. In various cases it has been found that the behaviour is linear at high temperatures. This means that the relaxation is only due to cubic depopulation processes. In other instances, however, it has been found that deviations from linearity occur. This is an indication for the onset of higher order or dephasing processes.

For the internal vibrations in crystals it has been found more difficult to establish a correlation between the linewidths and the excess vibrational energy [31]. An interesting case is that of the benzene crystal. The experimental results [37] are summarized in Table 5. It has been found that the relaxation times for the natural and isotopically pure crystal can be quite different. There are different mechanism by which impurities can affect the linewidth, like the reduction of symmetry, the relaxation of the selection rules, the creation of traps for the vibrational energy. The absence of a clear correlation with excess vibrational energy depends from two factors. First it is possible that the coupling coefficients for some of the internal phonons are particularly favorable. Secondly in the systems

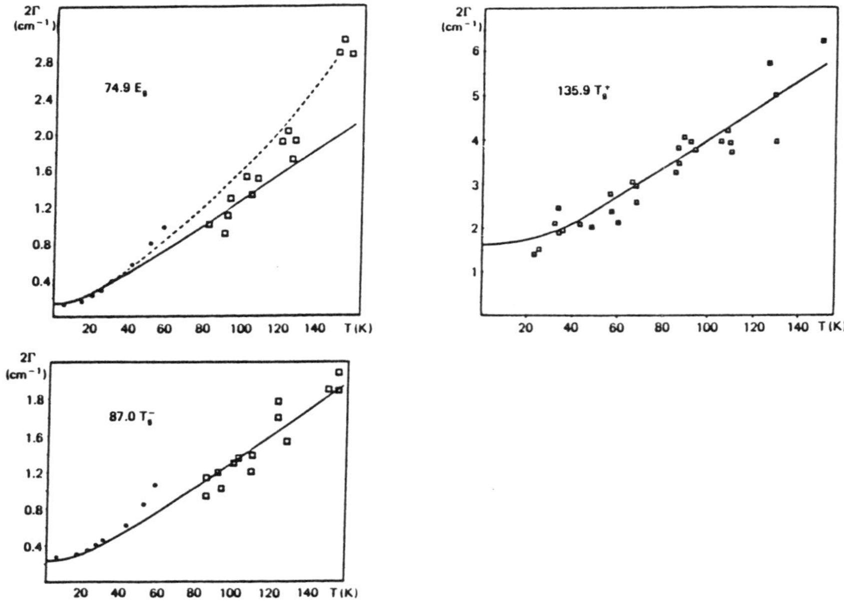

Fig. 12. Temperature dependence of phonon linewidths in crystal CO_2.

investigated so far the energy level diagram in the region of the internal vibrations is rather sparse as it is the case for the benzene crystal [31,37].

Finally, it should be mentioned that in several cases there have been indications that the perturbation treatment of the anharmonic interaction with the expansion limited to the cubic and quartic terms can be unsatisfactory. This is due to the fact that the perturbation expansion converges too slowly. In this respect the results of molecular dynamics simulation [38] have been found very useful. In molecular dynamics simulation the crystal is considerd classically and quantum effects, important at low temperature, are neglected. However, no expansion of the potential is performed and the anharmonicity of the intermolecular potential is fully considered. Comparison of results from lattice and molecular dynamics calculations give very useful information on the validity of the perturbation expansion.

10. REFERENCES

1 . Born, M. and Huang K. (1954) Dynamical Theory of Crystal Lattices, Clarendon Press, London

2 . Maradudin, A. A. ,Montroll, E. W., Weiss, G. H. ans Ipatova P. (1971) Theory of Lattice Dynamics in the Harmonic Approximation, Academic Press, New York

3 . Wilson, A. B., Decius, J. C. and Cross, P. C. (1955) Molecular Vibrations, McGraw Hill, New York

4 . Califano, S. (1976) Vibrational States, John Wiley, London

5 . Califano, S., Schettino, V. and Neto, N. (1981) Lattice Dynamics of Molecular Crystals, Springer Verlag, Berlin

6 . Califano, S. (Ed.) (1975) Lattice Dynamics and Intermolecular Forces, Academic Press, London

7 . Bratos, S. and Pick, R. (Eds) (1980) Vibrational Spectroscopy of Molecular Liquids and Solids, Plenum Press, New York

8 . Hirschfelder, J.O., Curtiss, C.F. and Bird, R.B. (1967) Molecular Theory of Gases and Liquids, John Wiley, New York

9 . Margenau, H. and Kestner, N. R. (1969) Theory of Intermolecular Forces, Pergamon Press, Oxford

10. Kihara, T. (1977) Intermolecular Forces,John Wiley, New York

11. Neto, N., Oehler, O. and Hexter, R., M. (1973) 'Vibrational Spectroscopy of Crystalline Iodoform', 58, 5661-5672

12. Davydov, D. A. (1962) Theory of Molecular Excitons, McGraw Hill, New York

13. Decius, J. C. and Hexter, R. M. (1977) Molecular Vibrations in Solids, McGraw Hill, New York

14. Castellucci, E. and Schettino, V. (1980) 'Lattice Dynamics of Ferro-electric Sodium Nitrite', J. Mol. Structure, 61, 191-194

15. Bokhenkov, E.L., Sheka, E.F., Dorner, B. and Natkaniec, I. (1977) 'Dispersion of Low Frequency Phonons in Deuterated Naphthalene Crystal', Solid State Commun., 23, 89-93

16. Righini, R., Califano.,S. and Walmsley, S.H. (1980) 'Calculated

Phonon Dispersion Curves for Fully Deuterated Naphthalene Crystal at Low Temperature', Chem. Phys., 50, 113-117
17. Lyubarskii, G. Ya. (1960) The Application of Group Theory in Physics, Pergamon Press, London
18. Bradley, C. J. and Cracknell, A. P. (1972) The Mathematical Theory of Symmetry in Solids, Pargamon Press, Oxford
19. Kovalev, O. V. (1965) Irreducible Representations of Space Groups, Academy of Sciences USSR, Kiev
20. Zak, J., Casher, A., Gluck, M. and Gur, Y., (1969) The Irreducible Representations of Space Groups, Benjamin, New York
21. Streitwolf (1971) Group Theory in Solid State Physics, McDonald, London
22. Schettino, V. and Califano, S. (1983) 'Infrared and Raman Spectra of Molecular Crystals', in Clark, R. J. H. and Hester, R. E. Advances in Infrared and Raman Spectroscopy, Heyden, London pp. 219-276
23. Hayes, W. and Loudon, R. (1978) Scattering of Light by Crystals, John Wiley, New York
24. Suzuki, M., Yokoyama, T. and Ito, M. (1968) 'Polarized Raman Spectra of Naphthalene and Anthracene Single Crystals', Spectrochim. Acta, 24A, 1091-1107
25. Marshall, W. and Lowesey, S.W. (1971) Theory of Thermal Neutron Scattering, Pergamon Press, Oxford
26. Schettino, V. and Califano, S. (1983) 'Lattice Dynamics and Interaction Potentials in Molecular Crystals', J. Mol. Structure, 100, 459-483
27. Righini, R., Neto, N., Califano, S. and Walmsley, S. H. (1978) 'Lattice Dynamics of Crystalline Ammonia and Deutero-ammonia', Chem. Phys., 33, 345-353
28. Burgos, E. and Righini, R. (1983) 'The Effect of Anisotropic atom-atom Interactions on the Crystal Structure and Lattice Dynamics of Solid CS_2', Chem. Phys. Letters, 96, 584-590
29. Velsko, S. and Hochstrasser, R. M. (1985) 'Studies of Vibrational Relaxation in Low-Temperature Molecular Crystals Using Coherent Raman Spectroscopy' J. Phys. Chem., 89, 2240-2253
30. Dlott, D. D. (1986) 'Optical Phonon Dynamics in Molecular Crystals' Ann. Rev. Phys. Chem., 37, 157-187
31. Califano, S. and Schettino, V. (1988) 'Vibrational Relaxation in Molecular Crystals' Int. Rev. Phys. Chem., 7, 19-57
32. Ranson, P., Ouillon, R. and Califano, S. (1984) 'Vibrational Relaxation in Molecular Crystals. High Resolution Raman Band Profiles of Some Naphthalene and Naphthalene-d_8 Phonons at Low Temperature', Chem. Phys., 86, 115-125
33. Angeloni, L., Righini, R., Castellucci, E., Foggi, P. and Califano, S. (1988) 'Temperature-Dependent Decay of Vibrational Excitons in K_2SO_4 Crystal Measured by Picosecond Time-Resolved CARS', Chem. Phys., 92, 983-988
34. Kosic, T. J., Cline Jr.,R. E. and Dlott, D. D. (1984) 'Picosecond Coherent Raman Investigation of the Relaxation of Low Frequency Vibrational Modes in Amino Acids and Peptides' J. Chem. Phys., 81, 4932-4949

35. Della Valle, R. G., Fracassi, P. F., Righini, R. and Califano, S. (1983) 'Anharmonic Processes in Molecular Crystals. Calculation of the Anharmonic Shifts, Bandwidths and Energy Decay Processes in Crystalline Naphthalene', Chem. Phys., 74, 179-195
36. Procacci, P., Righini, R. and Califano, S. (1987) 'Anharmonic Calculation of Bandwidths and Frequency Shifts in Crystalline CO_2' Chem.Phys., 116, 171-186
37. Trout, J., Velsko, S., Bozio, R., De Cola, P. L. and Hochstrasser, R. M. (1984) 'Nonlinear Raman Study of Line Shapes and Relaxation of Vibrational States of Isotopically Pure and Mixed Crystals of Benzene', J. Chem. Phys., 81, 4746-4759
38. Cardini, C., Procacci, P. and Righini, R. (1987) 'Molecular Dynamics and Anharmonic Effects in the Phonon Spectra of Solid Carbon Dioxide', Chem. Phys., 117, 355-366

A PLAIN INTRODUCTION TO JAHN–TELLER COUPLINGS

C.J. BALLHAUSEN
University of Copenhagen
Institute for Physical Chemistry
Universitetsparken 5
DK–2100 Copenhagen Ø
Denmark

ABSTRACT. The basic ideas, mathematical structure and manifestations of the first order Jahn–Teller couplings are introduced with special emphasize on inorganic complexes.

The fundamentals of Jahn–Teller Couplings.

The Jahn–Teller couplings are an outcome of the approximations introduced by the Born–Oppenheimer separation of the electronic (el) and nuclear (nuc) motions in a polyatomic molecule. For a non–degenerate electronic state the so–called crude Born–Oppenheimer wavefunctions take the form $\Psi = \psi_{el}(q,0)\chi_{nuc}(Q)$ where q are the electronic coordinates, Q the nuclear coordinates and $Q = 0$ indicates that the electronic wavefuntion is evaluated at the stable nuclear equilibrium conformation.

The electronic energies at $Q = 0$ are given by

$$\mathcal{H}^{\circ}\psi_{el}(q,0) = W^{\circ}\psi_{el}(q,0) \qquad (1)$$

with

$$\mathcal{H}^{\circ} = \hat{T}_{el} + V(q,0) \qquad (2)$$

Let the Q's stand for nuclear displacement coordinates away from the equilibrium positions of the N nuclei. Expanding the potential V(q,Q) in a Taylor Series around $Q = 0$ leads to first order in Q to

$$\mathcal{H} = \mathcal{H}^{\circ} + \sum_{i=1}^{3N} \left[\frac{\partial V}{\partial Q_i}\right]_{o} Q_i. \qquad (3)$$

We now introduce the nuclear symmetry coordinates ξ_j of the molecule, that is such linear combinations of the Q's which span the irreducible representations of the molecular point group peculiar to $Q = 0$. With $\xi_j = \sum_{i=1}^{3N} C_{ij}Q_i$ and $\mathcal{H} = \mathcal{H}^{\circ} + \mathcal{H}^{(1)}$,

$$\mathcal{H}^{(1)} = \sum_{j=1}^{3N} \left[\frac{\partial V}{\partial \xi_j}\right]_{o} \xi_j \qquad (4)$$

C. D. Flint (ed.), Vibronic Processes in Inorganic Chemistry, 53–78.
© 1989 by Kluwer Academic Publishers.

We observe that $\left[\frac{\partial V}{\partial \xi_j}\right]_0$ is solely a function of the electronic coordinates. For a non–degenerate electronic state we have then using first order perturbation theory

$$W_{el}^{(1)} = \int \psi_{el}^{o}(q)\mathcal{H}^{(1)}\psi_{el}^{o}(q)dq$$

$$= \sum_{j=1}^{3N} \xi_j \int \psi_{el}^{o}(q) \left[\frac{\partial V}{\partial \xi_j}\right]_0 \psi_{el}^{o}(q)dq$$

$$= \sum_{j=1}^{3N} c_j \xi_j$$

The c_j numbers are however all zero because at $Q = 0$ we are at the equilibrium point, and the electronic potential energy cannot therefore contain terms being linear in ξ_j.

Consider now the case in which the nuclear conformation point $Q = 0$ lead to electronic degeneracy. In other words $\psi_1(q,0)$, $\psi_2(q,0) \cdots \psi_k(q,0)$ all have the same energy W_{el}^{o}. Clearly this degeneracy is symmetry determined, being associated with the special nuclear arrangement $Q = 0$. In order to calculate the first order electronic perturbation energy we have therefore to use degenerate perturbation theory, that is, evaluate matrix elements exemplified by

$$\langle \psi_m^{o}(q) \mid \mathcal{H}^{(1)} \mid \psi_n^{o}(q) \rangle = \sum_{j=1}^{3N} \xi_j \sum_{m,n} c_{jmn} \, .$$

$$c_{jmn} = \int \psi_m^{o}(q) \left[\frac{\partial V}{\partial \xi_j}\right]_0 \psi_n^{o}(q)\, dq \tag{5}$$

However, in contrast to before Jahn and Teller proved the theorem that in the degenerate case all the c_{jmn} constants will not be zero, and the system will contain terms in the electronic energy which are linear in one or more non–symmetry preserving displacement coordinates ξ_j. We can therefore minimize the electronic energy with respect to ξ_j and obtain a new nuclear conformation having lower electronic energy. "Moving down" along the non–symmetry preserving nuclear coordinate has therefore taken the system away from the $Q = 0$ conformation; the symmetry which determined the electronic degeneracy has gone, and the electronic degeneracy is done away with. The content of the Jahn–Teller theorem is therefore: Electronic degeneracy in a molecule destroys the symmetry on which it is based. The only exemptions are degeneracies in linear molecules and Kramers degeneracies.

Physically one can understand the reason for the unstable nuclear configuration–points by reflecting that in an electronically degenerate state, the electronic charge cloud has different orientations for the different components. The linear terms in the nuclear displacement coordinates cannot, therefore, be eliminated simultaneously for all of the degenerate electronic components.

The Active Vibrations.

The question which non–symmetry preserving nuclear vibrations are active in a Jahn–Teller coupling can be answered by means of group–theory. The perturbation operator $\mathcal{H}^{(1)}$ must transform like a totally symmetric function because the energy must be independent of the chosen coordinate system. Only if ξ_k transform like $\left[\frac{\partial V}{\partial \xi_k}\right]_o$ will the product in eq. (4) be totally symmetric. But the matrix element $\langle \psi_m \mid \left[\frac{\partial V}{\partial \xi_k}\right]_o \psi_n \rangle$ is also only different from zero provided the function $\psi_m \left[\frac{\partial V}{\partial \xi_k}\right]_o \psi_n$ is a totally symmetric function. This is only the case when $\left[\frac{\partial V}{\partial \xi_k}\right]_o$ transform like $\psi_m(q)\psi_n(q)$. Hence the active vibrations can be found by looking at the representations the symmetric direct products $\psi_m(q)\psi_n(q)$ span.

In the octahedral and tetrahedral groups the degenerate representatives are of E, T_1 and T_2 type. Take the E representation spaned by $(3z^2-r^2)$ and (x^2-y^2). The three products $(x^2-y^2)(3z^2-r^2)$, $(3z^2-r^2)(3z^2-r^2)$ and $(3z^2-r^2)(x^2-y^2)$ span a reducible representation which can be broken down in the irreducible representations of E and A_1. Hence only an ϵ vibration is Jahn–Teller active to first order in an E state.The α_1 vibration is of course symmetry preserving, and only interferes with an ϵ vibration in second and higher order couplings. In a similar way we find that for the T_1 and T_2 states the Jahn–Teller active vibrations are ϵ and τ_2.

The Static $E \otimes \epsilon$ Jahn–Teller Coupling.

In this case we take the doubly degenerate electronic set spanning E as $\overset{o}{\psi}_1$ and $\overset{o}{\psi}_2$ and orient them after the molecular symmetry operation $\hat{C}_2'$:

$$\hat{C}_2' \begin{bmatrix} \overset{o}{\psi}_1 \\ \overset{o}{\psi}_2 \end{bmatrix} = \begin{bmatrix} 1 & 0 \\ 0 & -1 \end{bmatrix} \begin{bmatrix} \overset{o}{\psi}_1 \\ \overset{o}{\psi}_2 \end{bmatrix} \tag{6}$$

The doubly degenerate ϵ vibration spaned by the coordinates $\xi_{1\epsilon}$ and $\xi_{2\epsilon}$ are oriented in the same way

$$\hat{C}_2' \begin{bmatrix} \xi_{1\epsilon} \\ \xi_{2\epsilon} \end{bmatrix} = \begin{bmatrix} 1 & 0 \\ 0 & -1 \end{bmatrix} \begin{bmatrix} \xi_{1\epsilon} \\ \xi_{2\epsilon} \end{bmatrix} \tag{7}$$

With $$\mathcal{H}^{(1)} = \left[\frac{\partial V}{\partial \xi_{1\epsilon}}\right]_o \xi_{1\epsilon} + \left[\frac{\partial V}{\partial \xi_{2\epsilon}}\right]_o \xi_{2\epsilon} \tag{8}$$

we have for instance

$$\langle \psi_1^o \mid \mathcal{H}^{(1)} \mid \psi_1^o\rangle = a\, \xi_{1\epsilon} + b\, \xi_{2\epsilon} \tag{9}$$

From

$$C_2'\langle \psi_1^o \mid \mathcal{H}^{(1)} \mid \psi_1^o\rangle = a\, \xi_{1\epsilon} - b\, \xi_{2\epsilon} \tag{10}$$

we get, however, comparing (9) and (10) that b = 0. Similarly we have

$$\langle \psi_2^o \mid \mathcal{H}^{(1)} \mid \psi_2^o\rangle = c\, \xi_{1\epsilon} \tag{11}$$

and
$$\langle \psi_2^o \mid \mathcal{H}^{(1)} \mid \psi_1^o\rangle = \langle \psi_1^o \mid \mathcal{H}^{(1)} \mid \psi_2^o\rangle = d\, \xi_{2\epsilon}$$

From the fact that we also have three–fold symmetry $\hat{C}_3\langle \psi_2^o \mid \mathcal{H}^{(1)} \mid \psi_1^o\rangle = \langle -\tfrac{1}{2}\psi_2^o + \tfrac{\sqrt{3}}{2}\psi_1^o \mid \mathcal{H}^{(1)} \mid -\tfrac{1}{2}\psi_1^o - \tfrac{\sqrt{3}}{2}\psi_1^o\rangle \equiv d\,(-\tfrac{1}{2}\xi_{2\epsilon} + \tfrac{\sqrt{3}}{2}\xi_{1\epsilon})$ we get $a = -c = d$. Treating $\xi_{1\epsilon}$ and $\xi_{2\epsilon}$ as parameters the secular equation becomes then including the harmonic potential term $\tfrac{1}{2}k_\epsilon(\xi_{1\epsilon}^2 + \xi_{2\epsilon}^2)$ of $\mathcal{H}^o$

$$\begin{vmatrix} W^o + \tfrac{1}{2}k_\epsilon\left[\xi_{1\epsilon}^2 + \xi_{2\epsilon}^2\right] - c\xi_{1\epsilon} - V & c\xi_{2\epsilon} \\ c\xi_{2\epsilon} & W^o + \tfrac{1}{2}k_\epsilon\left[\xi_{1\epsilon}^2 + \epsilon_{2\epsilon}^2\right] + c\xi_{1\epsilon} - V \end{vmatrix} = 0 \tag{12}$$

from which we find the potential sheets

$$V = W^o + \tfrac{1}{2}k_\epsilon(\xi_{1\epsilon}^2 + \epsilon_{2\epsilon}^2) \pm c\sqrt{\xi_{1\epsilon}^2 + \xi_{2\epsilon}^2} \tag{13}$$

Changing to cylindrical coordinates $\xi_{1\epsilon} = \rho\cos\varphi$, $\xi_{2\epsilon} = \rho\sin\varphi$

$$V = W^o + \tfrac{1}{2}k_\epsilon\rho^2 \pm c\rho \tag{14}$$

Minimizing the lower potential surface with respect to ρ gives $\rho_o = \frac{c_\epsilon}{k_\epsilon}$.

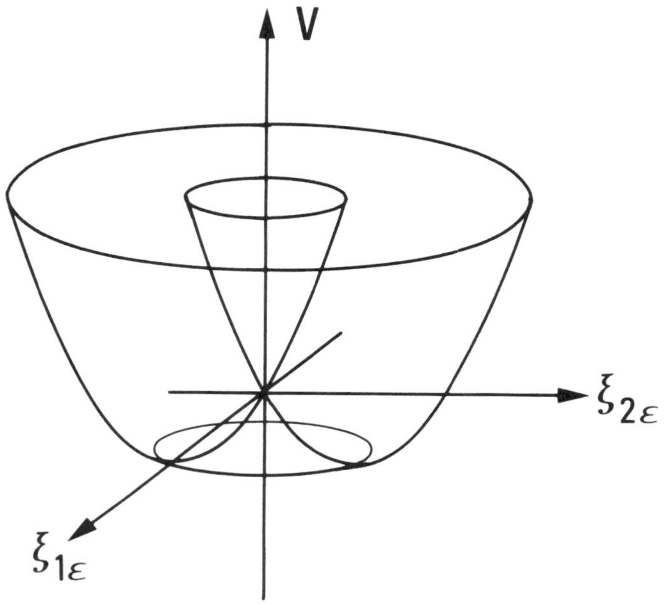

Figure 1. The "Mexican Hat" potential for the first order $E \otimes \epsilon$ coupling.

The lower potential is therefore stabilized with

$$\Delta V_{J-T} = -\frac{c_\epsilon^2}{2k_\epsilon} \tag{15}$$

from which we define the dimensionless Jahn–Teller $E \otimes \epsilon$ coupling factor D, being measured in harmonic vibrational quanta $\hbar\omega$, $\omega = \sqrt{k_\epsilon}$ in mass–normalized coordinates

$$D = \frac{c_\epsilon^2}{2k_\epsilon \hbar\omega} \tag{16}$$

The D factor is the same as the Huang–Rhys factor S used in bandshape analysis. The electronic energy has therefore gone down by ΔV_{J-T}, and the electronic degeneracy done away with. It is clear that for $D<1$ the structural effects will be small because then the distortion ρ_0 is of the same order of magnitude as the vibrational amplitude of the nuclei.

This static Jahn–Teller coupling calculation has then used the vibrational coordinates as parameters and not as dynamical variables. The static Jahn–Teller calculations therefore give us the potential surface for the systems; the energy levels are calculated by treating the vibrational coordinates as dynamical variables.

The dynamic $E \otimes \epsilon$ Coupling.

Including the kinetic energy of the nuclei in the molecular Hamiltonian we have

$$\mathcal{H} = \overset{o}{\mathcal{H}}_{el} - \frac{\hbar^2}{2}\left[\frac{\partial^2}{\partial\xi^2_{1\epsilon}} + \frac{\partial^2}{\partial\xi^2_{2\epsilon}}\right] + \tfrac{1}{2}k_\epsilon(\xi^2_{1\epsilon} + \xi^2_{2\epsilon}) + \mathcal{H}^{(1)} \tag{17}$$

$$\mathcal{H}^{(1)} = \left[\frac{\partial V}{\partial \xi_{1\epsilon}}\right]_0 \xi_{1\epsilon} + \left[\frac{\partial V}{\partial \xi_{2\epsilon}}\right]_0 \xi_{2\epsilon} \tag{18}$$

As a variational wavefunction we take in the spirit of the crude Born–Oppenheimer approximation

$$\Psi = \overset{o}{\psi}_1(q)\chi_1(\xi_{1\epsilon},\xi_{2\epsilon}) + \overset{o}{\psi}_2(q)\chi_2(\xi_{1\epsilon},\xi_{2\epsilon}) \tag{19}$$

where in a certain sense we may look at the vibrational wavefunctions χ_1 and χ_2 as the variational "parameters" to be determined. From the Schrödinger equation $\mathcal{H}\,\Psi = W\,\Psi$ we get by insertion of (17), (18) and (19) and integrating up over the electronic coordinates

$$\begin{bmatrix} \overset{o}{W} + \overset{o}{\mathcal{H}}_{vib} - c\xi_1 & c\xi_2 \\ c\xi_2 & \overset{o}{W} + \overset{o}{\mathcal{H}}_{vib} + c\xi_1 \end{bmatrix}\begin{bmatrix}\chi_1 \\ \chi_2\end{bmatrix} = W\begin{bmatrix}\chi_1 \\ \chi_2\end{bmatrix} \tag{20}$$

where we have droped the subscript ϵ and c as before, eq. (11), is the Jahn–Teller coupling constant. Applying the unitary transformation

$$U = \sqrt{\frac{1}{2}}\begin{bmatrix} e^{-\frac{i\pi}{4}} & -e^{-\frac{i\pi}{4}} \\ -e^{\frac{i\pi}{4}} & -e^{\frac{i\pi}{4}} \end{bmatrix}, \quad UU^* = 1$$

we get putting $\overset{o}{W} = 0$

$$\begin{bmatrix} \overset{o}{\mathcal{H}}_{vib} & c(\xi_1 - i\xi_2) \\ c(\xi_1 + i\xi_2) & \overset{o}{\mathcal{H}}_{vib} \end{bmatrix}\begin{bmatrix}\chi_1 \\ \chi_2\end{bmatrix} = W\begin{bmatrix}\chi_1 \\ \chi_2\end{bmatrix} \tag{21}$$

or changing to cylindrical coordinates $\xi_1 = \rho\cos\varphi$ and $\xi_2 = \rho\sin\varphi$

$$\begin{bmatrix} \mathcal{H}^{o}_{vib} & c\rho e^{-i\varphi} \\ c\rho e^{i\varphi} & \mathcal{H}^{o}_{vib} \end{bmatrix} \begin{bmatrix} \chi_1 \\ \chi_2 \end{bmatrix} = W \begin{bmatrix} \chi_1 \\ \chi_2 \end{bmatrix} \tag{22}$$

where in dimensionless units

$$\mathcal{H}^{o}_{vib} = -\tfrac{1}{2}\frac{1}{\rho}\frac{\partial}{\partial\rho}\left(\rho\frac{\partial}{\partial\rho}\right) - \tfrac{1}{2}\frac{1}{\rho^2}\frac{\partial^2}{\partial\varphi^2} + \tfrac{1}{2}\rho^2 \tag{23}$$

We can write $\mathcal{H}$ as

$$\mathcal{H} = \mathcal{H}^{o}_{vib}\begin{bmatrix} 1 & 0 \\ 0 & 1 \end{bmatrix} + c\rho\begin{bmatrix} 0 & e^{-i\varphi} \\ e^{i\varphi} & 0 \end{bmatrix}. \tag{24}$$

with eigenvalues of $\mathcal{H}^{o}_{vib}$ being $W_v = (v + 1)\hbar\omega$. $v = 0, 1, 2\cdots$. It is not difficult to prove that the operator $\hat{j}_z$

$$\hat{j}_z = -i\frac{\partial}{\partial\varphi}\begin{bmatrix} 1 & 0 \\ 0 & 1 \end{bmatrix} + \tfrac{1}{2}\begin{bmatrix} 1 & 0 \\ 0 & -1 \end{bmatrix} \tag{25}$$

commutes with $\mathcal{H}$:

$$[\mathcal{H}, \hat{j}_z] = 0 \tag{26}$$

$\hat{j}_z$ is therefore for the first order $E \otimes \epsilon$ Jahn–Teller coupling a constant of the motion. The eigenvalues of $\hat{j}_z$ are easily found

$$\hat{j}_z\phi_j(\varphi) = (m \pm \tfrac{1}{2})\,\phi_j(\varphi) \qquad m = 0, \pm1, \pm2\cdots \tag{27}$$

or $$j = m \pm \tfrac{1}{2} = \pm\tfrac{1}{2}, \pm\tfrac{3}{2}, \pm\tfrac{5}{2}\cdots$$

each level being twofold degenerate. j therefore measures the total angular momentum (vibrational plus electronic, unit $\hbar$) of the system. In the first order Jahn–Teller $E \otimes \epsilon$ coupling j is therefore a <u>good quantum number</u>.

The vibronic wavefunctions are then

$$\Psi = |\,\psi_+\rangle\, f_+(\rho)\, e^{i(j-\frac{1}{2})\varphi} + |\,\psi_-\rangle\, f_-(\rho)\, e^{i(j+\frac{1}{2})\varphi}$$

$$|\,\psi_+\rangle = -\frac{i}{\sqrt{2}}(\psi^{o}_1 + i\psi^{o}_2)$$
$$|\,\psi_-\rangle = \frac{i}{\sqrt{2}}(\psi^{o}_1 - i\psi^{o}_2)$$

With $\Psi(\rho,\varphi) = R(\rho)e^{ij\varphi}$, j being a half–integer, a rotation of $\varphi = 4\pi$ is needed before the phase of $\phi_j(\varphi)$ is +1 again. A "physical" rotation by $2\pi/3$ around the three–fold axis $\hat{C}_3$ therefore corresponds to $\varphi \to \varphi + \frac{4\pi}{3}$ and

$$\hat{C}_3 e^{ij\varphi} = \hat{C}_3 \left[e^{\frac{i\varphi}{2}}\right]^{2j} = \left[e^{\frac{2\pi i}{3}}\right]^{2j} e^{ij\varphi}$$

With $(e^{ij\varphi})^* = e^{-ij\varphi}$ the trace under $\hat{C}_3$ of the complex conjugated pair of wavefunctions is $2\cos\frac{2\pi}{3}2j$. The $2j = \text{mod. } 3$ levels therefore transform like (a_1, a_2), the others like e.

In order to calculate the energies of the vibronic levels we can, for <u>D small</u>,use second order perturbatian theory in the following way. Introducing an exponential transformation of the Hamiltonian $\mathcal{H} = \mathcal{H}^o + \mathcal{H}^{(1)}$

$$\tilde{\mathcal{H}} = e^{i\hat{S}}\,\mathcal{H}e^{-i\hat{S}} = \mathcal{H} + i[\hat{S},\mathcal{H}] + \tfrac{1}{2}\, i^2[\hat{S},[\hat{S},\mathcal{H}]] + \cdots \tag{28}$$

we choose $\hat{S}$ such that

$$\mathcal{H}^{(1)} + i[\hat{S},\mathcal{H}^o] = 0 \tag{29}$$

Therefore

$$\tilde{\mathcal{H}} = \mathcal{H}^o + \tfrac{1}{2}[\hat{S},[\hat{S},\mathcal{H}^o]] \tag{30}$$

Making use of the Pauli–matrices

$\sigma_1 = \begin{bmatrix}0 & 1\\ 1 & 0\end{bmatrix}$, $\sigma_2 = \begin{bmatrix}0 & -i\\ i & 0\end{bmatrix}$, $\sigma_3 = \begin{bmatrix}1 & 0\\ 0 & -1\end{bmatrix}$ and $I = \begin{bmatrix}1 & 0\\ 0 & 1\end{bmatrix}$ we have in dimensionless units

$$\mathcal{H}^o = \left[-\tfrac{1}{2}\left[\frac{\partial^2}{\partial\xi_2^2} + \frac{\partial^2}{\partial\xi_2^2}\right] + \tfrac{1}{2}\left[\xi_1^2 + \xi_2^2\right]\right]\hbar\omega \tag{31}$$

and from (21)

$$\mathcal{H}^{(1)} = (2D)^{\frac{1}{2}}\,(\xi_1\sigma_1 + \xi_2\sigma_2) \tag{32}$$

From (29), (31) and (32) we find

$$\hat{S} = i(2D)^{\frac{1}{2}}\left[\frac{\partial}{\partial\xi_1}\sigma_1 + \frac{\partial}{\partial\xi_2}\sigma_2\right] \tag{33}$$

leading to

$$\tilde{\mathcal{H}} = \mathcal{H}^o - 2D(I - i\sigma_3\frac{\partial}{\partial\varphi}) + \cdots \tag{34}$$

With the eigenvalues of $\mathcal{H}^{o}$ being $(v + 1)\hbar\omega$ we get

$$W = (v + 1 - 2D(1 - m))\hbar\omega \tag{35}$$

The motions for weak coupling can be described as those of a two–dimensional harmonic oscillator in the bottom of a two–dimensional, nearly harmonic potential. As the coupling increases the motion on the lower surface of figure 1 is that of a "radial" vibration together with a rotation around the cusp.

For strong coupling, D>>1, it is convenient to adopt a representation in which the potential is diagonal. Using an unitary transformation

$$U = \frac{1}{\sqrt{2}}\begin{bmatrix} e^{\frac{i\varphi}{2}} & e^{-\frac{i\varphi}{2}} \\ e^{\frac{i\varphi}{2}} & -e^{-\frac{i\varphi}{2}} \end{bmatrix} \tag{36}$$

The Hamiltonian (20) is transformed into

$$\mathcal{H} = \begin{bmatrix} -\frac{1}{2}\left[\frac{\partial^2}{\partial\rho^2} + \frac{1}{\rho}\frac{\partial}{\partial\rho} + \frac{1}{\rho^2}\frac{\partial^2}{\partial\varphi^2}\right] + \frac{1}{8\rho^2} + \frac{1}{2}\rho^2 + \sqrt{2D}\rho & \frac{i}{2\rho^2}\frac{\partial}{\partial\varphi} \\ \frac{i}{2\rho^2}\frac{\partial}{\partial\varphi} & -\frac{1}{2}\left[\frac{\partial^2}{\partial\rho^2} + \frac{1}{\rho}\frac{\partial}{\partial\rho} + \frac{1}{\rho^2}\frac{\partial^2}{\partial\varphi^2}\right] + \frac{1}{8\rho^2} + \frac{1}{2}\rho^2 - \sqrt{2D}\rho \end{bmatrix} \tag{37}$$

substituting

$$\begin{bmatrix} \chi_+ \\ \chi_- \end{bmatrix} = e^{ij\varphi}\rho^{-\frac{1}{2}}\begin{bmatrix} \eta_+(\rho) \\ \eta_-(\rho) \end{bmatrix} \tag{38}$$

$$\begin{bmatrix} -\frac{1}{2}\frac{\partial^2}{\partial\rho^2} + \frac{j^2}{2\rho^2} + \frac{1}{2}\rho^2 + \sqrt{2D}\rho - W & -\frac{j}{2\rho^2} \\ -\frac{j}{2\rho^2} & -\frac{1}{2}\frac{\partial^2}{\partial\rho^2} + \frac{j^2}{2\rho^2} + \frac{1}{2}\rho^2 - \sqrt{2D}\rho - W \end{bmatrix}\begin{bmatrix} \eta_+ \\ \eta_- \end{bmatrix} = 0 \tag{39}$$

For small values of j the lower surface potential V_- has its minimum at $\rho_0 \approx \sqrt{2D+j}$. With D large the coupling term for that value of ρ is $\approx j/4D$ which is very small. The low energy levels on the lower potential sheet can then be found by shifting the origin of ρ from zero to ρ_0, and expanding V_- around ρ_0. Retaining only terms to order D^{-1} we get

$$W_- = \left(-D + v + \tfrac{1}{2} + \frac{j^2}{4D}\right)(\hbar\omega) \quad v = 0, 1, 2, \cdots \tag{40}$$

Each vibrationel level on the lower branch therefore appear with "rotational fine structure".

The potential for the upper branch is given by

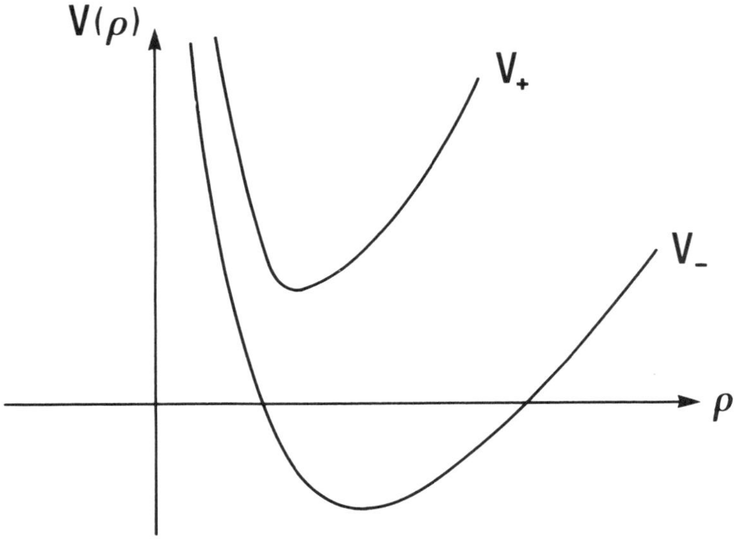

Figure 2. Upper and lower potential curves for D>>1.

$$V_+ = \frac{j^2}{2\rho^2} + \tfrac{1}{2}\rho^2 + \sqrt{2D}\rho \tag{41}$$

which for D large has its minimum at $\rho \approx \left[\frac{j^2}{\sqrt{2D}}\right]^{1/3}$. Expanding V_+ from that point in a Taylor series leads assuming $\left[\frac{4D^2}{j^2}\right]^{1/3} >> 1$ after a lot of plain tedious algebra to

$$W_+ = (2D)^{1/3}\left[\frac{3}{2}\,|j|^{2/3} + (v + \tfrac{1}{2})\,\frac{\sqrt{3}}{|j|^{1/3}}\right](\hbar\omega) \tag{42}$$

$$j = \pm\tfrac{1}{2}, \pm\tfrac{3}{2} \pm \tfrac{5}{2} \cdots$$
$$v = 0, 1, 2, \cdots$$

The full energy diagram for $E \otimes \epsilon$, obtained by numerical calculations are given in Figure 3 as a function of D. Great efforts have gone into obtaining analytical solutions, and these seem now to have succeeded.

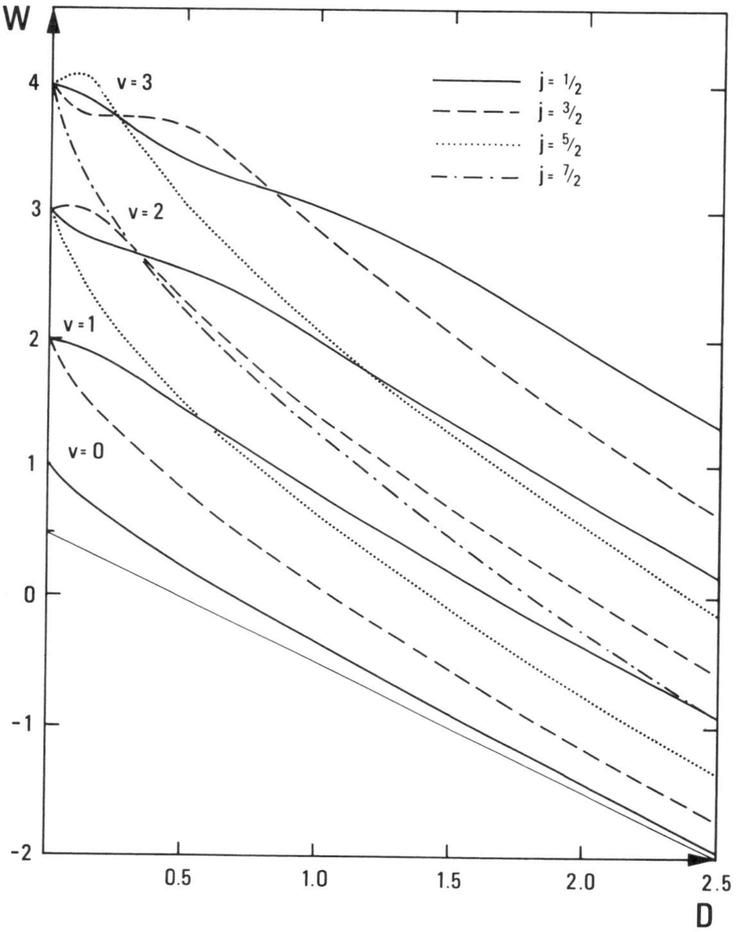

Figure 3. The complete $E \otimes \epsilon$ energy diagram as a function of D.

The $T \otimes \epsilon$ coupling.

For a three–fold degenerate electronic state be it a T_1 or T_2 representation interacting with a doubly degenerate ϵ vibrator, we have again

$$\mathcal{H}^{(1)} = \left[\frac{\partial V}{\partial \xi_1}\right]_o \xi_1 + \left[\frac{\partial V}{\partial \xi_2}\right]_o \xi_2$$

Demanding invariance of the matrix–elements under the symmetry operations of the group, the $\hat{C}_3$ operator gives easily with c being the Jahn–Teller coupling coefficient inside the T manifold

$$\mathcal{H}^{(1)} = c \begin{bmatrix} -\frac{1}{2}\xi_1 - \frac{\sqrt{3}}{2}\xi_2 & 0 & 0 \\ 0 & -\frac{1}{2}\xi_1 + \frac{\sqrt{3}}{2}\xi_2 & 0 \\ 0 & 0 & \xi_1 \end{bmatrix} \tag{43}$$

Adding $\mathcal{H}^{\text{o}} = \hat{T}_{\text{nuc}} + \frac{1}{2}k(\xi_1^2 + \xi_2^2)$ to the diagonal and closing the squares lead to

$$\mathcal{H}_1 = \hat{T}_{\text{nuc}} + \tfrac{1}{2}k\left[\left(\xi_1 - \frac{c}{2k}\right)^2 + \left(\xi_2 - \frac{\sqrt{3}c}{2k}\right)^2\right] - \frac{c^2}{2k} \tag{44a}$$

$$\mathcal{H}_2 = \hat{T}_{\text{nuc}} + \tfrac{1}{2}k\left[\left(\xi_1 - \frac{c}{2k}\right)^2 + \left(\xi_2 + \frac{\sqrt{3}c}{2k}\right)^2\right] - \frac{c^2}{2k} \tag{44b}$$

$$\mathcal{H}_3 = \hat{T}_{\text{nuc}} + \tfrac{1}{2}k\left[\left(\xi_1 + \frac{c}{k}\right)^2 + \xi_2^2\right] - \frac{c^2}{2k} \tag{44c}$$

Changing the origin of the coordinate system to the minima of the three potential surfaces we have

$$\mathcal{H}_i = \hat{T}_{i\ \text{nuc}} + \tfrac{1}{2}k(Q_{i1}^2 + Q_{i2}^2) - \frac{c^2}{2k} \quad i = 1, 2, 3, \tag{45}$$

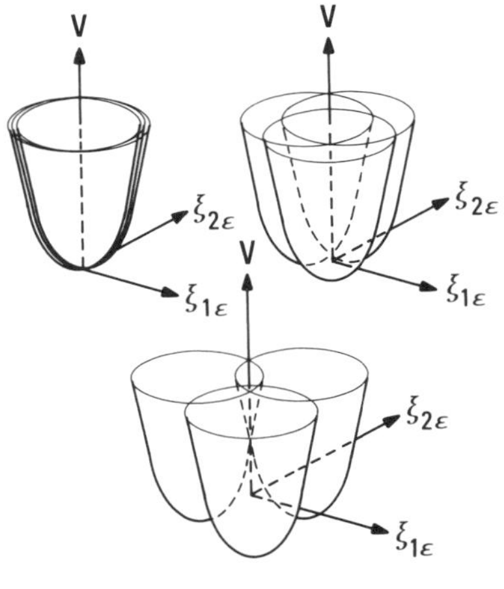

Figure 4. The three $T \otimes \epsilon$ potential surfaces.

The three non–coupled harmonic potential surfaces are pictured in figure 4. The energy levels are those of a two–fold degenerate harmonic oscillator

$$W_v = (v + 1)\hbar\omega \,, \; v = 0, 1, 2 \cdots \tag{46}$$

The vibrational manifold is seen to be $3(v + 1)$–fold degenerate.

The Ham effect.

An interesting and general phenomenum occurs when adding an additional non–diagonal perturbation to eq (43). This perturbatian may be "internal" as e.g. a spin–orbit coupling or it may be "external" as e.g. a strain along a C_3 axis lowering the symmetry of the system. This will cause a splitting of the vibrational manifold, and the splitting will be governed by the matrix elements

$$\langle \psi_i(q)\chi_i(Q) | V^{(1)}(q) | \psi_j(q)\chi_j(Q)\rangle = D\sigma\langle \chi_i(Q) | \chi_j(Q)\rangle \tag{47}$$

The electronic matrix element $D\sigma$ is multiplied by a vibrational overlap between two oscillators centered on different potential sheets. The distance between the minima of the three potential sheets is $\sqrt{3}c/k$ and the vibrational overlap S_{oo} between χ_{io} and χ_{jo} is calculated to be $S_{oo} = \exp(-\frac{3}{2}D)$. A strong Jahn–Teller coupling will therefore quench the expected splitting of $3D\sigma$. Such quenchings are called Ham effects; they may cause experimental undetectability of expected effects.

The $T \otimes \tau_2$ coupling.

Regardless of whether T is a T_1 or T_2 representation the symmetric product spans A, E and T_2. The perturbatian operator for a first order Jahn–Teller coupling to a τ_2 vibrator is therefore

$$\mathcal{H}^{(1)} = \left[\frac{\partial V}{\partial \xi_{1\tau}}\right]_o \xi_{1\tau} + \left[\frac{\partial V}{\partial \xi_{2\tau}}\right]_o \xi_{2\tau} + \left[\frac{\partial V}{\partial \xi_{3\tau}}\right]_o \xi_{3\tau} \tag{48}$$

where (dropping the τ) the three vibrational symmetry coordinates which span the three–fold degenerate vibration are oriented

$$\hat{C}_3 \begin{bmatrix} \xi_1 \\ \xi_2 \\ \xi_3 \end{bmatrix} = \begin{bmatrix} 0 & 1 & 0 \\ 0 & 0 & 1 \\ 1 & 0 & 0 \end{bmatrix} \begin{bmatrix} \xi_1 \\ \xi_2 \\ \xi_3 \end{bmatrix} \tag{49}$$

Using the symmetry operators $\hat{C}_3$ and $\hat{C}_2'$ we find easily that

$$\chi^{(1)} = c_\tau \begin{bmatrix} 0 & \xi_3 & \xi_2 \\ \xi_3 & 0 & \xi_1 \\ \xi_2 & \xi_1 & 0 \end{bmatrix} \tag{50}$$

Introducing the three matrices

$$\tau_1 = \begin{bmatrix} 0 & 0 & 0 \\ 0 & 0 & 1 \\ 0 & 1 & 0 \end{bmatrix} \quad \tau_2 = \begin{bmatrix} 0 & 0 & 1 \\ 0 & 0 & 0 \\ 1 & 0 & 0 \end{bmatrix} \quad \tau_3 = \begin{bmatrix} 0 & 1 & 0 \\ 1 & 0 & 0 \\ 0 & 0 & 0 \end{bmatrix} \tag{51}$$

for which

$$\tau_1^2 + \tau_2^2 + \tau_3^2 = 2 \begin{bmatrix} 1 & 0 & 0 \\ 0 & 1 & 0 \\ 0 & 0 & 1 \end{bmatrix} \tag{52}$$

we can write $\chi^{(1)}$

$$\chi^{(1)} = c_\tau (\xi_1 \tau_1 + \xi_2 \tau_2 + \xi_3 \tau_3) \tag{53}$$

With

$$V^\circ = \tfrac{1}{2} k_\tau (\xi_1^2 + \xi_2^2 + \xi_3^2) \tag{54}$$

the potential sheets V_i, i = 1, 2, 3 are found by solving the secular equation

$$\text{det.}\ | V^\circ + \chi^{(1)} - V | = 0 \tag{55}$$

Introducing the polar coordinates

$$\begin{aligned} \xi_1 &= r \sin \Theta \cos \varphi \\ \xi_2 &= r \sin \Theta \sin \varphi \\ \xi_3 &= r \cos \Theta \end{aligned} \tag{56}$$

we get expanding eq. (55)

$$(\tfrac{1}{2} k_\tau r^2 - V)^3 - (\tfrac{1}{2} k_\tau r^2 - V) c_\tau^2 r^2 + r^3 c_\tau^3 \sin^2\Theta \cos\Theta \sin 2\varphi = 0 \tag{57}$$

Clearly the extremal points of V occurs when the angular function has its minima and maxima. These occurs at $\varphi = \frac{\pi}{4}$ and $\cos \Theta = \pm \sqrt{\frac{1}{3}}$. The extremal points are therefore placed at the corners of a tetrahedron in ξ space, and the four minima correspond to a molecule of D_{3d} symmetry. Putting these values of φ and Θ back in (57) we get

$$(\tfrac{1}{2} k_\tau r^2 - V)^3 - (\tfrac{1}{2} k_\tau r^2 - V) c_\tau^2 r^2 \pm \frac{2}{3\sqrt{3}} r^3 c_\tau^3 = 0 \tag{58}$$

or

$$V = \tfrac{1}{2}k_\tau r^2 + \begin{cases} \pm \frac{1}{\sqrt{3}} c_\tau r & \text{2 fold degenerate} \\ \mp \frac{2}{\sqrt{3}} c_\tau r & \text{1 fold degenerate} \end{cases}$$

Taking $\frac{dV}{dr} = 0$ we get for the maximum stabilization $k_\tau r_0 - \frac{2}{\sqrt{3}} c_\tau = 0$, leading to

$$r_0 = \frac{2c_\tau}{k_\tau\sqrt{3}} \text{ and } V_{min} = \frac{-2c_\tau^2}{3k_\tau} \text{ with } |\xi_1| = |\xi_2| = |\xi_3| = r_0\sqrt{\frac{1}{3}} = \frac{2}{3}\frac{c_\tau}{k_\tau}$$

Comparing this result with the Jahn–Teller stabilization for $T \otimes \epsilon$, where the tetragonal coupling gave rize to a stabilization of $\frac{c_\epsilon^2}{2k_\epsilon}$, we observe that if

$$\frac{c_\epsilon^2}{2k_\epsilon} > \frac{2c_\tau^2}{3k_\tau} \tag{59}$$

it is the tetragonal configuration which will be stable; the reverse lead to the opposite. If the two quantities are comparable the dynamic situation is very complicated.

The dynamic $T \otimes \tau_2$ Jahn–Teller coupling is for small values of the coupling strenght parameter c_τ treated analogous to the $E \otimes \epsilon$ case. The transposed Hamiltonian $\tilde{\mathcal{H}}$ is given by $\tilde{\mathcal{H}} = e^{i\hat{S}} \mathcal{H} e^{-i\hat{S}}$ with

$$\hat{S} = \frac{c_\tau i}{k_\tau}\left(\frac{\partial}{\partial\xi_1}\tau_1 + \frac{\partial}{\partial\xi_2}\tau_2 + \frac{\partial}{\partial\xi_3}\tau_3\right) \tag{60}$$

and therefore as before, eq. (30), to second order in c_τ

$$\tilde{\mathcal{H}} = \mathcal{H}^o + \tfrac{1}{2}[\hat{S}, [\hat{S}, \mathcal{H}^o]] + \cdots \tag{61}$$

With

$$[\hat{S}, \mathcal{H}^o] = c_\tau i\,(\xi_1\tau_1 + \xi_2\tau_2 + \xi_3\tau_3)$$

and

$$\tfrac{1}{2}[\hat{S},[\hat{S}, \mathcal{H}^o]] = \frac{c_\tau^2}{2k}\{2 + (\xi_1\frac{\partial}{\partial\xi_2} - \xi_2\frac{\partial}{\partial\xi_1})\lambda_1 + (\xi_3\frac{\partial}{\partial\xi_1} - \xi_1\frac{\partial}{\partial\xi_3})\lambda_2 + (\xi_2\frac{\partial}{\partial\xi_3} - \xi_3\frac{\partial}{\partial\xi_2})\lambda_3\} \tag{62}$$

where the three components of the vector $\vec{\lambda}$ are given by $\vec{\tau} \times \vec{\tau} = -i\vec{\lambda}$. $\vec{\lambda}$ evidently represents an "electronic angular momentum" with the components

$$\lambda_1 = \begin{bmatrix} 0 & 0 & 0 \\ 0 & 0 & -i \\ 0 & i & 0 \end{bmatrix} \quad \lambda_2 = \begin{bmatrix} 0 & 0 & i \\ 0 & 0 & 0 \\ -i & 0 & 0 \end{bmatrix} \quad \lambda_3 = \begin{bmatrix} 0 & -i & 0 \\ i & 0 & 0 \\ 0 & 0 & 0 \end{bmatrix} \tag{63}$$

and $\vec{\lambda}^2 = \lambda_1^2 + \lambda_2^2 + \lambda_3^2 = 2.$

Defining the components of the vibrational "angular momentum" $\vec{M}$ (in units of $\hbar$)

$$\begin{aligned} \hat{M}_3 &= -i\,(\xi_1 \frac{\partial}{\partial \xi_2} - \xi_2 \frac{\partial}{\partial \xi_1}) \\ \hat{M}_2 &= -i\,(\xi_3 \frac{\partial}{\partial \xi_1} - \xi_1 \frac{\partial}{\partial \xi_3}) \\ \hat{M}_1 &= -i\,(\xi_2 \frac{\partial}{\partial \xi_3} - \xi_3 \frac{\partial}{\partial \xi_2}) \end{aligned} \tag{64}$$

we can write

$$\mathcal{H} = \mathcal{H}^{0} + \frac{c_\tau^2}{2k_\tau} \{ \vec{M} \cdot \vec{\lambda} - 2 \} \tag{65}$$

We have also $[\mathcal{H}^{0}, \hat{M}^2]$, $[\mathcal{H}^{0}, \hat{\lambda}^2]$ and $[\mathcal{H}^{0}, \hat{M}_3 + \hat{\lambda}_3] = 0$. Hence defining $\vec{L} = \vec{M} + \vec{\lambda}$

$$\vec{L}^2 = (\vec{M} + \vec{\lambda})^2 = \hat{M}^2 + \hat{\lambda}^2 + 2\vec{M} \cdot \vec{\lambda} \tag{66}$$

or

$$\mathcal{H} = \mathcal{H}^{0} + \frac{c_\tau^2}{4k_\tau} \{ L(L+1) - M(M+1) - 6 \} \tag{67}$$

leading to

$$W = (v + \tfrac{3}{2}) + \frac{c_\tau^2}{4k_\tau} \{ L(L+1) - M(M+1) - 6 \} \tag{68}$$

in units of $\hbar\omega$. For given v, M goes from $v, v-2 \cdots 0$ or 1, and L takes on the values $M+1$, M and $M-1$ when $M \geq 1$, but $L = 1$ when $M = 0$.

In the situation when the coupling constant c_τ is large, stabilization in one of the four equivalent trigonally distorted D_{3d} configurations occur. We saw that all displacement coordinates were numerically equal to $2c_\tau/3k_\tau = \Delta$. We consequently displace the coordinate system to the new equilibrium point, and define new vibrational coordinates D_i, $i = 1, 2, 3$ relative to the equilibrium point

$$\begin{aligned} D_1 &= \xi_1 + \Delta \\ D_2 &= \xi_2 + \Delta \\ D_3 &= \xi_3 + \Delta \end{aligned} \tag{69}$$

Then

$$\mathcal{H}^{\circ} = T_N + \tfrac{1}{2}k_\tau(D_1^2 + D_2^2 + D_3^2 + 3\Delta^2 - 2D_1\Delta - 2D_2\Delta - 2D_3\Delta) \tag{70}$$

and

$$\mathcal{H}^{(1)} = c_\tau(D_1\tau_1 + D_2\tau_2 + D_3\tau_3) - c_\tau\Delta(\tau_1 + \tau_2 + \tau_3) \tag{71}$$

or

$$\mathcal{H}^{\circ} = T_N + \tfrac{1}{2}k_\tau(D_1^2 + D_2^2 + D_3^2) + \frac{2}{3}\frac{c_\tau^2}{k_\tau} - \frac{2}{3}c_\tau(D_1 + D_2 + D_3) \tag{72}$$

$$\mathcal{H}^{(1)} = c_\tau\begin{bmatrix} 0 & D_3 & D_2 \\ D_3 & 0 & D_1 \\ D_2 & D_1 & 0 \end{bmatrix} - \frac{2}{3}\frac{c_\tau^2}{k_\tau}\begin{bmatrix} 0 & 1 & 1 \\ 1 & 0 & 1 \\ 1 & 1 & 0 \end{bmatrix} \tag{73}$$

The Hamiltonian is now transformed by a unitary transformation

$$U = \frac{1}{\sqrt{3}}\begin{bmatrix} 1 & 1 & 1 \\ 1 & \omega & \omega^2 \\ 1 & \omega^2 & \omega \end{bmatrix} \qquad \omega = \exp.(-\frac{2\pi}{3}i) \tag{74}$$

and new symmetry coordinates with respect to the three–fold symmetry axis is defined transforming like α_1 and ϵ

$$Q_3 = \sqrt{\frac{1}{3}}(D_1 + D_2 + D_3) \qquad (\alpha_1)$$

$$\left.\begin{aligned} Q_2 &= \sqrt{\frac{1}{2}}(D_3 - D_2) \\ Q_3 &= \sqrt{\frac{1}{6}}(-2D_1 + D_2 + D_3) \end{aligned}\right\}(\epsilon) \tag{75}$$

We get

$$\mathcal{H}_{\circ} = T_N + \tfrac{1}{2}k_\tau(Q_1^2 + Q_2^2 + Q_3^2) + \frac{2}{3}\frac{c_\tau^2}{k_\tau} - \sqrt{\frac{4}{3}}c_\tau Q_3 \tag{76}$$

$$\mathcal{H}^{(1)} = \begin{bmatrix} \sqrt{\frac{4}{3}}c_\tau Q_3 - \frac{4c_\tau^2}{3k_\tau} & \sqrt{\frac{1}{6}}c_\tau(Q_1 - iQ_2) & \sqrt{\frac{1}{6}}c_\tau(Q_1 + iQ_2) \\ \sqrt{\frac{1}{6}}c_\tau(Q_1 + iQ_2) & -\sqrt{\frac{1}{3}}c_\tau Q_3 + \frac{2}{3}\frac{c_\tau^2}{k_\tau} & -\sqrt{\frac{4}{6}}c_\tau(Q_1 - iQ_2) \\ \sqrt{\frac{1}{6}}c_\tau(Q_1 - iQ_2) & -\sqrt{\frac{4}{6}}c_\tau(Q_1 + iQ_2) & -\sqrt{\frac{1}{3}}c_\tau Q_3 + \frac{2c_\tau^2}{3k_\tau} \end{bmatrix} \tag{77}$$

Adding $\frac{c_\tau^2}{3k_\tau}$ to the diagonal elements we have finally dropping the subscript τ

$$\mathcal{H}=\begin{bmatrix} T_N+\frac{1}{2}k(Q_1^2+Q_2^2+Q_3^2)-\frac{1}{3}\frac{c^2}{k} & \sqrt{\frac{1}{6}}c(Q_1-iQ_2) & \sqrt{\frac{1}{6}}c(Q_1+iQ_2) \\ \sqrt{\frac{1}{6}}c(Q_1+iQ_2) & T_N+\frac{1}{2}k(Q_1^2+Q_2^2)+\frac{1}{6}\frac{c^2}{k} + \frac{1}{2}k(Q_3-\sqrt{3}\frac{c}{k})^2 & -\sqrt{\frac{2}{3}}c(Q_1-iQ_2) \\ \sqrt{\frac{1}{6}}c(Q_1-iQ_2) & -\sqrt{\frac{2}{3}}c(Q_1+iQ_2) & T_N+\frac{1}{2}k(Q_1^2+Q_2^2)+\frac{1}{6}\frac{c^2}{k} + \frac{1}{2}k(Q_3-\sqrt{3}\frac{c}{k})^2 \end{bmatrix} \quad (78)$$

Notice that there are no linear coupling terms in Q_3. On the other hand the low state is coupled to the doubly degenerate high state by the coupling terms $\sqrt{\frac{1}{6}}\,c\,(Q_1 \pm iQ_2)$. The doubly degenerate high state is also seen to experience a Jahn–Teller coupling in the doubly degenerate vibration (Q_1, Q_2), precisely as we have seen before.

The vibrational levels of the low state are therefore

$$W = -\frac{1}{3}c^2/k + \hbar\omega_\parallel(v_\parallel + \tfrac{1}{2}) + \hbar\omega_\perp(v_\perp + 1) \quad (79)$$

$v_\parallel = 0, 1, 2 \cdots \quad v_\perp = 0, 1, 2 \cdots\cdots.$

From $\mathcal{H}^{(1)}$ we find using second order perturbation theory

$$V_\perp^{(2)} = -\frac{\frac{1}{6}\,c^2\,(Q_1^2 + Q_2^2)\cdot 2}{2\,c^2/k} = -\frac{1}{6}k\,(Q_1^2 + Q_2^2) \quad (80)$$

To second order we have therefore for $V_\perp$

$$V_\perp = \tfrac{1}{2}\left(\tfrac{2}{3}k\right)(Q_1^2 + Q_2^2) \quad (81)$$

and therefore $\omega_\perp = \sqrt{\frac{2}{3}}\,\omega_\parallel$.

Cooperative Jahn–Teller couplings.

The elucidations of crystal structures in terms of local distortions of the molecular entities which make up the crystal are a separate field. Her I shall only illustrate the approach to the problem by means of a very simple example.

Interactions between unit cells in a crystal involve relative motions of groups. This give rize to three types of lattice modes, namely a bulk strain of the crystal as well as <u>acoustic and optic phonons</u>, that is the collective lattice vibrations as functions of the wave vector $\vec{k}\hbar = \vec{p}$, where $\vec{p}$ is the linear momentum. The strain has even parity and corresponds to a macroscopic change of the shape of the sample. With $Q(k)$ being a normal coordinate associated with an acoustic phonon we have $\lim_{k=0} k\, Q(k) = $ strain. The acoustic phonons evidently takes care of local fluctuations in the strain.

The potential function for a crystal with uniform strain can be written

$$V_s = \sum_n c_s\, \xi_\gamma \hat{O}_{\gamma n} + \tfrac{1}{2} N \kappa_\gamma\, \xi_\gamma^2 \tag{82}$$

where κ_γ is the elastic force constant corresponding to the strain ξ_γ, c_s the strength of the strain coupling, $\hat{O}_{n\gamma}$ an electronic operator in the n'th unit cell, transforming like ϵ_γ and N the number of unit cells. The strain is a macroscopic quantity which can be found by minimizing V_s:

$$\frac{dV_s}{d\xi_\gamma} = 0 = \sum_n c_s\, \hat{O}_{\gamma n} + N \kappa_\gamma \xi_\gamma \tag{83}$$

Substituting back in (82) for ξ_γ we can evidently take for the first order perturbation Hamiltonian associated with the strain

$$\mathcal{H}_s^{(1)} = -\frac{c_s^2}{2\kappa N} \sum_{n,m} \hat{O}_{\gamma n}\, \hat{O}_{\gamma m} \tag{84}$$

The strain therefore gives rise to an infinite–range interaction between all the unit cells.

The cells will further interact by the exchange of a virtual acoustic phonon, corresponding to an electronic exchange interaction. From general considerations one can take

$$\mathcal{H}_{n,m}^{(1)} = -\sum_k f(k)\, e^{i\vec{k}\cdot(\vec{R}_n - \vec{R}_m)}\, \hat{O}_{\gamma n}\, \hat{O}_{\gamma m} \tag{85}$$

as being the interaction between two ions in the unit cells n and m, located at $\vec{R}_n$ and $\vec{R}_m$ respectively. Summing over all the pairs in the crystal

$$\mathcal{H}^{(1)} = -\tfrac{1}{2} \sum_{n,m} J_{n,m}\, \hat{O}_{\gamma n}\, \hat{O}_{\gamma m} \tag{86}$$

For a two–level system the $\hat{O}_\gamma$ operators can be replaced by a pseudo spin operator $\hat{s}_z$ with the two eigenvalues ± 1. The total interaction, tending to make the distortions line up the same way, is called a <u>molecular field</u>; it is opposed by the thermal agitations.

We shall now treat the simple case of tetragonal symmetry with a degenerate electronic state E and a one–dimensional distortion. Using the pseudo spin formalism the Hamiltonian is the sum of eqs. (84) and (86)

$$\mathcal{H} = -\tfrac{1}{2} \sum_{n,m} \left(J_{n,m} + \frac{c_S^2}{\kappa N} \right) \hat{s}_{zn} \hat{s}_{zm} \tag{87}$$

We now consider eq. (87), the molecular field, the equivalent of an effective spin $\bar{S}_z$ acting on the local spins

$$\mathcal{H} = -\frac{1}{N} \left(\sum_{n} J_{n,m} + \frac{c_S^2}{\kappa} \right) \bar{S}_z \sum_{n} \hat{s}_{zn} \tag{88}$$

Introducing the order parameter $\langle S_z \rangle = N^{-1} |\sum_n \hat{s}_{zn}|$ the eigenvalues of (88) are

$$W = \pm \left(\sum_{n} J_{n,m} + \frac{c_S^2}{\kappa} \right) \langle S_z \rangle \tag{89}$$

We observe that the larger the order parameter $\langle S_z \rangle$ the greater the energy difference, 2W, between the two crystal phases. A selfconsistent solution for $\langle S_z \rangle$ may be obtained using the Boltzmann equation

$$\langle S_z \rangle = \frac{e^{W/k_B T} - e^{-W/k_B T}}{e^{W/k_B T} + e^{-W/k_B T}} = \tanh W/k_B T. \tag{90}$$

Substituting eq. (90) into eq. (89)

$$W = \left(\frac{c_S^2}{\kappa} + \sum_{n} J_{n,m} \right) \tanh W/k_B T \tag{91}$$

and the phase transition temperatur T_D is found by expanding eq. (91)

$$k_B T_D = \left(\frac{c_S^2}{\kappa} + \sum_{n} J_{n,m} \right) \tag{92}$$

Had we considered the case of a two–fold degenerate level interacting with a two–dimensional vibration, the first order interaction would have lead to a "Mexican Hat" potential, and the direction of the strain coordinate would be unspecified. This is unphysical for a crystal, and a second order Jahn–Teller coupling, introducing a three–fold modulation of the potential, would have to be invoked before meaningful results could be obtained.

LITTERATURE.

No litterature has been quoted in the above sections. The reason for this is the overwhelming mass of original papers. For text books I can point to R. Englman: "The Jahn–Teller effect in Molecules and Crystals". Wiley – Interscience 1972 and I.B. Bersukers books (though extremely expensive!) "The Jahn–Teller Effect and Vibronic Interactions in Modern Chemistry" Plenum Press 1984 and the bibliographic Review "The Jahn–Teller effect" Plenum Press 1984. An excellent review of the dynamic Jahn–Teller couplings has been written by H.C. Longuet–Higgins in "Advances in Spectroscopy" 2 (1961) 429. The static Jahn–Teller couplings have been exhaustively dealt with by A.D. Liehr: Progress in Inorganic Chemistry 3 (1962) 281. ibid 4 (1962) 455, ibid 5 (1963) 385 and J. Phys. Chem. 67 (1963) 389. The treatment of the dynamic Jahn–Teller couplings given here build upon W. Moffitts and W. Thorsons two papers in "Calcul des Fonctions D'onde Moleculaire" Edition du Centre National de la Recherche Scientifique 82, Paris 1958 and Phys. Rev. 108 (1957), 1251, respectively. A review, seen from the view point of solid state physicists, of the co–operative Jahn–Teller couplings has been written by G.A. Gehring and K.A. Gehring in "Reports on Progress in Physics" 38 (1975) 1.

The Experimental Disclosures.

The Jahn–Teller couplings manifest themselves in the following ways. Molecular shapes and crystal structures may be rationalized invoking a static Jahn–Teller coupling. Rotational and vibrational fine structure in absorption bands may be explained surmising a dynamic Jahn–Teller coupling, as may the values of the g–factors in Electron Spin Reasonans experiments. Here I can only discuss a few characteristic papers – a whole year would not suffice if I were to cover the litterature in depth as witnessed by Behrsukers bibliography, quoted in the last papagraph. It is also only inorganic molecules and ions I shall deal with.

a. THE GROUND STATES.

The VCl_4 molecule has a tetrahedral structure, with a ^2E groundstate, subject to a Jahn–Teller coupling. This will in first order cause the four chlorine atoms to perform a coupled rotation about the regular tetrahedral bond angles, each one rolling upon a cone with an apex angle of a few degrees. Figure 5. At the points marked 1, 2 and 3 the VCl_4 will have the structure of an elongated tetrahedron, corresponding to the weak mimima of the three–fold barrier introduced by the second–order Jahn–Teller coupling[(1)].

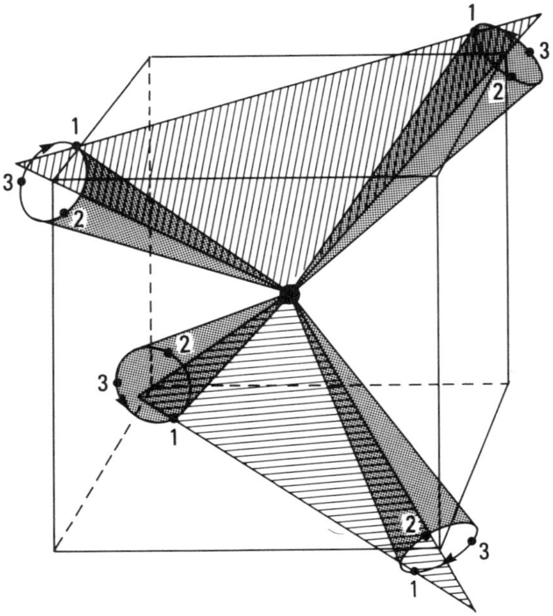

Figure 5. The movements of the chlorine atoms in VCl_4 due to the Jahn–Teller coupling.

In $TiCl_4$, which do not exibit a Jahn–Teller coupling, the mean square Cl–Cl distance will therefore be smaller than the mean square Cl–Cl distance in VCl_4, and this is presicely what is observed experimentally(2).

If Jahn–Teller active ions are doped into a cubic latice the ligand displacements will interact via the elastic strain of the crystal, and the entire crystal become unstable. For instance if the concentration x of Cu^{+2} ions in the cubic host lattice $Ba_2Zn_{1-x}WO_6$ exceeds 0.23 a phase transition from cubic to tetragonal is observed(3). The CuO_6 octahedra have a ^2E_g ground state, subject to a Jahn–Teller coupling with an ϵ vibration. The g–factors then reveal that the CuO_6 octahedra are tetragonally elongated, corresponding to the mimimum of the potential surface accountable for when second order Jahn–Teller couplings are introduced(4). That the Jahn–Teller coupling is responsible for distortions from cubic symmetry which occur in transition–metal oxides having spinel structure was first recognized by Dunitz and Orgel(5) and independently by McClure(6) in 1957.

The g–values from ESR measurements are very telling. Consider e.g. the ground term $^2T_2$ of an octahedrally coordinated $3d^1$ impurity ion(7). Only by the combined efforts of a trigonal crystal field, spin–orbit coupling, Jahn–Teller couplings, Ham effect and Zeeman splittings was it possible to account for the g–values of Ti^{+3} in Al_2O_3. From spectroscopic measurements of Ti^{+3} doped into Al_2O_3 it is known that the ground state is the ^2E component of $^2T_2$ of cubic parantage. Calculating the vibronic levels with both a Jahn–Teller coupling and a spin–orbit coupling present, results in the level pattern pictured in Figure 6.

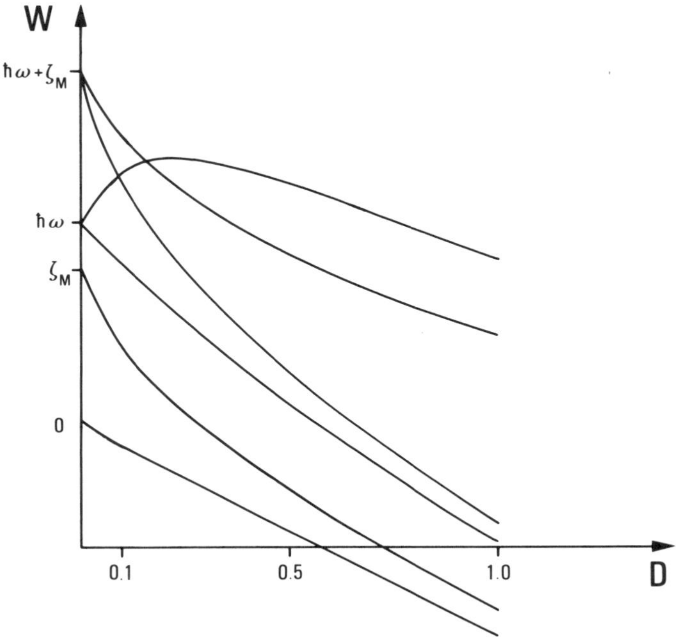

Figure 6. The six lowest vibronic levels if a ^2E state experiencing both a Jahn–Teller and a spin–orbit coupling with $\zeta_M = 1/3\hbar\omega$. Notice how the operation of the Ham effect diminishing the apparent spin–orbit coupling.

The Zeeman splittings of the lowest vibronic level of E symmetry then determines the g–values. With an effective ϵ mode of 200 cm^{-1}, D is about one for this system.

b. THE EXCITED STATES.

Assuming a D_{3h} symmetry the ground state of H_3 is ^2E, ripe for a Jahn–Teller coupling. As is well known H_3 is unstable, dissociating into $H_2 + H$. Whether this unstability can be associated with a Jahn–Teller coupling is not a question of physics – but of metaphysics. However, some excited ^2E Rydberg states of H_3 are stable and the spectra have been measured by Herzberg and co–workers[8, 9]. The results for the Rydberg state of $(1s)^2(3p)^1$, ^2E were that $D(H_3) = 0.0301$ and $D(D_3) = 0.0424$, corresponding to a very weak coupling. This is not unexpected: the 3p electron is so far removed from the nuclear framework that it hardly "feels" the vibrations.

The Cu^+ ion can be doped into a NaF crystal. Using two–photon spectroscopy McClure and co–workers[10] looked at the g→g $(3d)^{10} \rightarrow (3d)^9(4s)$ transitions. The upper configuration give rise to ^1E, ^3E, $^1T_2$ and $^3T_2$ states. The transitions occur in the 3000 – 1700 Å range, showing strong temperature dependence, indicating vibronic intensity. The Cu^+ ion is therefore located at the center of an octahedron of F^- ions. The ^1E_g spectrum shows the double humped feature expected for a Jahn–Teller coupling to an ϵ vibration. With the volume

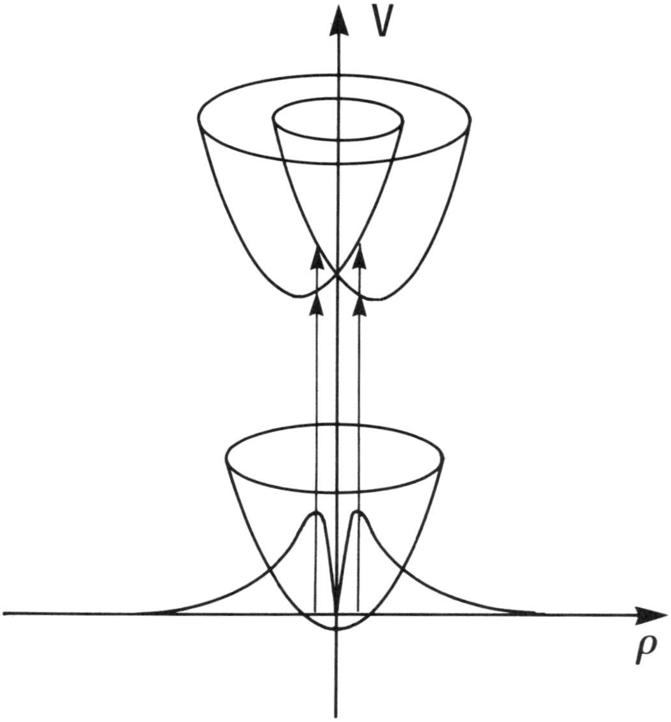

Figure 7. The Franck–Condon transition from an A state to $E \otimes \epsilon$.

element being $d\tau = \rho d\rho d\varphi$ the probability $|\chi_o|^2 d\tau$ of finding a doubly degenerate ϵ vibration has its maximum at $\rho = \sqrt{\frac{1}{2}}$. The Franck–Condon principle thenaccounts for the double humped feature in the spectrum. Using an effective ϵ vibration ≈ 300 cm^{-1}, and ΔV_{J-T} = 2250 cm^{-1} we get $D \simeq 7.5$, showing a strong coupling of the excited ^1E state to the localized vibrations of the NaF crystal lattice.

A dynamic Jahn–Teller coupling can be observed[11] in the spectrum of MnO_4^{-3} doped into $Sr_5(PO_4)_3Cl$. A single progression with 18 components and a double humped distribution confirms the identification of a $^3A_2 \rightarrow {}^3E$, parantage

$^3T_2$, band system. The average frequency in the progression of 95 cm^{-1} is associated with a much reduced ground–state ϵ vibration of 324 cm^{-1}. One of the most characteristic features in a Jahn–Teller coupled system is indeed a progression in single quanta in that non–totally symmetric mode which distorts the nuclear framework by coupling to the electronic state, thereby in its own turn becoming totally symmetric.

As a final example of a dynamic Jahn–Teller coupling we take the excited $^3T_2$ state of V^{+3} doped into Al_2O_3. From spectral measurements of the $^3T_2(^3A) \rightarrow {}^3T_2$ transition McClure observed(12) both the absence of any appreciable trigonal splitting and a relatively sharp line of about 15880 cm^{-1}, appearing equally strongly both in σ– and π–polarizations but at slightly different positions. McClure showed that the sharp line was to be identified with the no–phonon $^3A \rightarrow {}^3T_2$ electronic origin. Furthermore if there were no Jahn–Teller coupling the trigonal field would split the σ–polarized origin from the π–polarized origin by some hundred of wave numbers.

The detailed measurements of Scott and Sturge(13) confirmed that the splitting due to the trigonal field is partially quenched but still resolvable. With the spin being quantized along a tetragonal distortion the σ and π polarized lines should have equal intensity. (Remember σ and π refer to the trigonal axis of Al_2O_3). The Jahn–Teller coupling evidently enforce a tetragonal representation upon the system which cause both the trigonal field and the spin–orbit coupling operators to be off–diagonal. According to Ham(14) their effects are then partially quenched, and Scott and Sturge's result, that the first–order trigonal splitting is reduced by a factor of 45, is a fine demonstration of the workings of the Jahn–Teller coupling.

CONCLUSION

We have here considered some few experimental manifestations of Jahn–Teller couplings. With tongue in cheek I have availed myself of the strongest theorem of convergence known to chemical physicists: an infinite series converge if the first term of it does. The troubles we are faced with are that the second and higher order Jahn–Teller coupling terms cannot be treated matematically as simple as can the first order terms and that more parameters are introduced. However, as I have indicated above, second order couplings may be of major importance when it comes to the interpretations of experiments. Here one is always between the devil and the deep sea. The more accurate the experiments, the more sophisticated the interpretations need to be. Yet onesided overrefinements of a theory are of little value. Therefore when it comes to an understanding of the phenomena, one last word: be sober minded and remember to doubt.

REFERENCES.

1) C.J. Ballhausen and A.D. Liehr: Acta Chem. Scand. 15 (1961) 775.

2) Y. Morino and H. Uehara: J. Chem. Phys. 45 (1966) 4543.

3) D. Reinen and C. Friebel: Structure and Bonding 37 (1979) 1.

4) A.D. Liehr and C.J. Ballhausen: Ann. of Physics [N.Y.] 3 (1958) 304.

5) J.D. Dunitz and L.E. Orgel: J. Phys. Chem. Solids. 3 (1957) 20.

6) D.S. McClure: J. Phys. Chem. Solids. 3 (1957) 311.

7) R.M. MacFarlane, J.Y. Wong and M.D. Sturge: Phys. Rev. 166 (1968) 250.

8) G. Herzberg and J.K.G. Watson: Can. J. Phys. 58 (1980) 1250.

9) G. Herzberg, H. Lew, J.J. Sloan and J.K.G. Watson: Can. J. Phys. 59 (1981) 428.

10) S.A. Payne, A.B. Goldberg and D.S. McClure: J. Chem. Phys. 78 (1983) 3688.

11) P. Day, R. Borromei and L. Oleari: Chem. Phys. Lett. 77 (1981) 214.

12) D.S. McClure: J. Chem. Phys. 36 (1962) 2757.

13) W.C. Scott and M.D. Sturge: Phys. Rev. 146 (1966) 262.

14) F.S. Ham: Phys. Rev. 138 (1965) A 1727

THE THEORY OF THE JAHN-TELLER EFFECT

B. R. Judd
Department of Physics & Astronomy
The Johns Hopkins University
Baltimore, Maryland 21218
U. S. A.

ABSTRACT. After a brief historical introduction, the derivation of the so-called Jahn-Teller theorem is discussed. Group-theoretical concepts are introduced and the octahedral group is chosen as an example. The static distortion of a system comprising a T_{2g} electronic state coupled linearly to normal modes of the same type provides an illustration of the calculation of Clebsch-Gordan coefficients for the octahedral group. Annihilation and creation operators are introduced to discuss the vibronic Hamiltonian, and octahedral tensors are brought into play to make the analysis more concise. The differences that occur when electron spin is taken into account are pointed out. The advantages and drawbacks of using Glauber (coherent) states in the strong-coupling limit are briefly mentioned. Attention is paid to the Jahn-Teller system in which a T electronic state (such as a p electron) is equally coupled to octahedral t and e modes, these modes possessing equal frequencies and thereby leading to a Hamiltonian with spherical symmetry. Phase changes associated with the unequal rotational motion of the T state and the displaced ligands for this system are studied in the strong-coupling limit. The connection with Berry's phase is discussed.

1. Introduction

At the 1936 Spring meeting of the American Physical Society, which was held as tradition dictated (until very recently) in Washington D. C., Jahn and Teller (1936) announced the effect that bears their name. The abstract runs as follows:

As a general rule the electronic state of a polyatomic molecule can be degenerate only if the atomic configuration has a sufficiently high degree of symmetry. If the atomic nuclei are displaced the degenerate state may split up and if the splitting is a linear function of the displacement the original symmetrical configuration, and with it the

C. D. Flint (ed.), Vibronic Processes in Inorganic Chemistry, 79–101.
© 1989 by Kluwer Academic Publishers.

original degenerate state, does not correspond to an equilibrium state of the molecule. A group-theoretical investigation shows that except for molecules in which all atoms lie on a straight line only undegenerate states or the doubly degenerate states of molecules with an odd number of electrons can correspond to stable configurations. However, in the case of spin degeneracy and small spin-orbit interaction a nearly symmetrical configuration might be stable and the electronic state is then split to a very small extent. The same is true for degenerate orbits of electrons in inner incomplete shells if their interaction with the other atoms is small.

In his preface to the book by Englman (1972), Teller gave an account of the genesis of what we now refer to as the Jahn-Teller effect. It was Landau who expressed his concern to Teller in Copenhagen on the stability of molecules possessing degenerate electronic states, and it was later, in London, that Teller discussed the general problem in detail with Jahn. The time they spent together seems to have been quite short. In the afore-mentioned preface, Teller referred to Jahn, a native of Colchester (England), as a refugee from a German university, perhaps misled by Jahn's thesis work in Leipzig on the rotations and vibrations of methane (Jahn 1935). The overall tone of the Washington abstract is remarkably tentative. The opening phrase "As a general rule," strikes a curiously ambiguous note. In the second sentence, we read that the degenerate state "may" split up. Matters were soon tightened up. Jahn and Teller (1937) presented a formal article in which the orbital degeneracy and associated stability were studied, and Jahn (1938) tackled the problem of spin degeneracy. It is in the second of these two articles that concise expressions for what has become known as the Jahn-Teller effect were given: (A) A configuration of a polyatomic molecule for an electronic state having orbital degeneracy cannot be stable with respect to all displacements of the nuclei unless in the original configuration the nuclei all lie on a straight line; (B) A polyatomic molecule cannot possess a stable non-linear nuclear configuration in an electronic state having spin degeneracy unless this degeneracy is the special twofold one which can occur only when the molecule contains an odd number of electrons. Many years later, Teller (1982) discussed his work with Jahn and mused on the way nature so often provides deviations from perfect symmetry -- as in the case of elementary particles. Even though the Jahn-Teller effect might have implications for all of physics, our present interest takes molecules and clusters of atoms in solids for its focus. The extensive bibliography of Bersuker (1984a) gives a good impression of the vast extent of this limited field.

1.1 Derivation

To prove the statement (A) above, Jahn and Teller (1937) first considered the electronic wavefunctions a_i $(i = 1, 2, \ldots, n)$ that correspond to an n-fold degenerate level for some symmetric configuration Q_0 of the nuclei. To every normal coordinate Q_r that describes displacements of the nuclei there is an associated amplitude

d_r. Once we admit distortions from Q_0, the electronic Hamiltonian H_0 becomes augmented with terms of the type $V_r(q)d_r$, where $V_r(q)$ is a function of the electronic coordinates q. The perturbation matrix M thus contains entries of the following form:

$$M_{ab} = \sum_r d_r \langle a|V_r|b\rangle. \tag{1}$$

If E is a characteristic value of this matrix (that is, a solution to the determinantal equation $|M_{ab} - E\,\delta(a, b)| = 0$) for a given set of values of d_r, then - E is a characteristic value of the same matrix but with all the signs of the d_r reversed. As Jahn and Teller (1937) say, "Thus unless the characteristic values of this matrix all vanish (in which case the matrix itself must vanish) the configuration Q_0 cannot possibly be a stable one."

The question of the vanishing of the matrix was examined by using what today is called the Wigner-Eckart theorem. Let G be the symmetry group of the Hamiltonian H_0. If a, b and V_r are labelled by the irreducible representations R_a, R_b and R_r, then the matrix element appearing in eq.(1) necessarily vanishes unless the triple Kronecker product $R_a \times R_b \times R_r$ contains the identity representation R_1. Jahn and Teller (1937) found that R_1 occurs in all triple products except those appropriate to the symmetry group G for a linear molecule. For this case we need to examine terms in the Hamiltonian that are quadratic in the displacements d_r. Being quadratic, they do not change their signs when the displacements are reversed. Thus we cannot pursue the argument to decide whether a distortion will take place. A linear molecule may or may not buckle, depending on its actual physical characteristics. For all other types of molecules the possibility exists that at least one of the matrix elements appearing in (1) is non-zero. In fact, if we can be sure that G is the full symmetry group of the Hamiltonian and no other larger group G′ exists that contains G, then the vagaries of the physical nature of an actual molecule will guarantee that a distortion will take place. Its size will be determined by the terms in the total Hamiltonian quadratic in the displacements.

Teller, in his preface for the book by Englman (1972), recognized that the separate calculation of each triple Kronecker product made the proof of the Jahn-Teller theorem rather unwieldy. By that time, Ruch and Schoenhofer (1965) had given a derivation on general grounds, though at the price of a substantial digression into group theory. For other discussions, the reader is referred to Blount (1971) and Raghavacharyulu (1973).

The sketch of the proof of statement (A) given above has overlooked the fact that H_0 separates the electronic states into sets that, in the absence of accidental degeneracy, each span an irreducible representation of G. Thus we have $R_b = R_a$. The triple Kronecker product thus contains the part R_a^2. However, not all the parts of this Kronecker square are useful. The matrix M must be Hermitian, and this condition implies that only those irreducible representations occurring in the symmetric part of the square should be combined with R_r. This rather simple point has an interesting elaboration when we come to statement (B). When dealing with electron spin we must take into account the result of Kramers (1930) that, in the absence of a magnetic field (or, in general, any term in the Hamiltonian that does not possess time-reversal symmetry), a molecule with an odd number of electrons exhibits at least a residual two-fold degeneracy. This comes about because, as Wigner (1932) showed, the action of the time-reversal operator T on one of the components of the doublet cannot produce a multiple of itself, and hence must yield a second state, namely, the second component of the doublet. As a result of a detailed study, Jahn (1938) was able to deduce that, for molecules with an odd number of electrons, the antisymmetric rather than the symmetric part of the square $R_a{}^2$ must be taken. This point will be re-examined later. For the moment, we simply observe that, with this change, and the use of double groups, Jahn (1938) found that a non-linear molecule with a degenerate electronic state and an odd number of electrons is unstable unless the degeneracy is of the Kramers type.

1.2 An Example

Shortly after the initial work of Jahn and Teller, a detailed example was worked out by Van Vleck (1939). He chose a central ion surrounded by a regular octahedron of ligands, a configuration that remains of great interest today. The character table of the octahedral group, which comprises 24 elements, is as follows:

Irrep	Basis	C_1	C_2	C_3	C_4	C_5
A_1	$x^2+y^2+z^2$	1	1	1	1	1
A_2	xyz	1	-1	1	1	-1
E	$3z^2-r^2$, x^2-y^2	2	0	2	-1	0
T_1	x, y, z	3	-1	-1	0	1
T_2	yz, zx, xy	3	1	-1	0	-1

The five classes C_i correspond to similar types of rotations that send the octahedron into itself, and the characters specify the traces of the matrices that describe the transformations of the basis functions. The x, y and z axes define the positions of the ligands with respect to the central ion.

The first step is to work out the normal modes. A simple procedure for doing this has been described elsewhere (Judd 1984). The idea is to set up six similarly oriented coordinate axes at the undisplaced ligand sites (labelled 1, 2, 3, 4, 5 and 6 for the ligands along the x, y, z, -x, -y and -z axes). The 18 coordinates (x_j, y_j, z_j) describing the displacements are now subjected to sample transformations drawn from each of the five classes, in which the entire octahedron is rotated. It often happens that ligand sites are interchanged and the

18 transformations contribute nothing to the trace of the transformation matrix. We find that the representation whose basis is the 18 ligand displacements decomposes into the irreducible parts $A_1 + E + 3T_1 + 2T_2$. The linear combination of displacements belonging to a particular irreducible representation can be found by choosing a displacement coordinate and generating acceptable x_i, y_j and z_k to add to it by applying the five sample transformations and testing for consistency. The appearance of repeated representations can be partially eliminated by including the inversion operation in the octahedral group, which becomes O_h. The additional labels g (for gerade, or even) and u (for ungerade, or odd) are attached to the irreducible representations, and the string describing the 18 ligand displacements becomes

$$A_{1g} + E_g + T_{1g} + 2T_{1u} + T_{2g} + T_{2u}. \qquad (2)$$

Suppose, now, that the electronic state of the central ion is of the type T_{2g}. This could be provided by three linear combinations of the five spherical harmonics defining the wavefunctions of a d electron. According to the arguments given in section 1.1, we need to work out the symmetrical part of the Kronecker square $T_{2g}{}^2$. We can easily find $T_2{}^2$ from the character table set out above: it is $A_1 + E + T_1 + T_2$. The symmetrical part must have a dimension of 6 (that is, one half of 3 times 4), but this number can be constructed from $A_1 + E + T_1$, or $A_1 + E + T_2$, or $T_1 + T_2$. Which combination is the correct one?

Some ingenuity is called for here, since we would prefer to evade having to form the linear combinations of the nine products of the type $y_j z_j x_k y_k$ with assigned irreducible representations. The basis functions (x_j, y_j, z_j) for T_1 are simpler to deal with: A_1 obviously belongs to the symmetric form $x_1x_2 + y_1y_2 + z_1z_2$, and so E must also belong to the symmetric part of the Kronecker square (to make up the dimension of 6 from the available irreducible representations). Under the five sample transformations, the three symmetric combinations $y_1z_2 + z_1y_2$, $z_1x_2 + x_1z_2$ and $x_1y_2 + y_1x_2$ are found to yield the same characters as yz, zx and xy, which implies that they belong to T_2. Thus $T_1{}^2$ yields the symmetric part $A_1 + E + T_2$. Since A_2, when combined with T_1, yields T_2, the same sequence of three representations is the symmetric part of $(A_2T_2)^2$. However, $A_2{}^2 = A_1$, so the symmetric part of $T_2{}^2$ is $A_1 + E + T_2$. Since the Kronecker squares g^2 and u^2 are both g, the only representations appearing in (2) that can combine with an electronic state T_2 are A_{1g}, E_g or T_{2g}.

Suppose we pick the T_{2g} normal mode for study. We have now to construct the matrix **M**, that is, the 27 entries $\langle a|V_r|b\rangle$, where a, b and r run each run over the three basis functions for T_{2g}. It is a remarkable fact that these 27 numbers involve only one unknown, namely, the strength of the interaction between the electronic state and the normal mode. This is a direct consequence of the fact that $T_{2g}{}^3$ contains the identity representation A_1 only once. By the Wigner-Eckart theorem, the dependence of the entries of **M** on the components a, b and r are given by the Clebsch-Gordan coefficients

$$(T_{2g}\ a|T_{2g}\ r,\ T_{2g}\ b) \tag{3}$$

for the octahedral group. The reader might be tempted to turn to Table C2.2 of Griffith (1962) to find them. Unfortunately, that writer did not choose yz, zx or xy as a basis for T_2, but rather a set of functions in which expressions such as yz + izx appear. Nor are the extensive tables of Butler (1981) of any use (for rather more complicated reasons). We are thrown back on our own resources.

1.3 Clebsch-Gordan Coefficients for the Octahedral Group

The basic fact to use is that every matrix element $\langle a|r|b\rangle$ is a number and thus invariant with respect to any operation of the octahedral group that acts simultaneously on a, r and b. For the five sample transformations, take

$$(x,\ y,\ z) \rightarrow (x,\ y,\ z),\ (y,\ x,\ -z),\ (x,\ -y,\ -z),\ (z,\ x,\ y),\ (x,\ -z,\ y),$$

corresponding to operations from each of the five classes $C_1,\ldots,C_5$. Under the action of the third transformation we have

$$\langle zx|zx|zx\rangle \;=\; \langle(-zx)|(-zx)|(-zx)\rangle \;=\; -\ \langle zx|zx|zx\rangle \;=\; 0,$$

since a number that is the negative of itself is necessarily zero. Cyclic permutations yield two more zeros. Similarly,

$$\langle zx|zx|xy\rangle \;=\; \langle(-zx)|(-zx)|(-xy)\rangle \;=\; -\ \langle zx|zx|xy\rangle \;=\; 0,$$

and we see that $\langle zx|xy|zx\rangle$ must be zero too. Cyclic permutations and Hermiticity conditions yield 16 more zeros. There remain the six matrix elements of the type $\langle yz|zx|xy\rangle$, which can easily be shown to be identical to one another. Thus all Clebsch-Gordan coefficients (3) vanish unless a, r and b are distinct, in which case they are equal to one another.

1.4 The Secular Matrix

To prevent arbitrarily large distortions, the perturbation matrix M must be augmented by a potential term U of some kind. The most obvious expression to include is the quadratic form

$$U \;=\; (1/2)m\omega^2(Q_{yx}^2 + Q_{zx}^2 + Q_{xy}^2), \tag{4}$$

which is scalar with respect to the operations of the octahedral group. The eigenvalue equation $H|\psi\rangle = E|\psi\rangle$ is set up with electronic kets of the form

$$|\psi\rangle \;=\; a_{yz}|yz\rangle \;+ a_{zx}|zx\rangle \;+ a_{xy}|xy\rangle,$$

and the coefficients a_{ij} as well as the energy E can be found by insisting that the derivatives of E with respect to the normal coordinaties Q_{ij} be equal to zero. The details can be found from Van Vleck's article or from other sources (Judd 1974). Not surprisingly, the three normal coordinates Q_{yz}, Q_{zx} and Q_{xy} satisfy similar equations, namely $Q_{zx}Q_{xy}/Q_{yz} = Q$ (say) and its cyclic permutations. These lead to four solutions:

$$(Q_{yz}, Q_{zx}, Q_{xy}) = (Q, Q, Q), (Q, -Q, -Q), (-Q, -Q, Q), (-Q, Q, -Q). \quad (5)$$

Interestingly enough, it is only at this point, when we ask what the displacements actually are, that we need to know the linear combinations of the displacement coordinates (x_j, y_j, z_j) that correspond to the normal coordinates. The T_{2g} label turns out to specify angular displacements (that is, displacements which, to the first order of smallness, do not change the radial distance of the ligands from the central ion). Furthermore, the four solutions (5) preserve the trigonal symmetry of the octahedron: there are just four three-fold axes of symmetry for a regular octahedron, and the two pairs of equilateral triangles centered on each of these four axes either contract or expand, depending on the sign of the entries in **M**.

There are two points worth discussing here. First of all, any of the four displacements is as likely as the three others, so, in a sense, the octahedral symmetry of the problem is retained. This is only to be expected, because the total Hamiltonian possesses octahedral symmetry. From the point of view of quantum mechanics, we can imagine one of the distortions tunnelling to another, so that, on a long time scale, all distortions are represented in a given octahedral complex. Classically, we would have to consider an ensemble of octahedra to get an equal distribution of the distortions.

Secondly, the existence of well-defined displacements is not as universal as one might think. An octahedral problem of great interest is that of an E electronic state coupled to the two normal modes that also span E. (We sometimes write E × e here: the corresponding notation for the previous example would be $T_{2g} \times t_{2g}$.) In the present case, the modes correspond to purely radial displacements. But an infinity of displacements exists at the potential minimum: so that, although displacements of the ligands occur and the energy is lowered, there is no finite set of solutions for the displacement coordinates. Some ligands move out, others move in. Evidently some additional symmetry is present in the problem. A classic analysis of this situation was provided by Longuet-Higgins et al. (1958). No detailed treatment of the E × e system will be given here because Professor C. J. Ballhausen is treating it in his own lectures. Another octahedral problem of considerable interest is that of the system T × (e + t_2), when the frequencies of the e and t_2 modes are equal, and when, in addition, their couplings to the electronic T state are also equal. In this case (which we shall return to later) there is an even higher level of ambiguity in the displacements, and the ligands float around on the surfaces of spheres (not centered on the undisplaced sites, however) in a way that strongly couples them not only to one another but also to the electronic T state (Judd 1978).

2. Vibrations

So far we have only considered the stability of molecular systems. In real physical situations it is necessary to consider the kinetic energy of the ligands. To do this it is convenient to use the notation of second quantization. In one dimension, the position and momentum of a particle can be written as

$$x = (\hbar/2\omega m)^{\frac{1}{2}}(a^{\dagger} + a), \qquad p = i(\hbar\omega m/2)^{\frac{1}{2}}(a^{\dagger} - a) \tag{6}$$

where $a^{\dagger}$ and a are operators for the creation and annihilation respectively of a quantum of oscillation. The coefficients in equations (6) ensure that these operators satisfy the commutation relations of bosons and also that the two equivalent forms for the Hamiltonian for an oscillator are connected by the equation

$$(1/2m)p^2 + (m\omega^2/2)x^2 = (\hbar\omega/2)(a^{\dagger}a + aa^{\dagger}). \tag{7}$$

The Jahn-Teller coupling that we have so far considered is linear in the coordinates. In the one-dimensional case under study we would need to add a term of the type $vf^{\dagger}f(a^{\dagger} + a)$, where v determines the strength of the coupling, and where $f^{\dagger}$ and f create and annihilate the relevant electronic state (supposed, in this overly simplified example, to be non-degenerate). Since we are only dealing with a single electronic state, the statistics associated with $f^{\dagger}$ and f are immaterial. However, we might as well assume that they satisfy the anticommutation relations of fermions to add some zest to the calculations. The Hamiltonian

$$(\hbar\omega/2)(a^{\dagger}a + aa^{\dagger}) + vf^{\dagger}f(a^{\dagger} + a) \tag{8}$$

is easy to handle: all we have to do is write

$$a^{\dagger} \longrightarrow a^{\dagger} + bf^{\dagger}f, \qquad a \longrightarrow a + bf^{\dagger}f \tag{9}$$

and choose b to eliminate the term involving v in (8). We thus obtain a displaced one-dimensional oscillator. This simple result gives no indication of the complexities that arise when the electronic state and the normal mode are degenerate. Although the interested listener (or, perhaps one should say, reader) will no doubt be engrossed in the examples currently being provided in his lectures by Carl Ballhausen, it seems worthwhile to sketch the general case and make a few observations on a particular Jahn-Teller system.

2.1 The Hamiltonian

The construction and study of the relevant Jahn-Teller Hamiltonian must be done within the context of a particular point group G. To begin with, each normal mode, whether degenerate or not, is treated separately. Suppose we pick a mode characterized by the irreducible representation R of G; and suppose further that it possesses n_R components (that is, it is described by n_R normal coordinates). The creation operator a† of the previous section must now be replaced by a collection of n_R operators a_i† ($i = 1, 2, \ldots, n_R$). They necessarily form a basis for the representation R. They can thus be regarded as forming a tensor **a**† with a rank R. In a similar way, a degenerate electronic state characterized by R′ can be formed by the $n_{R'}$ creation operators f_j† ($j = 1, 2, \ldots, n_{R'}$), which form a tensor **f**† with a rank R′. Analogous tensors **a** and **f** can be constructed from the annihilation operators. With these preliminaries taken care of, we can now write down the generalization of (8):

$$H = (\hbar\omega/2)(\mathbf{a}^\dagger.\mathbf{a} + \mathbf{a}.\mathbf{a}^\dagger) + v(\mathbf{f}^\dagger\mathbf{f})^{(R)}.(\mathbf{a}^\dagger + \mathbf{a}). \qquad (10)$$

The dots in this expression indicate scalars with respect to G. The two fermionic tensors (each of rank R′) must be coupled to a rank R in order to give a scalar contribution to H when the coupled product is combined with **a**† + **a**. Should we wish to consider the coupling of the electronic state to a second mode, two additional terms of the same type as those appearing on the right-hand side of equation (10) should be added to H.

Because of the similarity between (8) and (10), one's natural inclination is to try to generalize the substitutions (9); that is, to write

$$\mathbf{a}^\dagger \longrightarrow \mathbf{a}^\dagger + b(\mathbf{f}^\dagger\mathbf{f})^{(R)}, \qquad \mathbf{a} \longrightarrow \mathbf{a} + b(\mathbf{f}^\dagger\mathbf{f})^{(R)}. \qquad (11)$$

Although the coefficient b can be chosen to eliminate the term in v in (10), we see at once that the new operators **a**† and **a** contain terms involving the fermion operators and hence do not necessarily satisfy the commutation relations for bosons. We could only be sure that they do satisfy these relations when $n_R = 1$, that is, for the trivial case of a one-dimensional mode. Otherwise we would expect to find the components of various tensors $(\mathbf{f}^\dagger\mathbf{f})^{(R'')}$ in the evaluation of

$$[(\mathbf{f}^\dagger\mathbf{f})_i{}^{(R)}, (\mathbf{f}^\dagger\mathbf{f})_j{}^{(R)}]$$

rather than the delta function on i and j that we require for bosons. This is not to say that transformations of the Hamiltonian are useless for the non-trivial cases. Froehlich (1952) considered the effect of unitary transformations involving the operator exp(S), where S is necessarily antiunitary, and this approach has been developed by Wagner (1984). It can be shown, for example, that a transformed Hamiltonian can be found for which the term linear in v is eliminated and v first appears as v^2. This is useful in the weak-coupling limit.

2.2 Spin Representations for R

It was mentioned in section 1.1 that the antisymmetric part of R_a^2 must be taken when R_a describes an electronic state formed from an odd number of electrons. In the notation of section 2.1, this means that the representation R must occur in the antisymmetric part of R'^2 in order to get a Hermitian Hamiltonian for systems with an odd number of electrons. To see why this should be so, consider the simple example of a spin-½ state. In this case R′ is the two-dimensional irreducible representation $D_{\frac{1}{2}}$ of SO(3), for which we have $D_{\frac{1}{2}}^2 = D_0 + D_1$. On purely dimensional grounds we know that D_0 occurs in the antisymmetric part of the product and D_1 in the symmetric part, since from n objects we can form n(n + 1)/2 symmetric pairs and n(n - 1)/2 antisymmetric pairs, and these two expression evaluate to 3 and 1 for n = 2. But the scalar representation D_0 corresponds to the number operator

$$\sum_m f_m^\dagger f_m = f_{\frac{1}{2}}^\dagger f_{\frac{1}{2}} + f_{-\frac{1}{2}}^\dagger f_{-\frac{1}{2}},$$

and it is this expression, rather than any labelled by D_1, that is Hermitian. This result depends only on the transformation properties of the spin-½ state, and it thus remains valid when S is an effective spin, such as we would obtain when an ion is placed in a crystalline environment. It is the half-integral nature of S that leads to the crucial result. Of course, if we had started with an integral spin (corresponding to systems with an even number of electrons), we would have found that D_0 occurred in in the symmetric part of the Kronecker square, thus leading to the more natural result described in section 1.1.

2.3. The T × t Jahn-Teller System in the Strong-Coupling Limit

The Hamiltonian H consists of two parts, an oscillator Hamiltonian and an interaction term. When the latter is small, it is obvious that the oscillator part provides the appropriate zeroth-order eigenfunctions. What do we do, however, when the interaction term dominates? This is a highly important case, since it corresponds to vibrations whose amplitudes are small compared to the static displacements. The classic analysis of the strong-coupling limit for the T × t class of Jahn-Teller systems was carried out by Moffitt and Thorson (1957), who found that the inclusion of a small kinetic-energy term to the static analysis (essentially the same as that described above) leads to two sets of vibrational levels. One is characterized by the same angular frequency ω as appears in H, the other by $(2/3)^{\frac{1}{2}}\omega$. The difference between these two frequencies comes about because of a subtle coupling between the lower and upper electronic states as soon as the normal coordinates depart from their static equilibrium positions. Their relative irrationality is somewhat surprising, of course, but there can be no question of the validity of the analysis. Small departures from the strong-coupling limit lead to splittings in the vibrational level.

Even the ground level splits, since the four possible static distortions are degenerate in the absence of any kinetic energy, but there is no four-dimensional irreducible representation in the character table of the octahedral group. In fact, the static ground level yields $A_2 + T_1$, and the splitting of the A_2 and T_1 levels has attracted considerable attention over the years. It is not easy to extend the analysis of Moffitt and Thorson (1957) to cope with this problem. Caner and Englman (1966) used a large set of weak-coupling eigenfunctions and extended the analysis as far towards the strong-coupling limit as they could go. Many of the low-lying levels were seen to be tending towards their strong-coupling counterparts. Schultz and Silbey (1976) used the second-quantized form of H given in equation (10), and subjected it to two unitary transformations.

Perturbation theory was then applied. The irrational form $(2/3)^{\frac{1}{2}}$ (= 0.81649...) is approached by the sequence 1 - 1/6 - 1/72 - ... (= 0.81944...). The energy between the lowest A_2 and T_1 levels is given to second order in perturbation theory by

$$E(A_2) - E(T_1) = (272/243)k^2\exp(-8k^2/9), \tag{12}$$

where the interaction parameter k of Schultz and Silbey (1976) is related to the v of equation (10) by $v = (3/2)^{\frac{1}{2}}k\hbar\omega$. The formula that Caner and Englman (1966) gave on the basis of their numerical results some distance from the strong-coupling limit is

$$E(A_2) - E(T_1) = 0.88k^2\exp(-0.827k^2), \tag{13}$$

which exhibits a modest agreement with equation (12).

In the process of deriving equation (12), Schultz and Silbey (1976) occasionally refer to the unpublished work of Beers, who tackled the $T \times t$ problem from a different direction. By using a variant of Brillouin-Wigner perturbation theory, he obtained

$$E(A_2) - E(T_1) = 1.06k^2\exp(-0.827k^2), \tag{14}$$

thus obtaining a striking agreement with the exponential function found numerically by Caner and Englman (1966). The factor 0.827, according to Beers, appears in the form $(8/9)[3\omega'/(1 + 2\omega')]$, with $\omega' = (2/3)^{\frac{1}{2}}$. Beers had the bad luck to encounter an unappreciative referee when he sent his article to the Physical Review, and his work was never published. Because of this, it seems worthwhile to give an elementary derivation of the exponential factor found by Beers. This is set out in the appendix. We should also mention the formula

$$E(A_2) - E(T_1) = 1.33k^2\exp(-0.827k^2)$$

given (without a detailed derivation) by Bersuker (1984b). The recurrence of the exponential factor of Beers suggests that it was obtained by a method similar to that used in the appendix.

2.4. Glauber states

Rather than transforming the Hamiltonian, we can tackle the strong-coupling limit by asking for the eigenfunctions of the interaction term (the term involving v in equation (10)). The oscillator part of the total Hamiltonian can then be added as a perturbation. The combination $a^\dagger + a$ that occurs in the one-dimensional Hamiltonian (8) can be handled by taking the annihilation operator first. From the work of Glauber (1963) on coherent states of the radiation field, we can easily write down an unnormalized eigenfunction: it is simply $\exp(Ka^\dagger)|0\rangle$, where $|0\rangle$ is the vacuum state (that is, the state corresponding to zero oscillator quanta and which satisfies $a|0\rangle = 0$). To see this, we have only to expand the exponential function and use the commutation relations for bosons. Normalization is easy: we get

$$\exp[-(1/2)K^2]\exp(Ka^\dagger)|0\rangle \tag{15}$$

in all. To generalize this result we replace the number K in the expression (15) by a set of numbers that multiply the various components $a_i{}^\dagger$ that span the irreducible representation R of G. Thus $Ka^\dagger$ is replaced by $\mathbf{K}.\mathbf{a}^\dagger$ and K^2 by $\mathbf{K}^2$.

Of course, the annihilation operator in the interaction Hamiltonian has a creation operator as a companion. A state such as (15) is not an eigenfunction of $a^\dagger$ as well as a. However, we can largely circumvent that difficulty by letting the creation operator act to the left and use the fact that the adjoint of (14) is an eigenbra of $a^\dagger$.

Glauber states have been used to analyze several Jahn-Teller systems (Judd 1974; Judd and Vogel 1975). The principal complication in the method is the non-orthogonality of the states, which means that overlap integrals have to be continually evaluated. On the other hand, exponential functions such as that given in equation (12), appear naturally in the analysis. For example, the effect of the oscillator part of the $T \times t$ Hamiltonian on Glauber states describing the four possible static distortions gives the first-order result

$$E(A_2) - E(T_1) = (32/27)k^2\exp(-8k^2/9), \tag{16}$$

which is in remarkably good agreement with the splitting (12) determined by more sophisticated methods by Schultz and Silbey (1976). However, as indicated in the appendix, the exponential form is not quite right. Nor is it clear how the Rayleigh-Schroedinger approach used by Schultz and Silbey could ever correct the exponential form.

3. High Symmetries

Some Jahn-Teller systems exhibit symmetries higher than what we would expect from an examination of the Hamiltonian. The classic case is the cylindrical symmetry of the linear $E \times e$ system, which has already been briefly mentioned above in connection with the static distortions. The question of why such higher symmetries should exist is an intriguing one, not least because there are sometimes several explanations, each of which seems to provide an answer.

Rather than discuss questions of that kind, we turn to the Jahn-Teller system $T_1 \times (e + t_2)$ and force it to exhibit a symmetry higher than octahedral. This is done by insisting that the two modes e and t_2 possess equal frequencies and correspond to equal linear couplings to the T_1 state. In a different context, we know that the five states of a d electron in an octahedral crystal field break up into the two substates e and t_2 (see, for example, Ballhausen (1962)); so by combining the e and t_2 modes of vibration in an appropriately balanced way we are effectively creating a d phonon. The Hamiltonian (10) becomes

$$H = (\hbar\omega/2)(\mathbf{a}^\dagger.\mathbf{a} + \mathbf{a}\mathbf{a}^\dagger) + v(\mathbf{f}^\dagger\mathbf{f})^{(2)}.(\mathbf{a}^\dagger + \mathbf{a}), \qquad (17)$$

where the creation and annihilation operators are now spherical tensors of rank 2. The symmetry of H is that of the rotation group in 3 dimensions, SO(3).

3.1. Static distortions

The classic theoretical analysis of the Hamiltonian (17) is that of O'Brien (1971), who developed the theory to account for the absorption spectrum of an electron trapped in an oxygen vacancy in CaO. The excitation $s \rightarrow p$ produces a sharp zero-phonon line and, to the blue, an attendant broad band corresponding to the interaction of the p electron with the surrounding octahedron of calcium ions. In octahedral symmetry the states of the p electron yield the three-fold degenerate state T_1. For our purposes, we need not go over O'Brien's analysis in full. Instead, we shall consider the static Jahn-Teller effect and then add a small amount of kinetic energy. Our approach is modeled on that of sections 1.2 - 1.4. In some respects our task is easier; for example, the Clebsch-Gordan coefficients for SO(3) can be rapidly found from tables. Since the Hamiltonian possesses SO(3) symmetry there can be no preferred directions for the displacements of the ligands. Thus we can replace the displacement tensor $\mathbf{a}^\dagger + \mathbf{a}$ by the set of rank-2 spherical harmonics characterized by the polar vector $\mathbf{R}$. On ordering the p-electron states in the sequence m = 1, 0, -1, the interaction matrix $A_{mm'}$ turns out to possess the following entries:

$$A_{11} = A_{-1-1} = v(3Z^2 - R^2)/(24)^{\frac{1}{2}},$$

$$A_{10} = -A_{0-1} = v(3/4)^{\frac{1}{2}}Z(X - iY),$$

$$A_{1-1} = v(3/8)^{\frac{1}{2}}(X - iY)^2, \qquad (18)$$

$$A_{01} = - A_{-10} = v(3/4)^{\frac{1}{2}}Z(X + iY),$$

$$A_{00} = - v(3Z^2 - R^2)/(6)^{\frac{1}{2}},$$

$$A_{-11} = v(3/8)^{\frac{1}{2}}(X + iY)^2.$$

The eigenvalues E of the determinantal equation $|A_{mm'} - E\delta(m, m')| = 0$ are found to be $vR^2/(6)^{\frac{1}{2}}$ (twice) and $-vR^2(2/3)^{\frac{1}{2}}$. Without further analysis we cannot know which of these two solutions corresponds to the ground level: however, it turns out to be the latter. In terms of the states x, y and z of the p electron, the normalized eigenfunction of the ground level takes the simple form

$$(X/R)|x\rangle + (Y/R)|y\rangle + (Z/R)|z\rangle \qquad (19)$$

To find the displacements (x_j, y_j, z_j) of ligand j, we have to match the five spherical harmonics with the normal coordinates describing the e and t_2 modes. We must also insist that the 18 displacements do not represent any other modes. The details of this have been given elsewhere (Judd 1978). The results can be specified as follows:

$$\begin{aligned} x_1 &= - x_4 = 3X^2 - R^2, \\ y_1 &= - y_4 = x_2 = - x_5 = 3XY, \\ z_1 &= - z_4 = x_3 = - x_6 = 3ZX, \\ y_2 &= - y_5 = 3Y^2 - R^2, \\ z_2 &= - z_5 = y_3 = - y_6 = 3YZ, \\ z_3 &= - z_6 = 3Z^2 - R^2. \end{aligned} \qquad (20)$$

A typical set of displacements is shown for the ligands 1, 2 and 3 in figure 1.

3.2. Movement in the potential minimum

As is described in the caption for figure 1, the displaced ligands can assume an infinity of positions for the same minimum potential energy. The loci of these positions for ligand j is a sphere whose center lies on the axis leading out from the origin O to the ligand j. The six vectors describing the positions of the displaced ligands from the centers of their six associated spheres forms a regular octahedron. If a small amount of kinetic energy is added to the system, this octahedron rotates. Thus we can expect a series of rotational levels characterized by the moment of inertia I of this octahedron.

There is, however, a surprising feature of the rotational energy

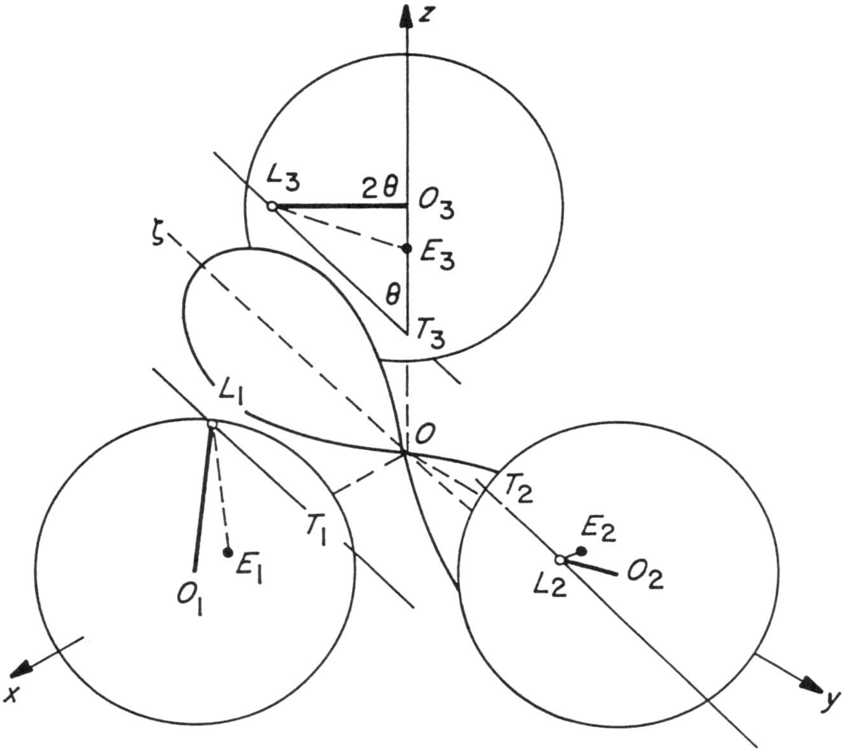

Figure 1. The Jahn-Teller system $T_1 \times (e + t_2)$ in the strong-coupling limit. The electronic T_1 state is represented by the lobes of a p electron and defines an axis ζ. The linear coupling between this electron and the e and t_2 modes shifts the octahedron of ligands from their original sites E_i to their displaced positions L_i. Of the six ligands, only those three lying initially on the positive x, y, and z axes are shown in the figure: the other three can be found by inversion in the origin O. Under conditions of equal frequencies and equal couplings, the equilibrium configuration corresponds to an infinity of static distortions, all possessing the same energy. The loci of the displaced positions L_i are the spheres centered at O_i and intersecting the axes at the interior points T_i. The lines L_iT_i are all parallel to the ζ axis of the p electron; moreover the vectors O_iL_i, if translated (without rotation) to unite the points O_i, define a regular octahedron. If a small amount of kinetic energy is added to the system, this octahedron turns in a rigid fashion to produce a series of rotational energy levels. Because of the way the different parts of this system are locked together, a rotation of the p electron that interchanges its two lobes corresponds to a complete rotation of the octahedron. Thus the ligands return to their positions L_i while the wavefunction of the p electron changes sign, thereby providing an example of Berry's phase.

levels: only those angular momenta L for which L is an odd integer occur. This was recognized twenty years ago by O'Brien (1969). The reason lies in the quadratic nature of the coordinates (20). If **R** is replaced by -**R**, the ligand coordinates are unchanged, while the eigenfunction (19) reverses sign. In fact, we can track the relative movements of the ligands and the orbit of the p electron from figure 1, and it is clear that a rotation that interchanges the positive and negative lobes of the p electron brings the ligands through a full circle. Thus the ligands return to their initial displaced positions while the wavefunction of the p electron changes sign. Since the total wavefunction of a system must be single-valued, we have to compensate for this sign change by multiplying (19) by functions of **R** that are also odd under inversion. This can be done by choosing the odd spherical harmonics Y_L to represent the rotational characteristics of the system. Thus odd L are allowed while even L are excluded.

We should also mention here a theorem of Longuet-Higgins (1975), which is stated in the following form: "If the wavefunction of a given electronic state changes sign when transported adiabatically round any loop in configuration space, then its potential energy surface must intersect that of another state at some point within the loop." For the example under study, the intersection takes place at **R** = 0, where the potential energy surfaces corresponding to the three states of the p electron intersect.

4. Berry's phase

M. Jourdain: Par ma foi! il y a plus de quarante ans que je dis de la prose sans que j'en susse rien.

Molière: Le Bourgeois Gentilhomme.

A few years ago, practitioners of the Jahn-Teller effect learned that they had unknowingly been using Berry's phase for many years -- perhaps not the forty of Molière's hero, but certainly for the thirty years since the article of Longuet-Higgins et al. (1958). In the abstract to his article, Berry (1984) wrote "A quantal system in an eigenstate, slowly transported round a circuit C by varying parameters **R** in its Hamiltonian H(**R**), will acquire a geometrical phase factor $\exp\{i\gamma(C)\}$ in addition to the familiar dynamical phase factor." As we shall see, the phase factor of -1 just obtained in section 3.2 is an example of this.

Of course, the achievement of Berry was to provide a general perspective on such phase factors and to show how they appear in many areas of physics. For an account of the wide scope of Berry's phase, as well as a detailed list of references to later work, the reader is referred to the review article of Jackiw (1988). Because of the enormous interest that Berry's work has engendered, it seems worthwhile to outline his discussion of the origin of his geometrical phase and to illustrate his approach by means of the octahedral Jahn-Teller system $T_1 \times (e + t_2)$.

4.1. Derivation

Suppose we are studying a general problem involving the electronic wavefunction of a molecule whose nuclei are fixed in some configuration defined by the coordinates $\mathbf{R}$. Different electronic states may be distinguished by the label n. The time dependence of the electronic wavefunction is removed by writing

$$|\psi(t)\rangle = \exp(-iE_n t/\hbar)\,|n(\mathbf{R})\rangle, \tag{21}$$

where E_n is the energy of the state n. Suppose, now, that the nuclear frame undergoes an adiabatic change so that $\mathbf{R}$ becomes a function of t. The electronic state in (21) becomes $|n(\mathbf{R}(t))\rangle$, with the result that $|\psi(t)\rangle$ no longer satisfies the Schroedinger equation

$$H|\psi(t)\rangle = i\hbar(\partial/\partial t)|\psi(t)\rangle.$$

We must evidently include a compensatory phase factor $\exp(i\gamma_n(t))$ on the right-hand side of equation (21), where $\gamma_n(t)$ satisfies

$$(\partial/\partial t)\exp(i\gamma_n(t))\,|n(\mathbf{R}(t))\rangle = 0. \tag{22}$$

That is,

$$d\gamma_n = \langle n(\mathbf{R})|\,i d\mathbf{R}.\nabla_R\,|n(\mathbf{R})\rangle. \tag{23}$$

On integrating round a circuit C, the phase change is given by

$$\gamma_n(C) = \int_C \mathbf{A}.d\mathbf{R}, \tag{24}$$

where the vector potential $\mathbf{A}$ satisfies

$$\mathbf{A} = i\langle n(\mathbf{R})|\nabla_R|n(\mathbf{R})\rangle. \tag{25}$$

4.2. Application to $T_1 \times (e + t_2)$

The naive substitution of the eigenfunction (19) into equation (25) yields the immediate result $\mathbf{A} = 0$. The reason for this is that (19) is not single-valued: as we have already seen, there are two forms of (19) (differing with respect to a sign) corresponding to the same set of ligand displacements $\mathbf{R}$. Berry (1984) makes the explicit statement that $|n(\mathbf{R})\rangle$ should be single-valued for his derivation to work. In the case of a double-valuedness, we would obviously have to take some linear combination of the two states and include separate phase factors for each, thus complicating the compensatory mechanism described by equation (22). The simplest way to make (19) single-valued (while preserving its normalization) is to multiply it by the factor $\exp(i\varphi)$, where φ is the second polar angle (as usual). This is the procedure of

Chancey and O'Brien (1988). In the Cartesian coordinate system that we are using, we thus replace (19) by

$$[(X + iY)/(X^2 + Y^2)^{\frac{1}{2}}][(X/R)|x\rangle + (Y/R)|y\rangle + (Z/R)|z\rangle]. \tag{26}$$

We can now bring equation (25) into play to find **A**. The nabla operator in (25) is interpreted as a vector operator with the partial derivatives with respect to X, Y and Z as its three components. We could, of course, work in polar coordinates, but that scheme possesses the complication that the term involving $\partial/\partial\varphi$ is prefaced by the function cosec θ that diverges when $\theta = 0$. The result of the calculation is

$$A_X = Y/(X^2 + Y^2), \quad A_Y = -X/(X^2 + Y^2), \quad A_Z = 0. \tag{27}$$

These equations show that the source for **A** can be thought of as a flux tube along the Z axis with a strength of -1, a result obtained by Chancey and O'Brien (1988). Provided we avoid the singular points for which X = Y = 0, equation (24) yields the result

$$\gamma(C) = -\int_C d\varphi \, . \tag{28}$$

Adiabatic motions of the ligands that return them to their original positions are of two types: $\mathbf{R} \to \mathbf{R}$ and $\mathbf{R} \to -\mathbf{R}$. In the first case the angle φ is increased or decreased by a multiple of 2π, and $\exp\{i\gamma(C)\} = 1$. In the second case the angle is increased or decreased by an odd multiple of π and so $\exp\{i\gamma(C)\} = -1$. Thus Berry's phase exactly matches the phase changes determined for our Jahn-Teller system in section 3.2. It is important to recognize that a complete encirclement of the flux tube produces no phase change: it is the half-turn that is crucial in producing the sign reversal.

4.3. Discussion

A number of points need to be made about the analysis above. In the first place, the flux tube is a mathematical construct. If we had made the wavefunction (19) single-valued by multiplying it by the expression $(Y + iZ)/(Y^2 + Z^2)^{\frac{1}{2}}$, we would have found that our flux tube coincided with the X axis. This possibility should come as no surprise, since our Jahn-Teller system is symmetrical with respect to X, Y and Z. However, it has an interesting consequence. Since the Berry phase is built up by integrating over φ, we might have thought that we could write $\gamma(\mathbf{R})$. But in taking the ligands from a starting position $\mathbf{R}_0$ to some other position **R** there is no reason why the angular change with respect to the Z axis should be the same as that with respect to the X axis. Thus, as Berry (1984) recognized, his phase is demonstrably not a function of **R**. It is only when the trajectories $\mathbf{R}_0 \to \mathbf{R}_0$ or $\mathbf{R}_0 \to -\mathbf{R}_0$ are completed that the two angular changes become identical.

One of the reasons that **A** is called a vector potential is to draw out the analogy with electromagnetic theory, where gauge transformations involving **A** are often considered. They amount to adding the gradient of a scalar function λ to **A**. From equations (27) we can see that for us

$$\mathbf{A} = -\nabla[\tan^{-1}(Y/X)] = -\nabla\varphi,$$

so a simple gauge change would correspond to a rotation of the X and Y axes about the Z axis. The shift in the flux line described in the preceding paragraph can be effected by taking

$$\lambda = \tan^{-1}(Y/X) - \tan^{-1}(Z/Y),$$

which annihilates the flux line on the Z axis and creates a new one on the X axis.

Berry (1984) was able to develop his formula (24) further by using Stokes' theorem to convert the line integral to a surface integral involving the curl of **A**. His new formula no longer requires single-valuedness for $|n(\mathbf{R})\rangle$; however we cannot immediately take advantage of this result for the Jahn-Teller system under study because our flux lines produce singularities in the surface being integrated over, thus complicating the application of Stokes' theorem. Furthermore, the transformation that carries **R** into -**R** takes the ligands through a complete cycle, but does not produce a closed loop in the space defined by the X, Y and Z axes. To utilize Stokes' theorem we would have to either complete the loop in some artificial way or else choose a coordinate system tied more directly to the positions of the ligands.

Chancey and O'Brien (1988) refer to the work of Aitchison (to appear in Physica Scripta) in which the Jahn-Teller system $E \times e$ is investigated from Berry's point of view. There are many parallels to the analysis for $T_1 \times (e + t_2)$. In particular, Aitchison has found that a flux line of strength $\frac{1}{2}$ appears, thus matching the flux line of strength -1 described above.

Jackiw (1988) and some of the writers he cites have pondered whether Berry's phase can be observed or not. To the extent that analyses of such Jahn-Teller systems as $E \times e$ and $T_1 \times (e + t_2)$ agree with experiment, while, at the same time, depend on sign reversals in the electronic wavefunction when the ligands are taken through a complete cycle, an affirmative answer can clearly be given. Chancey and O'Brien (1988) point out that the existence of a vibronic triplet for the lowest level in $T_1 \times (e + t_2)$ is directly related to Berry's phase, as we have seen in section 4.2.

Acknowledgements

Drs. C. C. Chancey and M. C. M. O'Brien are thanked for making available a preprint of their article on Berry's phase, and conversations with Dr. G. M. S. Lister proved useful in getting an understanding of it.

Appendix. Overlap in the T × t System

As can be seen from equations (13) and (14), the exponential factor $\exp(-0.827k^2)$ quoted by Caner and Englman (1966) is matched by Beers in his unpublished analysis. It also recurs in the formula of Bersuker (1984b). We can understand how the analytical form found by Beers for the factor 0.827 comes about by some elementary reasoning. To do this, we calculate the wavefunction overlap of two oscillators located in two of the four possible static potential minima of the T × t Jahn-Teller system. The principal complication in doing this is the necessity to allow for the two frequencies ω (for longitudinal vibrations) and ω' (for transverse vibrations) found by Moffitt and Thorson (1957) and described in section 2.3. A one-dimensional oscillator chracterized by a mass m and angular frequency ω has an unnormalized ground-state wavefunction given by $\exp(-\frac{1}{2}qx^2)$, where $q = m\omega/\hbar$. If we consider the first of the four displaced sites (5), corresponding to the coordinates (Q, Q, Q), we begin by writing

$$(x - Q)^2 + (y - Q)^2 + (z - Q)^2$$
$$= (2/3)(x^2 + y^2 + z^2 - yz - zx - xy) + (x + y + z - 3Q)^2/3,$$

thereby breaking the square of the radial distance up into a transverse part and a longitudinal part. The ground-state wavefunction of the three-dimensional oscillator is thus proportional to

$$\exp[-q(x + y + z - 3Q)^2/6 - (q'/3)(x^2 + y^2 + z^2 - yz - zx - xy)], \tag{A1}$$

where, following the result of Thorson and Moffitt (1957), we have $q' = (2/3)^{\frac{1}{2}}q$. A similar wavefunction centered at the displaced site $(-Q, -Q, Q)$ has the form

$$\exp[-q(z - x - y - 3Q)^2/6 - (q'/3)(x^2 + y^2 + z^2 + yz + zx - xy)]. \tag{A2}$$

The overlap between (A1) and (A2) is the integral of their product. The latter is

$$\exp[-(q + 2q')(x^2 + y^2 + z^2)/3 + 2(q' - q)xy/3 + 2qzQ - 3qQ^2]. \tag{A3}$$

By writing

$$z = z' + 3qQ/(q + 2q') \tag{A4}$$

the term $2qzQ$ in (A3) can be eliminated. The integration over z' proceeds without difficulty and yields a number. However, $3q^2Q^2/(q + 2q')$ remains in the exponential, and, when combined with the last term in (A3), yields

$$\exp[-\ 6qq'Q^2/(q + 2q')]. \tag{A5}$$

No additional terms come from manipulations with x and y, so the exponential function (A5) is the only exponential to be left after the integration has been carried out. Furthermore, all integrals of the Hamiltonian between (A1) and (A2) will involve the same exponential function, so (A5) must represent the exponential part of the energy separation between the A_2 and T_1 levels. If we set $q' = q$, thus ignoring the analysis of Thorson and Moffitt (1957), the expression (A5) becomes $\exp(-\ 2qQ^2)$, which, with a little manipulation, turns out to be $\exp(-\ 8k^2/9)$. This is the form of the exponential given by Schultz and Silbey (see equation (12)) and by the Glauber-state analysis (see equation (16)). However, on putting $q' = (2/3)^{\frac{1}{2}}q$ we find that (A5) becomes

$$\exp[-\ (8k^2/9)(3q'/(q + 2q'))] = \exp(-\ 0.827k^2).$$

This more accurate expression agrees not only with the unpublished result of Beers but also with the exponential function given by Caner and Englman (1966) on the basis of their numerical work. In fact, the agreement is so good that it seems likely that Caner or Englman carried out an analysis similar to that given here in order to find the right asymptotic form for the energy difference. Although they do not mention any such calculation, it is interesting to note that some years later Englman (1972) gave the slightly different form $\exp(-\ 0.81k^2)$ as the best approximation to the numerical results.

References

Ballhausen, C. J. (1962) Introduction to Ligand Field Theory, McGraw-Hill, New York.

Berry, M. V. (1984) Quantal phase factors accompanying adiabatic changes, Proc. Roy. Soc. (London) A 392, 45-57.

Bersuker, I. B. (1984a) The Jahn-Teller Effect: a Bibiographic Review, IFI/Plenum, New York.

Bersuker, I. B. (1984b) The Jahn-Teller Effect and Vibronic Interactions in Modern Chemistry, Plenum, New York.

Blount, E. I. (1971) The Jahn-Teller theorem, J. Math. Phys. 12, 1890-1896.

Butler, P. H. (1981) Point Group Symmetry Applications, Plenum, New York.

Caner, M. and Englman, R. (1966) Jahn-Teller effect on a triplet due to threefold degenerate vibrations, J. Chem. Phys. 44, 4054-4055.

Chancey, C. C. and O'Brien, M. C. M. (1988) Berry's geometric quantum phase and the $T_1 \times (e_g + t_{2g})$ Jahn-Teller effect, J. Phys. A: Math. Gen. 21, 3347-3353.
Englman, R. (1972) The Jahn-Teller Effect in Molecules and Crystals, Wiley-Interscience, London.
Froehlich, H. (1952) Interaction of electrons with lattice vibrations, Proc. Roy. Soc. (London) A 215, 291-298.
Glauber, R. J. (1963) Coherent and incoherent states of the radiation field, Phys. Rev. 131, 2766-2788.
Griffith, J. S. (1962) The Irreducible Tensor Method for Molecular Symmetry Groups, Prentice-Hall, Englewood Cliffs, New Jersey.
Jackiw, R. (1988) Berry's phase -- Topological ideas from Atomic, Molecular and Optical Physics, Comments At. Mol. Phys. 21, 71-82.
Jahn, H. (1935) Rotation und Schwingung des Methanmolekuels, Ann. Phys. 23, 529-556.
Jahn, H. (1938) Stability of polyatomic molecules in degenerate electronic states. II - Spin degeneracy, Proc. Roy. Soc. (London) A 164, 117-131.
Jahn, H. and Teller, E. (1936) Stability of degenerate electronic states in polyatomic molecules, Phys. Rev. 49, 874.
Jahn, H and Teller, E. (1937) Stability of polyatomic molecules in degenerate electronic states. I - Orbital degeneracy, Proc. Roy. Soc. (London) A 161, 220-235.
Judd, B. R. (1974) Lie groups and the Jahn-Teller effect, Can. J. Phys. 52, 999-1044.
Judd, B. R. (1978) Ligand trajectories for a degenerate Jahn-Teller system, J. Chem. Phys. 68, 5643-5646.
Judd, B. R. (1984) Jahn-Teller trajectories, in I. Prigogine and S. A. Rice (eds.), Advances in Chemical Physics, vol. LVII, John Wiley & Sons, New York.
Judd, B. R. and Vogel, E. E. (1975) Coherent states and the Jahn-Teller effect, Phys. Rev. B 11, 2427-2435.
Kramers, H. A. (1930) Théorie générale de la rotation paramagnétique dans les cristaux, Proc. Kon. Ned. Akad. Wet. (Amsterdam) 33, 959-972.
Longuet-Higgins, H. C. (1975) The intersection of potential energy surfaces in polyatomic molecules, Proc. Roy. Soc. (London) A 344, 147-156.
Longuet-Higgins, H. C., Oepik, U., Pryce, M. H. L. and Sack, R. A. (1958) Studies of the Jahn-Teller effect. II. The dynamical problem, Proc. Roy. Soc. (London) A 244, 1-16.
Moffitt, W. and Thorson, W. (1957) Vibronic states of octahedral complexes, Phys. Rev. 108, 1251-1255.
O'Brien, M. C. M. (1969) Dynamic Jahn-Teller effect in an orbital triplet state coupled to both E_g and T_{2g} vibrations, Phys. Rev. 187, 407-418.
O'Brien, M. C. M. (1971) The Jahn-Teller effect in a p state equally coupled to E_g and T_{2g} vibrations, J. Phys. C: Solid St. Phys. 4, 2524-2536.
Raghavacharyulu, I. V. V. (1973) Simple proof of the Jahn-Teller theorem, J. Phys. C: Solid St. Phys. 6, L455-457.

Ruch, E. and Schoenhofer, A (1965) Ein Beweis des Jahn-Teller-Theorems mit Hilfe eines Satzes ueber die Induktion von Darstellungen endlicher Gruppen, Theoret. Chim. Acta (Berlin) 3, 291-304.
Schultz, M. J. and Silbey, R. (1976) A theoretical study of the strongly coupled T × t Jahn-Teller system, J. Chem. Phys. 65, 4375-4383.
Teller, E. (1982) The Jahn-Teller effect - Its history and applicability, Physica 114A, 14-18.
Van Vleck, J. H. (1939) The Jahn-Teller effect and crystalline Stark splitting for clusters of the form XY_6, J. Chem. Phys. 7, 72-84.
Wagner, M. (1984) Unitary transformation methods in vibronic problems, in Yu. E. Perlin and M. Wagner (eds.), The Dynamical Jahn-Teller Effect in Localized Systems, North-Holland, Amsterdam.
Wigner, E. (1932) Ueber die Operation der Zeitumkehr in der Quantenmechanik, Nachr. Gesell. Wiss. (Math. Phys. Kl.) Goettingen 546-559.

VIBRONIC INTENSITIES

GAD FISCHER
Department of Chemistry
The Australian National University
GPO Box 4, Canberra ACT 2601
Australia

ABSTRACT. An expression for the vibronic transition moment is obtained from a semi-classical treatment of the matter-radiation interaction. The significance of the various terms in the expansion of the transition moment as a power series in the nuclear coordinates is discussed. Finally, a number of questions are raised regarding the validity and appropriateness of the usual Herzberg-Teller approach to the determination of the induced intensity. Some transition intensities in the benzene spectra are given as illustration.

For the absorption intensities of molecular transitions (f ← g), the extinction coefficient ε is commonly used as the experimental measure of absorption properties. The extinction coefficient is related to the incident and transmitted light intensities by the Beer-Lambert law. Now, it can be shown that the extinction coefficient is also related to the Einstein B-coefficient by

$$\int \varepsilon d\nu \;\alpha\; \nu B \tag{1}$$

and
$$B \;\alpha\; |\underline{M}|^2 \tag{2}$$

where ν is the frequency and $\underline{M}$ the transition moment Hence we have a relation between the experimentally determined extinction coefficient and the theoretically formulated transition moment. It is given by

$$\int \varepsilon d\nu \;\alpha\; \nu |M|^2 \tag{3}$$

An expression for the transition moment can be obtained from a consideration of the interaction of the radiation field with the molecular system.
The Hamiltonian for the interaction between the momentum of the particles and the radiation field is presented in the conventional form, (S.I. units)

$$H_{int} = \sum_{j=1}^{N} (-eZ_j/m_j)\underline{A}_j \cdot p'_j + (e^2 Z_j^2/2m_j)A_j^2 \tag{4}$$

C. D. Flint (ed.), Vibronic Processes in Inorganic Chemistry, 103–110.
© 1989 by Kluwer Academic Publishers.

where a system of N particles (electrons and nuclei) is considered with masses m_j, charges eZ_j, momentum vectors $\underline{p}'_j$ and position vectors $\underline{q}'_j$. The $\underline{A}_j$ are operators for the vector potential field where $\underline{A}_j$ means the vector potential at the position of particle j. The vector potential is given by

$$\underline{A}_j(\underline{q}'_j,t) = \sum_{\lambda}\{\underline{A}_{j\lambda}(t)\exp i\underline{k}_{\lambda}\bullet\underline{q}'_j + \underline{A}^*_{j\lambda}(t)\exp -i\underline{k}_{\lambda}\bullet\underline{q}'_j\} \tag{5}$$

where the $\underline{A}_{j\lambda}$ are described as mode amplitudes and k is the wave vector whose direction is the direction of propagation and $k/2\pi$ is the wavenumber

$$k/2\pi = \nu/c \tag{6}$$

The asterisk denotes the complex conjugate.

The mode amplitudes are transverse to the direction of propagation and are specified in terms of two mutually orthogonal directions denoted by the unit vectors $\underline{e}_{\lambda}$ that may be real or complex, hence

$$\underline{A}_j(\underline{q}',t) = \sum_{\lambda}\{\underline{e}_{\lambda}a_{\lambda}\exp i\underline{k}_{\lambda}\bullet\underline{q}'_j + \underline{e}^*_{\lambda}a^*_{\lambda}\exp -i\underline{k}_{\lambda}\bullet\underline{q}'_j\} \tag{7}$$

The amplitude of a transition between the states f and g deriving from the $\underline{A}\bullet\underline{p}'$ term is proportional to the transition matrix element,

$$<f(q')|\sum_j(eZ_j/m_j)\exp(i\underline{k}_{\lambda}\bullet\underline{q}'_j)\underline{e}_{\lambda}\bullet\underline{p}'_j|g(q')> \tag{8}$$

Only one-photon processes are considered so that the A^2 term in H_{int} (Eqn (4)) has been omitted.

In the approximation in which the variation of the vector potential $\underline{A}$ over the molecular dimensions is neglected (known as the dipole approximation) we have

$$\exp(i\underline{k}_{\lambda}\bullet\underline{q}'_j) = 1 \tag{9}$$

and the transition matrix element is given by

$$<f(q')|\sum_j(eZ_j/m_j)\underline{e}_{\lambda}\bullet\underline{p}'_j|g(q')> = \underline{e}_{\lambda}\bullet\underline{M}^p_{fg} \tag{10}$$

The transformation from momentum to dipole formulation makes use of the commutator relation

$$[\underline{q}'_j,H] = i\hbar\underline{p}'_j/m_j \tag{11}$$

giving

$$\langle f(q')|\sum_j (eZ_j/m_j)\underline{e}_\lambda \cdot \underline{p}'_j|g(q')\rangle = i\hbar^{-1}\Delta E_{fg}\langle f(q')|\sum_j eZ_j\underline{e}_\lambda \cdot \underline{q}'_j|g(q')\rangle \tag{12}$$

Hence for a transition between the vibronic states fv and gu the transition moment may be written

$$\underline{M}_{fv,gu} = \langle fv(q,Q)|\underline{\mu}|gu(q,Q))\rangle \tag{13}$$

where $\underline{\mu}$ is the electric-dipole operator and is given by the sum of the dipole operators of the electrons and of the nuclei,

$$\sum_j eZ_j\underline{q}'_j = \underline{\mu}(q) + \underline{\mu}(Q) \tag{14}$$

Within the BO approximation the transition moment, Eqn (13), may therefore be expressed as the sum of two terms[a]

$$\underline{M}_{fv,gu} = \langle f(q,Q)|g(q,Q)\rangle(v_f(Q)|\underline{\mu}(Q)|u_g(Q)) + (v_f(Q)|\langle f(q,Q)|\underline{\mu}(q)|g(q,Q)\rangle|u_g(Q)) \tag{15}$$

Now, because the electronic wavefunctions are orthogonal, the first term vanishes unless
f = g. It is the term which governs the intensities of vibrational transitions. The second term is of relevance when transitions between different electronic states are concerned. Generally, the electronic transition moment is separately defined as

$$\underline{M}_{fg}(Q) = \langle f(q,Q)|\underline{\mu}(q)|g(q,Q)\rangle \tag{16}$$

and since it is a function of the nuclear coordinates, the integration over the nuclear coordinates must take this fact into account. The usual procedure is to expand the electronic transition moment as a Taylor series in the nuclear coordinates about the equilibrium nuclear configuration, Q_0

$$\underline{M}(Q) = \underline{M}(Q_0) + \sum_r \left(\frac{\partial \underline{M}(Q)}{\partial Q_r}\right)_{Q_0} Q_r + \tfrac{1}{2}\sum_{r,s}\left(\frac{\partial^2 \underline{M}(Q)}{\partial Q_r \partial Q_s}\right)_{Q_0} Q_r Q_s + \ldots \tag{17}$$

It should be recognized that this expansion of the electronic transition moment parallels the Herzberg-Teller expansion, discussed earlier , in that it makes explicit the Q dependence. This Q dependence is absent if only the first term $\underline{M}(Q_0)$ of Eqn (17) is taken into consideration corresponding to the electronic transition moment at the fixed nuclear configuration Q_0. Clearly, the transition moment can then be interpreted as that between CA electronic states.

Substitution of Eqn (17) in the second term of Eqn (15) leads to the following well-known expression for the vibronic transition moment

---

[a] We have used round and angular brackets to designate integration over nuclear and electronic coordinates respectively.

$$\underline{M}_{fv,gu} = \underline{M}(Q_o)(v_f(Q)|u_g(Q)) + \sum_r\left(\frac{\partial \underline{M}(Q)}{\partial Q_r}\right)_{Q_o}(v_f(Q)|Q_r|u_g(Q))$$
$$+ \tfrac{1}{2}\sum_{r,s}\left(\frac{\partial^2 \underline{M}(Q)}{\partial Q_r \partial Q_s}\right)_{Q_o}(v_f(Q)|Q_rQ_s|u_g(Q)) + \ldots \qquad (18)$$

The Condon approximation consists in the neglect of all but the first term of this expression. The transition moment may then be readily expressed as a product of an electronic transition moment $\underline{M}(Q_o)$ which is independent of the nuclear coordinates and a vibrational overlap integral $(v_f(Q)|u_g(Q))$, whose absolute square is known as the Franck-Condon factor. In this limit the transition intensities are therefore proportional to the Franck-Condon factors.

TABLE 1. Measured intensities of some prominent second-order HT origins relative to the combination $\nu_4 + \nu_{10}$ in the electronic spectrum of benzene.

Band[a)]	Intensity
10^0_2	0.4
16^0_2	0.15
17^0_2	0.7
$4^0_1 10^0_1$	1.0
$5^0_1 10^0_1$	~0
$11^0_1 16^0_1$	0.5
$11^0_1 17^0_1$	~0
$16^0_1 17^0_1$	0.15
(6^0_1	~50)

a) Band designation and numbering follow Callomon, Dunn and Mills (1966) Phil. Trans. R. Soc. A259, 499.

In the breakdown of the Condon approximation, more generally called HT coupling, the Q dependence of the electronic transition moment is introduced. That is, the higher order terms of Eqn (18) are included. The term linear in Q corresponds to the HT induced intensity, the quadratic term to the second order HT induced intensity and so on.

The electronic spectrum of the first singlet system of benzene provides a very clear and well-documented illustration of first- and higher-order HT vibronic coupling. For the isolated (unperturbed) molecule, the vibrationless transition is forbidden on symmetry grounds and intensity is only induced by vibronic coupling. Although the higher-order vibronically induced intensities are one or more orders of magnitude smaller than those for the usual first-order, they nevertheless account for an appreciable fraction of the total transition intensity. The measured relative intensities of some prominent second-order HT origins are listed in Table 1.

A number of questions has been raised regarding the validity and appropriateness of the HT approach. One concerns a possible cancellation of HT induced contributions.

Although the standard HT approach has been very successful in explaining the vibrational structure in the electronic spectra of many molecules, the approach has nevertheless come under some criticism. It has been suggested that interference between HT terms that derive from the vibronic coupling of the excited and ground states may be effective enough to lead to the cancellation of the HT contribution to the transition intensity so that intensity induced through breakdown of the BO approximation (BO-B) becomes predominant. Recall that expression (18) has been obtained within the BO approximation and justified on the grounds that within the weak coupling domain the contribution to the transition intensity from BO breakdown is much smaller than that due to HT coupling.

One means of assessing the validity of this argument is through a measurement of the ratio of the transition intensities for different isotopes.

The HT induced transition moment is given by the second term of Eqn (18)

$$\underline{M}^{HT}_{fv,\,gu} = \sum_r \left(\frac{\partial \underline{M}(Q)}{\partial Q_r}\right)_{Q_0} (v_f|Q_r|u_g) \tag{19}$$

For the purposes of this discussion, only one of the HT terms is considered. The coupling of the ground state is neglected. This is generally a good assumption because its contribution is smaller due to the much larger energy separation of the electronic states. The HT induced moment is then given by

$$\underline{M}^{HT}_{fv,\,gu} \simeq \sum_r \sum_p (\langle f|\partial/\partial Q_r|p\rangle \langle p|\underline{\mu}(q)|g\rangle)_{Q_o} (v_f|Q_r|u_g) \tag{20}$$

It follows that the ratio of HT induced transition intensities for two different isotopes I_H/I_D varies as the squared ratio of the vibrational integrals,

$$\left| \frac{(v_f|Q_r|u_g)_H}{(v_f|Q_r|u_g)_D} \right|^2 \tag{21}$$

With the assumption that the inducing modes r can be treated as harmonic oscillators, and recalling that the vibrational coordinates are mass-weighted, we then have that

$$\frac{I_H}{I_D} \propto \frac{\nu_H}{\nu_D} \tag{22}$$

Expression (22) requires a further qualification. It has been assumed that the overlap integrals are equal, that is

$$(v_f|u_g)_H = (v_f|u_g)_D$$

For the non-BO mechanism (BO breakdown) the respective transition moment is given by

$$\underline{M}^{BO-B}_{fv,gu} = \sum_{p,w} (\varepsilon_{pw} - \varepsilon_{gu})^{-1}(\langle fv|\underline{\mu}(q)|pw\rangle)(\langle pw|T(Q)|gu\rangle)$$

$$+ (\varepsilon_{pw} - \varepsilon_{fv})^{-1}(\langle fv|T(Q)|rw\rangle)(\langle pw|\mu(Q)|gu\rangle) \quad (23)$$

where the summations extend over the complete BO sets. Making the same assumptions as above, and setting the vibronic energy gap

$$\varepsilon_{pw} - \varepsilon_{fv} = E_p - E_f, \quad (24)$$

which is justified if the energy gaps are much larger than the vibrational energy quanta (weak coupling domain), the approximate expression is obtained for the transition moment

$$\underline{M}^{BO-B}_{fv,gu} \simeq \sum_r \sum_p [\langle f|\partial/\partial Q_r|p\rangle\langle p|\underline{\mu}(q)|g\rangle]_{Q_o}$$

$$\times (E_p - E_g)^{-1}(v_f|\partial/\partial Q_r|u_g) \quad (25)$$

Therefore, here the ratio of transition intensities I_H/I_D, for two different isotopes varies approximately as

$$\left|\frac{(v_f|\partial/\partial Q_r|u_g)_H}{(v_f|\partial/\partial Q_r|u_g)_D}\right|^2 \alpha \left(\frac{\nu_H}{\nu_D}\right)^3 \quad (26)$$

Hence measurement of the transition intensities provides a diagnostic means of establishing which mechanism is responsible for inducing the intensity.

For second-order coupling, by the same reasoning, intensities jointly induced by vibrations r and s are proportional to $\nu_r\nu_s$ (HT) or $(\nu_r\nu_s)^3$ (BO-B), or for overtones to ν^2 (HT), ν_r^6(BO-B). In the spectra of benzene-h_6 and benzene-d_6 there are bands supposedly induced by second-order HT coupling whose intensities can be used to explore these alternatives. As an example the three bands 10^2_0, 16^1_1 and 16^2_2 are considered, (Fischer and Jakobson).

The intensity ratio for the two isotopes is given by

$$I(10^2_0)_H/I(10^2_0)_D = (\nu_H/\nu_D)^p_{10} \prod_{r\neq 10} (S_h/S_d)^2_r \quad (27)$$

where p=2 (HT) or 6(BO-B), and each term in the product is a quotient of zero-quantum Franck-Condon factors. $(S_H)_r$ is the overlap integral $(O_f/O_g)_r$ for benzene-h_6. The only term in the product which is significantly different from unity is that attributable to ν_1, the ring breathing mode. Its value may be determined from a consideration of the origin shifts for the two isotopes, which accompany expansion (1.05), or it may be extracted from experimentally measured intensities of other bands. From the intensities of the pair $(6^1_0)_H$ and $(6^1_0)_D$, and assuming these are induced by the HT mechanism, one obtains a value for the product of Franck-Condon factors of about 1.9.

Table 2 makes the comparisons using absorption data. Although agreement is not exact, the HT mechanism is clearly supported. It should be appreciated that the measured absorption intensities have substantial associated errors. Furthermore, a number of assumptions are made in applying Eqn (27).

TABLE 2. Isotope effects on intensities induced by second order vibronic coupling.

Observed band intensities[a)]			Intensity ratio		
Band	C_6H_6	C_6D_6	Obs.	calc(HT)[b)]	calc(BO-B)[b)]
10^2_0	0.45	0.22	2.04	1.72	4.46
16^1_1	1.00	0.58	1.72	1.40	2.48
16^2_2	3.7	2.7	1.37	1.40	2.48

a)These are absolute intensities, taken from absorption spectra, corrected for Boltzmann factors, and then expressed relative to the well-defined band 16^1_1 in the absorption of benzene-h_6.

b)From Eqn (27).

A general point concerns the question whether estimates of HT induced intensity are better carried out in the transition momentum representation, Eqn (10) rather than in the transition dipole representation, in view of the fact that complete summations over states are not taken. A full treatment of this question is not given here but the overall conclusion can be stated.

First, the two approaches are equivalent if complete summations are taken. However, the equivalence does not extend to the individual terms. Second, in the usual case where the ground state contributions are neglected and the energy gaps between the ground and excited states are large, the transition momentum provides a better estimate of the induced intensity.

A problem of a similar nature is encountered in the calculation of vibrational intensities. It concerns the order in which the transformation from momentum to length, and the BO approximation are applied.

BIBLIOGRAPHY

Blinder, S. (1974) Foundations of Quantum Dynamics, Academic Press, London.

Craig, D.P. and Thirunamachandran, T. (1984) Molecular Quantum Electrodynamics, Academic Press, London.

Cohan, N.V. and Hameka, H.F. (1966) Born-Oppenheimer Approximation and the Calculation of Infrared Intensities, J. Chem. Phys. 45, 4392-4399.

Fischer, G. (1984) Vibronic Coupling, Academic Press, London.

Fischer, G., Reimers, J.R. and Ross, I.G. (1981) CNDO-Calculation of Second Order Vibronic Coupling in the $^1B_{2u}$-$^1A_{1g}$ Transition of Benzene. Chem. Phys. 62, 187-193.
Fischer, G. and Jakobson, S. (1979) Band Intensities in the Absorption Spectrum of Benzene Vapour, Mol. Phys. 38, 299-308. Also (1987) Unpublished results.
Orlandi, G. and Siebrand, W. (1973) Theory of Vibronic Intensity Borrowing. Comparison of Herzberg-Teller and Born-Oppenheimer Coupling. J. Chem. Phys. 58, 4513-4523.
Sharf, B. (1973) Suggesting a Better Approximation to Compute the Herzberg-Teller Induced Intensity. Chem. Phys. Letts. 18, 132-134.

ELECTRONIC SPECTROSCOPY

R.G. DENNING
Inorganic Chemistry Laboratory,
South Parks Road,
Oxford, OX1 3QR,
United Kingdom

ABSTRACT. The factors controlling vibronic features in the electronic spectra of insulating solids are described. The bandshape, intensity mechanisms, phonon dispersion and width of the zero-phonon lines are discussed. Some examples from inorganic materials are used to illustrate the principles.

1. Introduction

This article describes the role of vibronic interactions in the optical electronic spectroscopy of insulating solids, with the emphasis on inorganic materials. In crystalline solids the atomic environment of the optical centres can be regarded as fixed in its absolute orientation on a time scale long compared to a vibrational period; there is no rotational freedom. {This is not necessarily the case in glassy solids where re-orientation may occur over a range of timescales, representing the adaptation of the structure towards the ordered equilibrium state, and can be extraordinarily slow.} This rigidity provides a powerful means of determining the nature of the vibronic wavefunctions by exploiting the polarisation properties of the optical spectrum, while the low temperature at which solids can be examined provides a major simplification of the spectral properties by limiting the thermal population of vibrational states. Important features of the spectra of crystals arise from the transfer of electronic and vibrational excitations between lattice sites and the requirements of translational symmetry. In what follows we shall see how the optical properties are controlled by vibronic interactions, and how the analysis of the optical spectrum can be used to quantify these interactions.

2. Theoretical Background

2.1 The Adiabatic Approximation

The Hamiltonian is divided into three components

$$H = H(r) + H_0(Q) + V(r,Q) \tag{1}$$

C. D. Flint (ed.), Vibronic Processes in Inorganic Chemistry, 111–137.
© 1989 by Kluwer Academic Publishers.

where r and Q represent electronic and nuclear coordinates respectively, H(r) includes the kinetic energy of the electrons and inter-electron repulsion, $H_0(Q)$ is the kinetic energy of the nucleii, and V(r,Q) includes both the internuclear electrostatic repulsion and the electron-nuclear attractions. To examine the role of the vibrations, which change Q and therefore V(r,Q), the interaction is expanded in a Taylor series in Q, where Q_i represents the displacement coordinate of an atom;

$$V(r,Q) = V(r,0) + V'(r,Q)$$

$$V(r,Q) = V(r,0) + \sum_i (\delta V/\delta Q_i)_{Q=0}\, Q_i + \\ + (1/2)\sum_{i,j}(\delta^2 V/\delta Q_i \delta Q_j)_{Q=0}\, Q_i Q_j + \dots \qquad (2)$$

with Q = 0 as the equilibrium nuclear position. The first term then represents the interactions which exist when the nucleii are fixed. The second and third terms are the linear and quadratic components of the electron-phonon (or vibronic) coupling, V'(r,Q). When Q = 0 we can find solutions to the electronic Hamiltonian such that

$$\{H(r) + V(r,0)\}\,\varphi_k(r) = \epsilon_k\, \varphi_k(r) \qquad (3)$$

When the second and third terms in equation (2) are included we define a wavefunction, $\Psi(r,Q)$, describing the situation when $Q \neq 0$, and assume that this can be expanded in terms of the stationary electronic functions $\varphi(r)$.

$$\Psi(r,Q) = \sum_k X_k(Q)\,\varphi_k(r) \qquad (4)$$

where $\Psi(r,Q)$ is an eigenfunction of a Schrodinger equation giving an energy E. Expressions for the X_k can be obtained by projecting out a single electronic function, φ_m, in the Hamiltonian of equation (1) ;

$$\sum_k X_k(Q)[< \varphi_m |H(r)| \varphi_k > + < \varphi_m |V(r,0)| \varphi_k > \\ + H_0(Q)< \varphi_m | \varphi_k > + < \varphi_m |V'(r,Q)| \varphi_k >] = E\, X_k(Q) \qquad (5)$$

This expression is simplified by using equation (3) to give

$$[H_0(Q) + V'_{kk}(Q) + \epsilon_k - E]X_k(Q) + \sum_{m \neq k} V'_{km}(Q)X_m(Q) = 0 \qquad (6)$$

where V'_{kk} and V'_{km} are the diagonal and off-diagonal matrix elements of the vibronic interaction.

The third term of equation (5) implies that the nuclear kinetic energy is independent of the electronic wavefunctions, which is an approximation. The equations in (6) are decoupled if we ignore the off-diagonal elements, V'_{km}, and become a set of eigenvalue equations

$$\{H_0(Q) + V'_{kk}(Q) + \epsilon_k\}X_k(Q) = E X_k(Q)$$

in which the X_k are the vibrational wavefunctions, and the nucleii move in an effective potential described by the vibronic interaction V'_{kk}, which is a characteristic of each electronic state. The energies are just the sum of the electronic energy ϵ_k, and the vibrational energy associated with the vibronic potential of the state.

If the Q_i are replaced by normal coordinates, we can recognize the quadratic vibronic coupling coefficient as equivalent to the force constant for the vibration. At equilibrium the vibronic potential is at a minimum and the linear coupling must be zero. The quadratic coupling defines an harmonic potential. We will need the linear coupling to describe the change in equilibrium distance which can occur as a consequence of the electronic excitation.

In this procedure both the dependence of the nuclear kinetic energy on the electronic wavefunctions, and also the elements of the vibronic interaction off-diagonal in the equilibrium electronic functions are ignored. This constitutes the Born-Oppenheimer or crude *adiabatic approximation.* The total wavefunction for each state is a simple product function or vibronic function, $\Psi_{kn} = \varphi_k X_n$ where X_n now stands for the product of all the vibrational wavefunctions describing all the normal modes of the system. The exclusion of off-diagonal elements becomes serious when the coupling approaches the magnitude of the separation between electronic states; the most spectacular consequences occur when electronic degeneracy gives rise to Jahn-Teller instability. These effects are dealt with in other chapters.

2.2. Electronic Excitation

Consider an optical transition between two electronic states which both obey the adiabatic approximation. Initially the radiation field is assumed to be weak so that its influence need only be taken to first order in a perturbation expansion. This will be the case if $\Omega = \mu.E/\hbar << \omega$, where E is the amplitude of the radiation field which interacts with a transition dipole μ, ω is a characteristic separation between electronic states, and Ω is called the Rabi frequency. In intense laser fields it will be necessary to consider higher terms in the perturbation expansion, to enable the discussion of two-photon absorption.

In a weak radiation field the probability of absorption is determined by the Einstein B coefficient, and as an example, for light propagating in the z direction, linearly polarised in the x direction, we have

$$B = 2\pi/\hbar^2|M_x|^2$$

$$M_x = < j\,|ex|\,a > + < j\,|\mu_B l_y|\,a > + (i2\pi^2\omega_{aj}/c)< j\,|exz|\,a > \quad (7)$$

where | a > and | j > are the ground and excited state wavefunctions, μ_B is the Bohr Magneton, l_y is the angular momentum operator, ω_{aj} is the angular frequency of the transition, and each matrix element, for compactness, implies a summation over all electrons.

M_x is called the transition moment and has three terms, the leading one arises from that component of the radiation field which is effectively constant over the dimensions of the chromophore and describes the induced electric-dipole moment. The other two terms come from the curvature of the radiation field in the region of the chromophore. The first of these corresponds to an interaction with the magnetic field while the second represents the quadrupolar component of the field. The approximate relative magnitude of these terms in the absence of any symmetry restrictions is $1 : 10^{-6} : 10^{-7}$.

Retaining only the electric-dipole term, the intensity for a transition between a ground vibronic state | an > and an excited vibronic state | jm > then has the form;

$$I \propto |\langle an|er|jm\rangle|^2 \tag{8}$$

where a and j are electronic functions and n and m represent the set of vibrational wavefunctions describing all the oscillators.

In an electronic transition the dipole operator only operates on electronic coordinates so equation (8) reduces to;

$$I \propto |\langle a|er|j\rangle\langle n|m\rangle|^2 \tag{9}$$

The assumption in equation (9) that the electronic transition moment is independent of the vibrational wavefunctions is called the Condon approximation. To obtain the intensity in any one vibronic transition both the electronic factor and a vibrational overlap (or Franck-Condon) factor must be evaluated. The electronic matrix element is often constrained by symmetry considerations and we shall see how this is useful in due course, but for the meantime we assume that it is non-zero and restrict attention to the Franck-Condon factors. Figure 1 represents the two potential surfaces with respect to a single normal coordinate.

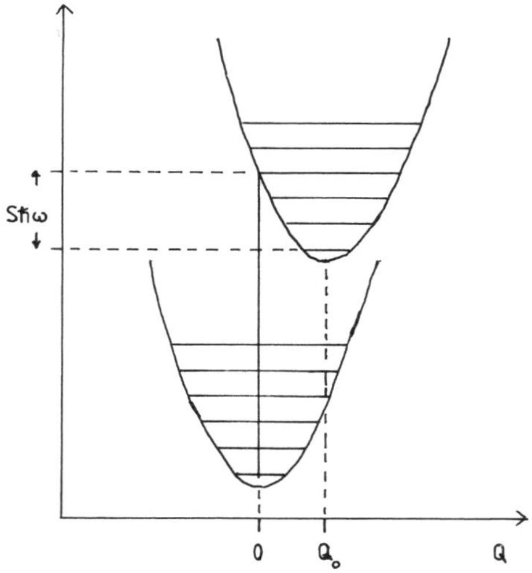

Figure 1. Vibronic Energies with respect to a single normal coordinate.

To relate the vibrational coordinates in the two electronic states the excited state potential is expressed as an expansion in the ground state normal coordinates;

$$V(Q) = \epsilon_0 + AQ + (W/2)Q^2 \tag{10}$$

where ϵ_0 is the pure electronic excitation energy, $A = -(\delta V/\delta Q)_{Q=0}$, is the linear vibronic coupling, $W = (\delta^2 V/\delta Q^2)$, is the quadratic coupling coefficient, and Q is a collective coordinate describing the deformation. The linear coupling describes a force which, when balanced by the elastic force in the bonds, determines the sign and magnitude of the energy displacement between the two potential minima relative to the pure electronic energy difference in the ground state equilibrium geometry. The quadratic coupling, W, can be seen as an additional force constant operating in the excited state that modifies the excited state vibrational frequencies in relation to those of the ground state. This constant can have either sign; if it is negative the mode becomes softer, with a lower limit where it is equal and opposite to the ground state force constant. Eqn. (10) makes the approximation that the same vibrational coordinates prevail in the ground and excited states.

The linear coupling plays a leading role in the analysis of the overall shape of an absorption band, while the quadratic coupling is important in determining the sensitivity of the so-called zero-phonon line, or pure electronic excitation, to temperature.

2.3. The Bandshape

Consider electronic absorption in a molecule at a sufficiently low temperature that the initial state is $| a0 >$. It can be excited to a series of vibronic states $| jn >$. The evaluation of the vibrational integrals $< 0 | n >$ in equation (9), is described in a number of sources [1,2]. The solution for the case where the ground and excited state frequencies are equal, gives;

$$I_n \propto |M_{aj}|^2 \exp(-S).(S^n/n!) \tag{11}$$

where I_n is the intensity of the nth vibrational quantum, M_{aj} is the electronic matrix element in (9) and

$$S = (1/2)M\omega/\hbar Q_0^2 \tag{12}$$

where ω is the angular frequency and M is the effective mass.

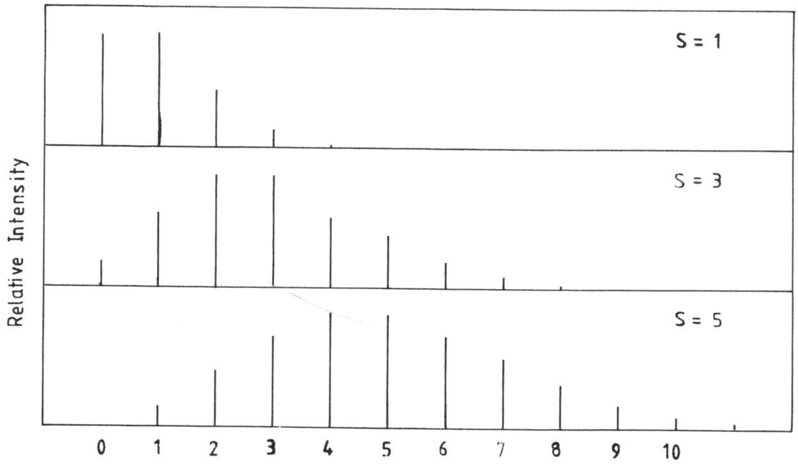

Figure 2. Intensity distribution as a function of the excited state vibrational quantum number for various values of S, the Huang-Rhys parameter.

S is dimensionless, and is called the *Huang-Rhys* parameter. It relates the equilibrium displacement in the coordinate Q_0 to the mean amplitude of the zero-point vibrations, $\tilde{q}$, through the expression $S = (Q_0/\tilde{q})^2$. Characteristic intensity patterns obtained from (11) are shown in Figure 2 for three values of S. When S is large the maximum intensity occurs when $n \sim S$. The semi-classical interpretation of S can be seen by noting that at equilibrium the deformation energy, ΔE, accompanying the distortion in the excited state is balanced by the harmonic restoring force such that

$$\Delta E = (1/2)kQ_0^2, \quad \text{with} \quad Q_0 = A/k \tag{13}$$

where k is the force constant with respect to Q. With $k = \omega^2 M$, (12) and (13) give the deformation energy

$$\Delta E = S\hbar\omega \tag{14}$$

So S, which is proportional to the square of the linear vibronic coupling constant, describes the mean number of quanta of the deformation mode created in the optical transition (see figure 1). S can be quickly estimated from the peak of the absorption band, or as half the difference between the absorption peak and luminescence peak.

2.3.1. *The Ham Effect.* If either the ground state or the excited state are degenerate (but not Kramers degenerate) the linear electron-phonon coupling can lead to a removal of degeneracy in some distorted nuclear configuration. This Jahn-Teller distortion, produces multiple potential surfaces, whose form is discussed in other sections of this book. Although the degeneracy is absent at any one configuration there still exists a degeneracy between states of different nuclear configuration. As a consequence operators which can be diagonal in the undistorted configuration, are forced to be off-diagonal between states whose potential minima differ spatially.

Important examples of operators in this class are the angular momentum and the trigonal field in octahedral environments. The evaluation of the matrix-elements of the operator then resembles the evaluation of the optical transition moments which we have just described. They are attenuated by a factor which is determined by the overlap of displaced vibrational wavefunctions which can be expressed in terms of the Huang-Rhys parameter. This quenching, known as the *Ham* effect, is most obvious in electronic spectroscopy as a result of the quenched orbital angular momentum with consequent reductions in the spin-orbit and Zeeman interactions. As an illustration consider a cubic point group with a T_1 or T_2 electronic state coupled to a mode of *e* symmetry. Electronic matrix-elements between the zeroth vibrational levels are quenched by a Ham reduction factor [1,3,4];

$$R = \exp(-(3/2)S) \tag{15}$$

The optical spectra frequently reveal the spin-orbit components and their energies can be used to assess the Huang-Rhys parameter associated with the *e* Jahn-Teller mode.

2.3.2. *The Effect of Temperature.* In any real solid above 0K some vibrational modes are thermally excited and so we need to examine the temperature dependence of the bandshape, considering all the possible transitions from $| \, an >$ to $| \, jm >$. This was the problem solved by Huang and Rhys and others[5], with the important assumption that all phonons have the same frequency, ω. We use the Bose-Einstein factor, or phonon occupation number, $\tilde{n}$, given by;

$$\tilde{n} = (\exp(\hbar\omega/k_BT)-1)^{-1} \quad (16)$$

to describe the thermal population of the levels. Then if p is the number of vibrational quanta absorbed or emitted in the optical transition it can be shown [1,5,6] that

$$I \propto |M_{aj}|^2 \exp\{-S(2\tilde{n}+1)\}\{(\tilde{n} + 1)/\tilde{n}\}^{p/2} \{[S\ \tilde{n}(\tilde{n} + 1)]^{2\tilde{n}+p}/\tilde{n}!(\tilde{n} + p)!\} \quad (17)$$

When $k_BT/\hbar\omega >> 1$ the Bose factor tends to $k_BT/\hbar\omega$ and, after some manipulation, equation (17) reduces to

$$I = I_{max} \exp[-h^2(\nu - \nu_{max})^2/4Sk_BT\hbar\omega] \quad (18)$$

where ν_{max} is the frequency of maximum intensity, I_{max}, in the band and ω is the phonon frequency. The line has a Gaussian lineshape, with a width which is proportional to the square root of both S and T. Broad bands are associated with large linear vibronic coupling and high temperatures.

The validity of this high temperature limit depends on the frequencies of the modes with large S. For most *molecular* solids, where the coupling is dominated by high frequency local modes, $\hbar\omega > k_BT$ at normal temperatures and so this is not a suitable approximation, on the other hand colour centres and ions in continuous lattices can be strongly coupled to the lattice modes, and in this case this description of the bandshape is useful, although approximate because of dispersion. In the low temperature limit equation (17) reduces to equation (11) as expected.

2.4. Intensity Mechanisms

2.4.1. *Symmetry Constraints.* In section 2.2. we noted that the electric dipole mechanism should provide many orders of magnitude more intensity than the magnetic-dipole or electric-quadrupole mechanisms. There are, however, many interesting cases where the electric-dipole mechanism should be zero, as we can show by symmetry arguments. The quantity of interest is the electronic matrix element of equation (7). In a molecule with symmetry any observable must be invariant to the symmetry operations of the group, with the implication that the direct product

$$\Gamma_a \times \Gamma_o \times \Gamma_j$$

be invariant, where Γ_a, Γ_j are the irreducible representations of the electronic wavefunctions, and Γ_o is the representation carried by the transition moment operator. Or equivalently

$$\Gamma_a \times \Gamma_o \text{ contains } \Gamma_j \quad (19)$$

Concentrating now on the electric-dipole part of the intensity, the most significant requirement occurs in centro-symmetric systems, where the *ungerade* symmetry of Γ_o and (19) give the well-known selection rule $g \longleftrightarrow u$, but $g \not\longleftrightarrow g$, and $u \not\longleftrightarrow u$. In centro-symmetric sites the valence intra-shell transitions of all transition metal (d-shell) and lanthanide (f-shell) ions conserve parity so that there should be no electric-dipole intensity in their optical transitions. If electric-dipole intensity is observed in such a site it signals another manifestation of vibronic coupling.

2.4.2. *Vibronic Intensity.* To obtain a source of electric-dipole intensity in centro-symmetric sites, we must abandon the simple adiabatic approximation, and allow a suitable vibrational deformation to modify the electronic wavefunctions. Assume that the parity selection rule forbids the pure electronic transition $| an > \rightarrow | jn >$, but that there exists a vibronic state $| kn >$ to which electric-dipole transitions are allowed from either state. We first consider a transition in which a quantum of a suitable vibration is created in the excited state, i.e. $| an > \rightarrow | jn{+}1 >$. Using only the linear vibronic coupling from equation (2), the off-diagonal elements in equation (6) give, to first-order in the perturbation;

$$| jn{+}1 > = | jn{+}1 >_0 + \sum_k c_{jk} \, | kn >$$

$$| an > = | an >_0 + \sum_k c_{ak} \, | kn{+}1 > \qquad (20)$$

Here $| an >_0$ and $| jn{+}1 >_0$ are the adiabatic vibronic wavefunctions and the coefficients are

$$c_{jk} = \Delta E_{jk}^{-1} < j \, |A| \, k >< n{+}1 \, |Q| \, n >$$

$$c_{ak} = \Delta E_{ak}^{-1} < k \, |A| \, a >< n{+}1 \, |Q| \, n > \qquad (21)$$

where ΔE_{jk} and ΔE_{ak} are the energy separations of the coupled states and Q is the normal coördinate of the vibration. Using (20), (21) and (8) gives two contributions to the intensity of the transition $| an > \rightarrow | jn{+}1>$;

$$I \propto \Big| \sum_k \Delta E_{jk}^{-1} < j \, |A| \, k >< k \, |er| \, a >< n{+}1 \, |Q| \, n > +$$

$$+ \sum_k \Delta E_{ak}^{-1} < j \, |er| \, k >< k \, |A| \, a >< n{+}1 \, |Q| \, n > \Big|^2 \qquad (22)$$

The optical spectrum is then modified by these *Herzberg-Teller* perturbations in both the ground and excited states. The symmetry requirements for the transition to be electric-dipole allowed can now be defined by the requirement that another electronic state exists such that the electronic matrix elements in (22) are non-zero. For either term in (22) this means that

$$\Gamma_a \times \Gamma_o \times \Gamma_Q \quad \text{contains} \quad \Gamma_j \qquad (23)$$

The mode that brings about this coupling is called the enabling mode. It is not possible to distinguish between the two mechanisms in (22) experimentally. An equivalent argument can be made for a transition in which a quantum of the enabling mode is destroyed in the transition, i.e. $| an > \rightarrow | jn{-}1 >$; the symmetry requirements are the same.

The magnitude of the coefficients in equation (21) can be estimated from the linear electron-phonon coupling due to the vibration, which is of the order of a vibrational quantum, say 200 cm^{-1}, while the separation between the states could be 20,000 cm^{-1}. The coefficients are then of the order of 10^{-2} and the intensities, which depend on their square, are approximately 10^{-4} times the magnitude of the intensity of the fully allowed transition from which the intensity is "borrowed" by the perturbation. Notice that the intensity gained by this mechanism still exceeds that due to the parity-conserving magnetic-dipole and electric-quadrupole mechanisms. Intensity derived in this way is described as *vibronic*. While this mechanism is highly significant for the optical spectrum, the vibronic coefficients are usually too small to effect the dynamics of the adiabatic state.

This new source of intensity has a characteristic temperature dependence. We now include the process in which a phonon is annihilated. Using (22) and collecting the nuclear integrals, the transition moment is

$$M = |\{< n+1 \,|Q|\, n > + < n-1 \,|Q|\, n >\} \; x$$

$$\sum\nolimits_k \{\Delta E_{ak}{}^{-1} < j \,|er|\, k >< k \,|A|\, a > + \, \Delta E_{jk}{}^{-1} < j \,|A|\, k >< k \,|er|\, a >\}|(24)$$

The first term in the nuclear coordinates describes the creation of a phonon of the enabling mode in the vibronic excited state while the second term leads to the loss of a phonon of this mode. At low temperatures, of course, only the first mechanism is possible, but at high temperatures the second is also important. The simplest way to evaluate the thermal average of this expression is to use the phonon creation and annihilation operators [7,8] (and mass weighted normal coordinates Q) which lead to the result that;

$$< \tilde{n}+1 \,|Q|\, \tilde{n} > = \{(\tilde{n}+1)\hbar/2\omega\}^{1/2}; \qquad < \tilde{n}-1 \,|Q|\, \tilde{n} > = \{\tilde{n}\hbar/2\omega\}^{1/2} \qquad (25)$$

where $\tilde{n}$ is the phonon occupation number, or the ensemble average vibrational quantum number, and is given by $\tilde{n} = (\exp(\hbar\omega/kT)-1)^{-1}$. The ensemble thermally averaged transition probability, using (24) then becomes

$$|M|^2 = [M_0{}^2\{\hbar/2\omega\}\{(\tilde{n}+1) + \tilde{n}\}] \qquad (26)$$

where M_0 is represents the electronic integrals in (24). The intensity of the transition is proportional to $2\tilde{n} + 1$. Notice that, if all other factors are equal, low frequency modes are the most effective. Using (16) this can be rearranged to

$$I = I_0 \coth(\hbar\omega/2k_BT) \qquad (27)$$

This expression can be used with considerable success to model the temperature dependence of vibronic intensity, or in circumstance where the identity of the enabling mode is unknown to determine ω. If more than one enabling mode is active the intensity is just the sum of contributions of the type given in (27). The source of the temperature dependence is seen from equation (25) to arise from the increasing amplitude of the vibrations with the increase in the phonon occupation number. This temperature dependence is diagnostic of the vibronic intensity mechanism, because it can be easily shown [1] that transitions allowed by the static mechanism, retain an unchanged integrated intensity as a function of temperature. This test can be a useful indication of symmetry where the geometry of the site is unknown.

2.4.3. *Intensity from Acoustic Modes.* In a continuous solid vibronic intensity can be drawn not only from the optical modes but also from the acoustic elastic waves of the lattice. These acoustic phonons introduce strain at the lattice site as a result of the *difference* in amplitude of the displacement at the optical centre and the atoms adjacent to it. Translation symmetry requires that these modes be characterised by a wave-vector **k**.

In the so-called *Debye* approximation it is assumed that the waves are dispersionless, i.e. there is a velocity of sound for transverse waves, v_t, and one for longditudinal waves, v_l which is independent of the frequency and which relates the wavelength to the frequency. The waves propagate as though they were in an isotropic elastic medium. This should be a reasonable description when the wavelengths are long compared with a unit cell dimension. It is just these low frequency modes which are significant at low temperature. In terms of the wavevector, $\mathbf{k} = 2\pi/\lambda$;

$$\omega_t = v_t \mathbf{k}_t \quad ; \quad \omega_l = v_l \mathbf{k}_l \tag{28}$$

In such a medium the Brillouin zone can be represented as a sphere in k-space. The number of states within a small range dk is proportional to the $4\pi k^2$, corresponding to the area of a sphere of radius k. The sphere is bounded by the condition that the total number of modes is 3N where N is the total of atoms in the lattice, leading to an upper limit for the frequency of these modes, ω_D, called the Debye frequency. Using equation (28) the number of phonon states per unit frequency range per unit volume of the crystal, in a crystal of volume V, is given by [8]

$$\rho(\omega) = (9N/V)(\omega^2/\omega_D^3); \tag{29}$$

Here $\rho(\omega)$ is called the phonon density of states, and influences the spectral intensity through the probability of creating a final state with this phonon energy.

The linear vibronic coupling depends on the deformations in the immediate vicinity of optical centre. When a phonon is created by the optical transition the amplitude is given by equation (25) (where Q is now a normal coordinate describing the lattice modes) and is inversely proportional to the square root of the frequency; i.e. low frequency phonons have larger amplitudes. However what matters in the vibronic interaction is not only the amplitude of the displacement but also the strain at the site, ϵ, which describes the relative displacement of adjacent atoms. For example, with a displacement in the x-direction the phonon amplitude is described by

$$u(x) = (\hbar/2M\omega_k)^{1/2}\{\exp(ikx).b_k + \exp(-ikx).b_k^+\} \tag{30}$$

where b_k and b_k^+ are the phonon creation and annihilation operators, where k is the wavevector. The strain in the x-direction is then

$$\epsilon = du/dx = i(\hbar\omega_k/2Mv^2)^{1/2}(b_k - b_k^+) \tag{31}$$

where we have used $\omega = vk$. Qualitatively this means that long wavelength phonons produce small amplitude differences at adjacent sites and are less effective in the vibronic coupling mechanism. The strain operator is now the effective displacement coordinate in equation (21). It follows from (31) that the intensity of a single vibronic transition in the acoustic phonon spectrum is directly proportional to ω_k.

Finally we observe that the transition probability, given by the Fermi golden rule, is weighted by the density of final states, so we have

$$I(\omega) = (C/v^2)|A|^2\omega\rho(\omega) \tag{32}$$

where C is a constant. If a Debye density of states is valid then $I(\omega) \propto \omega^3$. Alternatively the spectral intensity distribution can be used to determine $\rho(\omega)$.

It is interesting to compare this situation with that in a glassy solid, where the disorder destroys the translational symmetry. It is argued that this biases the density of vibrational states to low frequency where there are many modes with large amplitude involving the motions of small segments of the lattice. Experimental evidence from inelastic neutron scattering and Raman spectroscopy support this view. Since there are no phonons with the translational symmetry of the lattice all these vibrations can be considered as local modes, so that the vibronic coupling is related to the amplitude of the vibrations. Because this is inversely proportional to the square root of the frequency we have

$$I(\omega) = C\,|A|^2 \omega^{-1} \rho(\omega) \tag{33}$$

Figure 3 compares the phonon sideband structure on the ${}^4A_2 \rightarrow {}^2E$ transition of Cr^{3+} in a single crystal of MgO and in a silicate glass [9].

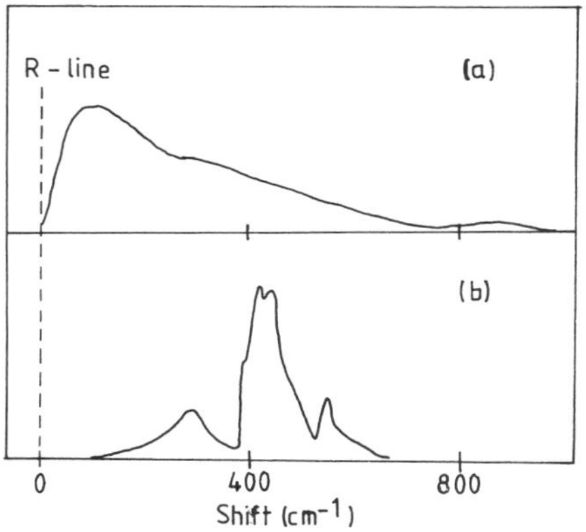

Figure 3. Vibronic structure in the spectrum of Cr^{3+} (a) in ED2 silicate glass, (b) in a crystal of MgO.

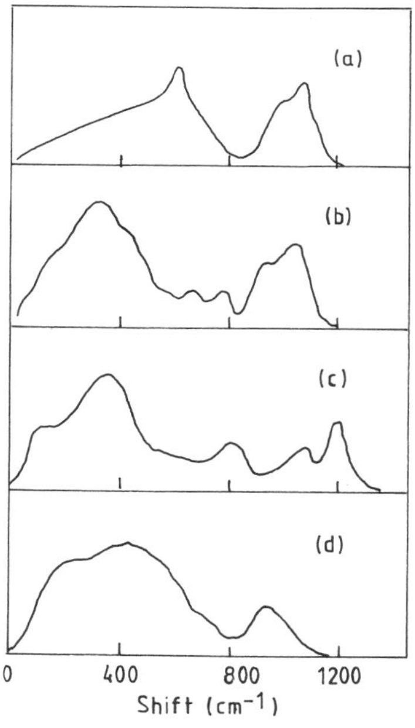

Figure 4. Phonon density of states for a silicate glass. (a),(b) from Raman spectroscopy, (c) from inelastic neutron scattering, (d) from Fig 3(a) and eqn. (33).

While the MgO crystal exhibits the monotonic rise of intensity associated with the Debye distribution, the glass has much greater intensity at low frequency. The density of states which can be inferred by using (33) is shown in Figure 4 to be close to that determined by other methods.

2.5. Zero-Phonon Line Shifts and Broadening

A prominent feature of vibronic spectra is the strong dependence of their sharpness on temperature. An obvious cause is the presence of transitions from thermally occupied vibronic states, which smear the whole spectral density at higher temperatures, but a more important phenomenon is the broadening of individual transitions as the temperature rises. The simplest case to analyse is that of the pure electronic or zero-phonon transition. This not only broadens with increasing temperature but also shifts in absolute energy. The temperature sensitivity is dependent on the structure of the solid. In measuring the linewidth it is important to remove the effects of site inhomogeneity in the lattice and to determine the true homogeneous width of a single optical site. It is only recently that a number of laser-based techniques for the proper determination of the homogeneous linewidth have become available[10].

The source of the broadening is the coherence lifetime of the electronic state. This has two components, population decay characterised by a relaxation time, T_1, and dephasing processes characterised by T_2. Here we are only interested in the latter and it is common to use luminescent systems for which T_1 is measurably long, say > 10^{-9}s, to be certain that relaxation is dominated by dephasing. A Lorentzian line with a full width at half maximum of $1\mathrm{cm}^{-1}$ corresponds to a lifetime of 5.3ps.

We can probe the excited state in the radiation field, creating a coherent superposition of two states. In the absence of any perturbations the phase of the superposition survives indefinitely, or in reality as long as the coherence lifetime of the probe beam. In this situation the measured linewidth is determined only by the coherence of the source. But, in the presence of a vibrational perturbation, there can be a time-dependent modulation of the electronic energy difference between the two-states if the vibronic coupling terms of equation (10) are non-zero. This energy modulation produces a progressive loss of phase in the time domain and it is this which introduces the line broadening.

The shift in the mean electronic excitation energy with temperature can be understood as follows. The zero-phonon excitations by definition conserve the phonon energy but when the quadratic vibronic coupling in (10) is not zero the ground and excited state phonons differ in their force constants and mean amplitudes, leading to a change in the average strain at the optical centre in the two sites and therefore to a shift in the pure electronic excitation.

The original treatment of this problem, pioneered by McCumber and Sturge[11], expresses the vibronic interaction as a perturbation expansion, taking the linear vibronic coupling to second order and the quadratic coupling to first order. Recent theoretical progress has shown that the theory does not need to be be perturbative, and that an analytical solution exists for both the line-broadening and the line shift[12,13], providing that the harmonic approximation is made, i.e. that the Taylor expansion is cut off as in equation (10). The temperature dependent parts are found to be independent of A, the linear coupling, and dependent only on the quadratic coupling constant W. The lineshift is linearly dependent on W, and the line-broadening, $\Delta\tilde{\nu}$, in cm^{-1}, is given by

$$\Delta\tilde{\nu} = (1/4\pi^2c)\int_0^\infty d\omega \ln\{1 + [4\tilde{n}(\omega)(\tilde{n}(\omega) + 1)W^2\, \Gamma_0(\omega)\, \Gamma_1(\omega)]\} \tag{34}$$

where $\Gamma_0(\omega)$ is the weighted phonon density of states in the ground state and $\Gamma_1(\omega)$ is the equivalent quantity for the excited state. Here the weighting refers to the influence of each normal mode on the local coordinate, Q, which appears in (10). In the case of an acoustic phonon the weighting is a function of the strain at the site, as discussed in Section 2.4.3. As before $\bar{n}(\omega)$ is the phonon occupation number given by eqn.(16). W is defined to be dimensionless [12]. With this definition we can have $-1 < W < \infty$. $W = 0$ if the ground state and excited state force constants are identical and tends to -1 as the excited state force-constant tends to zero. Positive values of W indicate a stiffer vibration in the excited state than in the ground state.

Equation (35) can be greatly simplified when $|W\, \Gamma_0(\omega)| << 1$, i.e., when there are only small differences in the quadratic coupling in spectral regions where the density of states is significant. Eqn.(34) then reduces to the same expression as that obtained by the perturbation method, namely;

$$\Delta\tilde{\nu} = (W^2/\pi^2 c)\int_0^\infty d\omega\,\bar{n}(\omega)\{\bar{n}(\omega)+1\}\,\Gamma_0(\omega)^2 \qquad (35)$$

There are two simple but important cases of equation (35). One uses the Debye density of states, with $\rho(\omega) \propto \omega^2$, which, when weighted by the strain gives $\Gamma_0 \propto \omega^3$. With $x = \hbar\omega/k_BT$, (35) gives the famous result that

$$\Delta\tilde{\nu} = \omega_D(9W^2/4c)(T/T_D)^7\int_0^{\omega_D} dx\,\{x^6 e^x/(e^x-1)^2\} \qquad (36)$$

where ω_D is the Debye frequency and $T_D = \hbar\omega_D/k_B$. At low temperatures the integral in (36) is almost constant and the linewidth depends on T^7.

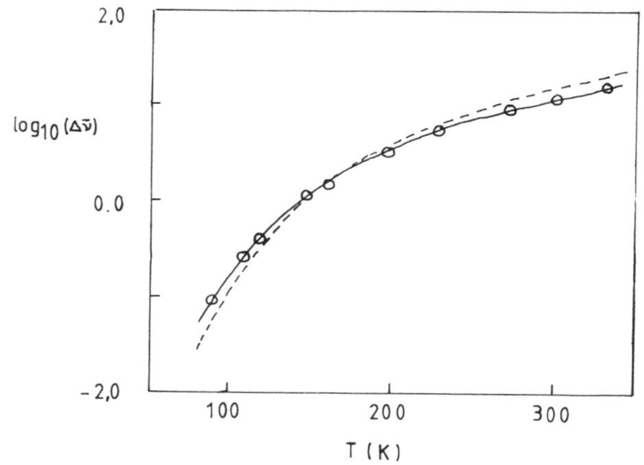

Figure 5. Homogeneous width, $\Delta\tilde{\nu}$, in cm^{-1} of the R_1 line of ruby. The experimental points are circles. The full line uses eqn. (34), the dashed line eqn. (36).

To illustrate the success of this theory we show in Figure 5 the results of applying a Debye density of states to the broadening of the R_1 line of ruby [12]. The full line is equation (34) with $W = -0.312$ and 3.01, and the dashed line is the approximate expression (35) with $W = |0.373|$. The fit is ambiguous as to the sign of W but it is more likely that the negative value is correct. The full theory is clearly more successful, but the main point of chemical interest is the large magnitude of W. Remembering the very small linear vibronic coupling for the doublet states of Cr^{3+} it is perhaps surprising that the excited state force constant is some 30% smaller than that of the ground state, and it is important to realise that magnitude of the two components of the coupling need not be correlated.

When an impurity is introduced into a lattice the perturbation produced by differences in its size and mass from the host partially decouple the local deformations from the lattice modes, creating a so-called pseudo-local mode. In molecular lattices these may be low frequency torsional modes of large molecules. Because they are localised these modes are particularly important in the line-broadening. The effect may be understood by assuming that $\Gamma_0(\omega)$ in (35) is only significant in a narrow range centred at a frequency ω_0. At this frequency, and at temperatures where $\hbar\omega_0 >> k_BT$

$$\tilde{n}(\omega)\{\tilde{n}(\omega) + 1\} = \exp(\hbar\,\omega_0/k_BT)/\{\exp(\hbar\,\omega_0/k_BT) - 1\}^2$$

$$\sim \exp(-\hbar\,\omega_0/k_BT) \qquad (37)$$

This result, which was first pointed out by Small [14], gives an Arrhenius temperature dependence and is found to apply to a number of experiments on mixed organic crystals. Equation (34) shows that a dispersionless optical mode cannot provide a line-broadening. In this case both $\Gamma_0(\omega)$ and $\Gamma_1(\omega)$ are delta functions. When W is non-zero they describe oscillators of different frequency, the density of states functions do not overlap, and eqn.(34) gives a zero linewidth.

When the pseudo-local mode is embedded in the acoustic phonon spectrum it is possible [15] to refine eqn.(37). The evaluation of (34) can be undertaken providing both density of states functions and W are known. The latter is derived directly from the difference in the observed ground and excited state local mode frequencies;

$$W = (k_1 - k_0)/k_0 = (\omega_1/\omega_0)^2 - 1 \qquad (38)$$

where k_1 and k_0 are the excited and ground state force constants. The density of states is obtained by fitting a Lorentzian to the observed width of the phonon sideband, with a half-width $\Delta\nu$. Then, with assumptions about the density of bath phonon states it can be shown that [15];

$$W = [2(\Delta\nu_1/\Delta\nu_0)^{1/2} - 1]^2 - 1 \qquad (39)$$

If W is obtained from (38), then (39) allows the determination of both density of states functions in (34) from the width, $\Delta\nu_0$ or $\Delta\nu_1$, of a single vibronic pseudolocal mode sideband.

A good example of a pseudo-local mode problem is found in the work of Pack and McClure[16] on Cu^+ in NaF. Here the modes which couple strongly are the t_{1u} asymmetric stretching of the fluorine atoms. In the A_{1g} ground state the frequency of this mode is 93cm^{-1}, while in the $^1T_{2g}$ excited state it is 32.7cm^{-1}. Figure 6 shows the experimental lineshape as a function of temperature for the zero-phonon line, determined by two-photon excitation spectroscopy. The phonon sideband widths are 0.31cm^{-1} in the ground state and 2.5cm^{-1} in the excited state. From (38) W = -0.876, and using the sideband widths to give the density of states functions in (34) the linewidth can be calculated with no adjustable parameters[15]. Agreement with experiment up to 40K is excellent, but fails at higher temperature. In view of the extreme anharmonicity of the oscillator [16] this is not surprising.

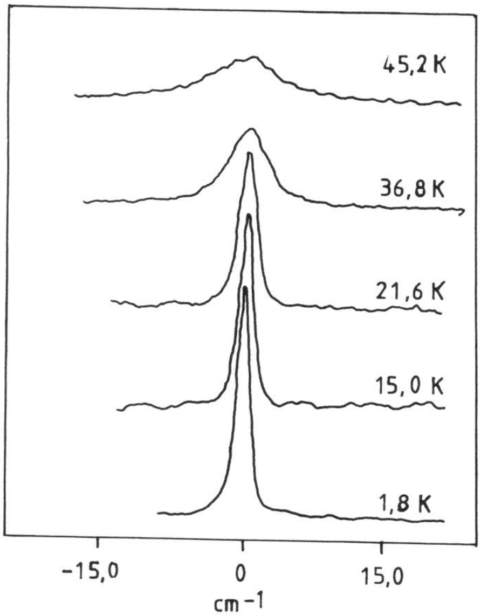

Figure 6. Two-Photon excitation spectrum of the ${}^1A_{1g} \rightarrow {}^1T_{2g}$ zero-phonon line of NaF:Cu^+, at 35,325.6cm^{-1} after reference [16].

Qualitatively we now see that the homogeneous broadening of the spectrum with increasing temperature has a number of causes. The zero-phonon line, on which all other features are superimposed is strongly dependent on temperature, and confers its width upon them. The relaxation of pseudo-local modes into the phonon bath creates a broadening associated with vibrational dephasing, which is also a strong function of temperature. This effect operates on modes lying in the acoustic spectral range. Local modes lying above the Debye frequency will however remain relatively sharp, because there decay requires the creation of more than one phonon. Finally we expect hot structure from the thermal population of phonons which enable vibronic intensity.

2.6. Vibronic Structure in Two-Photon Absorption

The theory of two-photon absorption is dealt with elsewhere in this volume. The transition probability depends on the presence of a non-stationary or virtual state, created in the radiation field, by the admixture of a set of intermediate stationary states $| k >$ such that, when only photons of a single energy, $\hbar\omega_0$, are present the transition probability is given by;

$$G_{aj} \propto \sum_k |< j |er| k >< k |er| a >|^2 / (\hbar\omega_0 - \hbar\omega_{ak})^2 \tag{40}$$

where $\hbar\omega_{ak}$ is the separation of the contributing intermediate states from the ground state. It is immediately obvious from the arguments of Section 2.4.2. that in a centro-symmetric site G_{aj} is non-zero when both $| a >$ and $| j >$ are *gerade*. So vibronic intensity will be confined to *gerade* modes. In an interesting application of this principle McClure [17] observed the even (*gerade*) harmonics of *ungerade* modes in an strongly anharmonic system.

The symmetry constraints on G_{aj} in a vibronic system follow the principles of section 2.4.2. with the overall requirement that

$$\Gamma_a \times \Gamma_p \times \Gamma_q \times \Gamma_Q \text{ contains } \Gamma_j \qquad (41)$$

where Γ_p and Γ_q are the representation of the transition moment vectors associated with the two photons and Γ_Q is the representation of the enabling mode. It is a feature of (40) that there are now three states susceptible to Herzberg-Teller coupling and it has been shown that in at least one case the mechanism is dominated by coupling in the intermediate state [18]. In the intra-shell transitions of the f and d shell ions, groups of states with the same parity cover much of the visible spectrum. The energy denominators in (22) are then typically 5000cm^{-1} but sometimes much less so that vibronic intensity in two-photon absorption is more significant relative to an allowed electronic transition, than in one-photon absorption. Some examples will be found in Section 3.

3. Examples of Vibronic Intensity

3.1. Use of Linear Polarisation

The identity of the enabling modes is most easily established if they are well resolved in the spectrum; then the full apparatus of isotopic labelling and optical polarisation, usually applied in pure vibrational spectroscopy, can be used to characterise them. We can illustrate the various factors that need to be taken into account by using specific examples, starting with the very detailed spectrum of the uranyl ion [19]. In this case the visible spectrum of a single crystal at low temperatures typically contains more than five hundred features.

The crystal structure has only one molecule per unit cell, which removes some of the potential complication of spectral features caused by the inter-chromophore coupling. The electronic excitation or exciton in a pure, as opposed to dilute, crystal must reflect the translational symmetry of the lattice and the electronic coupling between sites. But the long-wavelength of the radiation field (~500nm) compared to the distance between chromophores (~100pm), effectively gives access only to those components of the excitation wave close to the Brillouin zone centre, for which **k** = 0. Thus, although there can be dispersion in the exciton band, the pure electronic, or zero-phonon, transition does not reveal it.

A small portion of the polarised absorption spectra of $Cs_2UO_2Cl_4$ is shown in Figure 7; it covers 900 cm^{-1} in the region of the first electronic excited states, but similar structure spreads over the whole spectral region from 20,000 cm^{-1} to 29,000 cm^{-1}.

The optical chromophore in $Cs_2UO_2Cl_4$ is the complex anion. Its relation to the external habit of the single crystals in shown in Figure 8. The monoclinic crystal symmetry forces C_{2h} point symmetry (which implies a centre of symmetry) at the uranium atom, but the effective symmetry is almost D_{4h}. In figure 7 there are six experiments distinguished by the propagation axes (upper case) and electric vector polarisation axes (lower case) which refer to Figure 8. We can immediately detect that the features labelled I and II are not allowed by the electric-dipole mechanism. For example I is common to the X(z) experiment and the Z(x) experiment but is forbidden in all other cases. It follows that the transition is allowed by the magnetic-dipole mechanism.

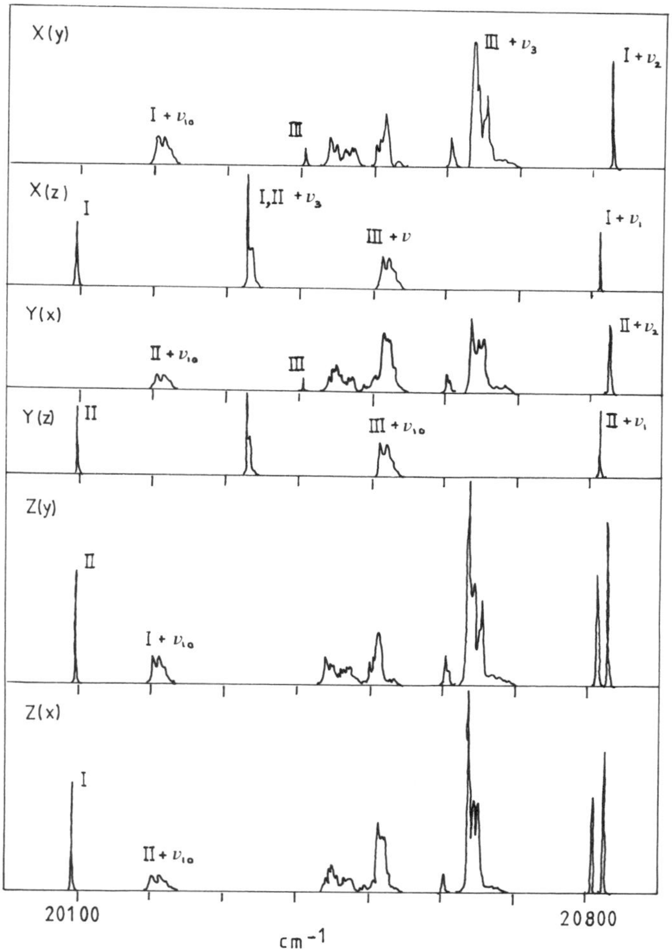

Figure 7. Single crystal absorption spectrum of $Cs_2UO_2Cl_4$ at 4.2K, in six different polarisations.

We now use the fact that the ground state of the uranyl ion is a closed shell species to assign the invariant symmetry representation to this state. It follows from (9) that the electronic excited states form a basis for the same irreducible representations as the angular-momentum operators which appear in equation (7) . Thus we assign E_g symmetry to this pair of states in D_{4h}, and B_{2g} and B_{3g} symmetry in D_{2h} (which is the lowest effective symmetry in which it is necessary to operate). The 1.6cm^{-1} splitting is a reflection of the small deviation from tetragonal symmetry. Near 20,810cm^{-1} we see the same polarisations repeated in features which mark the beginning of a progression with an interval of ~710cm^{-1}; this is the symmetric UO_2 stretching mode. We can immediately determine from figure 1 that the Huang-Rhys factor is close to 1. Using equation (12) it is easy to evaluate the change in bond length as 0.07A, as compared to a ground-state bond length of 1.76A.

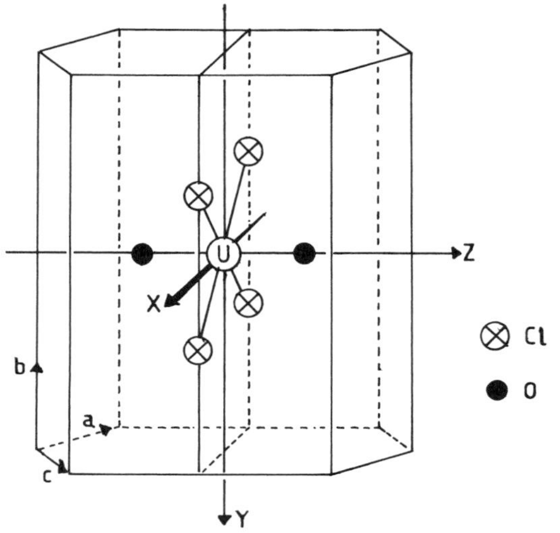

Figure 8. Crystallographic axes, crystal habit and molecular axes of $Cs_2UO_2Cl_4$.

Once the electronic excited state symmetry is assigned, the polarisation of the electric dipole vibronic structure can be interpreted. Intensity in the x and y polarisations is borrowed from states of $E_u(D_{4h})$ symmetry while that in the z polarisation is borrowed from states with $A_{2u}(D_{4h})$ symmetry. The vibronic coupling matrix element of an enabling vibration of symmetry Γ_v is then non-zero if

$$E_g \times \Gamma_v = E_u \text{ or } A_{2u}$$

so that a_{2u} and b_{1u} modes appear in the x and y polarisations and e_u modes appear in the z polarisation. It is then obvious that the electric-dipole feature near 20,830cm^{-1} is the UO_2 asymmetric stretching mode which has a_{2u} symmetry, and this is confirmed by its oxygen-18 isotope shift. Similarly the feature at 20,330cm^{-1} must have e_u vibrational symmetry, and has an isotopic shift characteristic of the UO_2 bending mode.

The broad structured features near 20,200cm^{-1} illustrate a further point. They occur about 105cm^{-1} above the electronic origins I and II. There are two local modes of the chlorine atoms which are expected near this frequency. Both are out-of-plane deformations of the four chlorine atoms. Turning to D_{2h} symmetry there is one deformation, of b_{1u} symmetry, in which the four chlorines deform to give a square pyramid, and another, of a_u symmetry, in which they deform towards a tetrahedral geometry. When these vibrations are coupled to origin I, with B_{2g} (D_{2h}) symmetry, the vibronic symmetry for the b_{1u} mode is B_{2u}, giving y polarised intensity, and that for the a_u mode is B_{3u}, giving x polarised intensity. When the same argument is extended to origin II we anticipate that if the b_{1u} mode is active then the feature will appear lower in energy in the y polarisation than in the x polarisation, but that if the a_u mode is active then the feature will be lower in energy in the x polarisation than in the y-polarisation. Careful measurements show that the latter assignment is the correct one, and that the a_u mode is active.

The shape of this feature illustrates a further point. There are no other local modes of the $UO_2Cl_4^{=}$ anion expected near this frequency, and yet there are sub-components to this band, separated by about 10cm^{-1}, giving a total width of ~25cm^{-1}. The pure vibrational spectrum of the compound suggests that the optical branches of the lattice modes lie below this frequency. We therefore look for the source of the structure in the vibrational coupling between adjacent chromophores. With only one molecule per unit cell, factor-group or Davydov splitting is excluded, and we infer that the structure reflects the dispersion of the vibrational frequency associated with the inter-ion coupling.

For a vibronic band the conservation of the wavevector in the optical excitation, $\mathbf{k} = 0$, is satisfied if $\mathbf{k_e} + \mathbf{k_v} = 0$. We can see the meaning of this by considering an exciton at the zone boundary in a one-dimensional system, with $k = \pi/a$, where a is the lattice parameter, and the enabling phonon is also at the zone boundary. The electronic wavefunction then changes sign at each lattice site but because the phonon also alternates in amplitude at adjacent sites, the electronic state to which the coupling occurs has the same phase factor at each site, i.e., $\mathbf{k} = 0$. Vibronic bands therefore reflect the combined dispersion of the exciton and the phonon. In most cases the exciton dispersion is negligible compared to the phonon dispersion; in the present case it is <0.1cm^{-1} [20]. The structure of the vibronic band then reflects the density of phonon states, which can be derived from lattice dynamical calculations. For acoustic phonons the amplitude of the deformation at any one lattice site is a strong function of the wave-vector so that the intensity should follow eqn.(32). In $Cs_2UO_2Cl_4$ the Debye frequency derived from (36) is ~60cm^{-1}, and the absorption of a thick crystal in the acoustic phonon region shows the expected form for $I(\omega)$ [21].

We notice that while the low-frequency chlorine modes of the ion exhibit a large dispersion the high frequency oxygen modes are sharp. We can understand this on a qualitative basis because the chloride ions together with the cesium ions together compose the lattice and as such are strongly coupled, whereas the covalently bound uranyl ions are effectively isolated from interaction with one another by heavy cesium ions, and can be treated as uncoupled molecular oscillators.

Returning to Figure 7 the same disperse vibronic feature is observed in the z polarisations, and is clearly associated with another pure electronic transition, which is not electric-dipole allowed, labelled III. The vibronic symmetry of this state must be $A_{2u}(D_{4h})$ and since we know that the vibrational symmetry is $B_{1u}(D_{4h})$ the electronic state must be assigned $B_{2g}(D_{4h})$ symmetry. Almost all the remaining electric-dipole structure in the figure can be assigned as vibronic structure on this origin. Because the intensity all appears in the x and y polarisations, these vibrations all must have $e_u(D_{4h})$ symmetry. Thus vibrations which provide electric-dipole intensity in the z polarisation on origins I and II should provide intensity on origin III in the x and y polarisations. The figure shows that many more vibrations of this type are effective in the latter case, and is a clear indication of the importance of the anisotropy in the linear coupling imposed by the electronic wavefunctions.

3.2. Vibronic Structure In Natural Circular Dichroism

The vibronic structure in a circular dichroism spectrum need have little relationship to the structure in an absorption spectrum. For any vibronic transition in an optically active molecule we can define a rotational strength

$$R_{anjm} = \mathbf{Im} < jm \mid e\mathbf{r} \mid an >< jm \mid (e/2mc)\mathbf{l} \mid an > \qquad (42)$$

where **Im** stands for the imaginary part and as before we imply a sum over all electrons.

In the spectroscopy of f-f and d-d transitions we can distinguish two sources of electric-dipole intensity, a static part, which is a property of the pure electronic wavefunctions and a vibronically induced part associated with vibrations which operate through the linear vibronic coupling. If the deviations from centrosymmetric equilibrium geometry are small, i.e. comparable to the magnitude of a vibrational displacement, then these contributions to the overall absorption intensity of the band can be similar. Furthermore, the intra-shell nature of the transitions means that they will often be magnetic-dipole allowed.

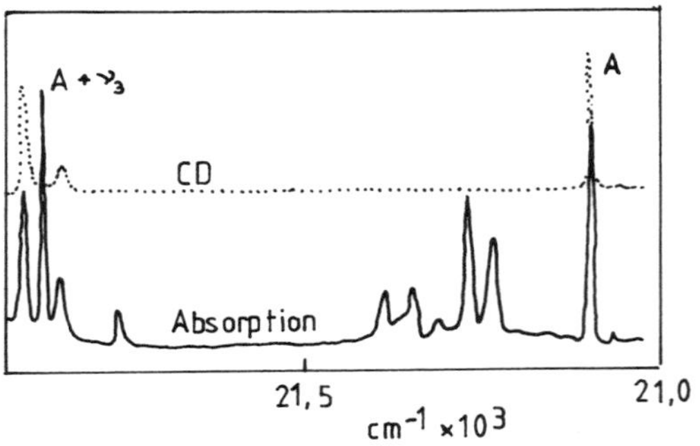

Figure 9. The Absorption and CD spectrum of $NaUO_2(CH_3COO)_3$ at 4.2K

It transpires that the vibronic intensity does not contribute to the circular dichroism in these circumstances, a point first realised by Moffitt and Moscowitz [22]. In the presence of the radiation field the coherent superposition of ground and excited states, generates electric and (if the transition is magnetic-dipole allowed) magnetic polarisations, oscillating at the radiation field frequency. These polarisations may be represented by vectors rotating in a complex coordinate frame. If we define new axes which rotate in phase with the these polarisation vectors (a rotating axis frame) they will, of course, appear static. This is an adequate description when the source of the electric-dipole transition moment is a static perturbation, but when the source is a vibronic perturbation the electric-dipole transition moment also oscillates in phase with the enabling vibration, relative to the rotating frame. Equation (42) shows that the rotational strength of the transition is a pseudo-scalar product of the electric-dipole transition moment, and the magnetic-dipole transition moment. The latter is static relative to the rotating frame so in the case of vibronic intensity the time average of the projection, over a period large compared with the vibrational oscillation, is zero.

A clear example of this selection rule is seen in the spectrum of another uranyl compound, $NaUO_2(CH_3COO)_3$, which crystallizes in a cubic enantiomeric space group. Single crystals are naturally optically active. Figure 9 shows a section of the spectrum which corresponds to the same electronic transitions as those presented in Figure 7 [23]. Remembering that the first electronic transitions, I and II, which are degenerate in the D_3 symmetry of this species, were magnetic-dipole allowed, it is not surprising to find strong circular dichroism in the origin band, labelled A. There is, however, no CD intensity in the vibronic sidebands based on this origin until the first members of two progressions in totally symmetric modes appear near 21,900 cm^{-1}. Notably the strong vibronic coupling of the asymmetric UO_2 stretching frequency, marked A + ν_3, is completely absent from the CD.

It is evident that this experiment has the potential for separating the static and vibronic intensity sources where both are present. The separation is not possible when the vibronic coupling provides not only the source of electric-dipole intensity but also the magnetic-dipole intensity. Then the fluctuations of both electric and magnetic polarisations will remain in phase, generating a source of vibronic CD. This mechanism has been explored by Weigang [24].

3.3. Electronic Spectroscopy of a Jahn-Teller Excited State

A well studied example is the Cr^{3+} ion in the octahedral field of six chloride ions, as a substitutional impurity in the *el Pasolite* lattices, $Cs_2NaInCl_6$ or Cs_2NaYCl_6 [25]. When the ion is in a cubic site, linearly polarised spectroscopy cannot provide any information, but a powerful method for the assignment of vibronic states comes from the magnetic circular dichroism (MCD) experiment. This technique has wide-ranging scope [26] in the analysis of electronic structural problems in high symmetry environments, but here we illustrate its use in the identification of the source of vibronic intensity.

3.3.1. *The Ham Effect.* In the first spin allowed excited state, $^4T_{2g}$ near 12,500cm^{-1}, we expect to find four electronic origins, with, in order of increasing energy, E", U', U' and E' symmetry (in the double group O^*) arising from the spin-orbit interaction. These origins are all observed in absorption [25] and span 33cm^{-1}. When the spin-orbit coupling parameter is chosen to fit the relative energies of all the other states in the spectrum it is predicted that these origins should span 105cm^{-1}. The most natural explanation of this reduction is the Ham effect. Using (15) gives a Huang-Rhys factor for the e_g mode of 1.3.

3.3.2. *The Vibronic Intensity Mechanism.* Sharp features at 14,600cm^{-1} are the familiar $^4A_{2g} \longrightarrow {}^2E_g$ transitions. Because of the spin-selection rule this region is supposed to draw its intensity, via the spin-orbit interaction, from the $^4A_{2g} \longrightarrow {}^4T_{2g}$ transition. It is assumed that the vibronic coupling intensity mechanism observed in the doublet states reflects the mechanism in the quartet. The ^2E_g region is shown in Figure 10. Here the weak magnetic-dipole allowed electronic origin at 14,430cm^{-1} is accompanied by sidebands at 117, 189, and 309cm^{-1}. These features are the t_{2u} and t_{1u} bending modes and the t_{1u} stretching modes respectively. This assignment is confirmed by the sign of the MCD, which symmetry requires to be opposite for the t_{2u} and t_{1u} modes.

Because there are four electronic states in a small energy range the vibronic structure in the absorption spectrum of the $^4T_{2g}$ state is too complex to interpret, but if the temperature is low enough the emission spectrum comes only from the lowest E" component. In a magnetic field the circularly polarised luminescence reveals that the same three modes observed in the doublet state are the source of the vibronic intensity in the quartet. Figure 11 shows the application of eqn.(27) to the temperature dependence of the integrated band area of the quartet state. Here the observed frequencies and relative intensities of the enabling modes in Figure 10 are used as the parameters in eqn.(27). The agreement with experiment is excellent.

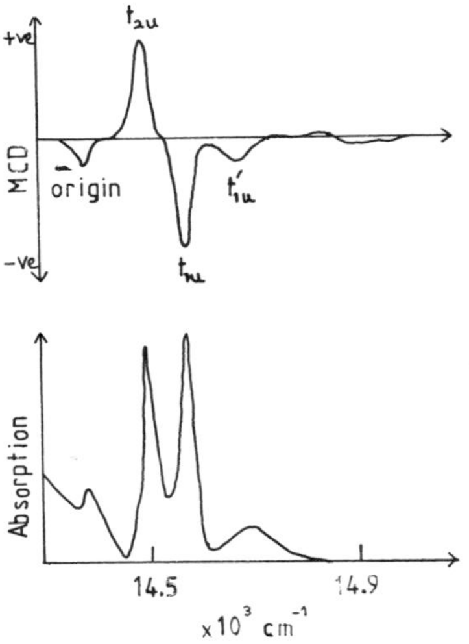

Figure 10. Absorption and MCD of the $^4A_{2g} \rightarrow {}^2E_g$ transition of Cr^{3+} in $Cs_2NaInCl_6$ at 4.2K.

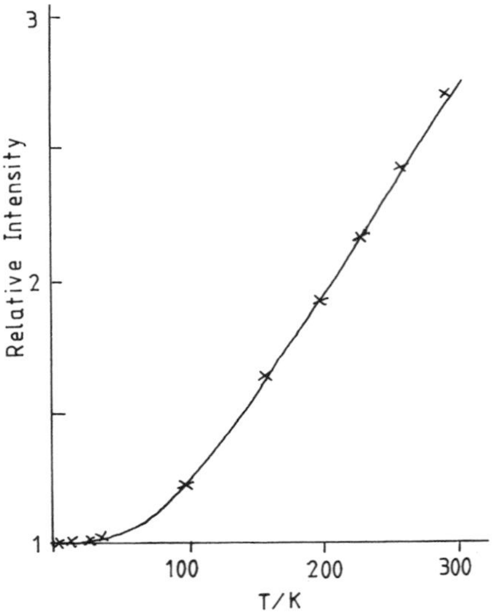

Figure 11. Temperature dependence of the intensity of the $^4A_{2g} \rightarrow {}^4T_{2g}$ transition.

The most obvious feature of Figure 10 is the much stronger intensity in the bending modes than in the stretching mode. This feature is common to many octahedral ions with this electronic configuration [27]. Superficially one might expect a stronger perturbation from the stretching modes. Ceulemans [28] first realised that this effect has a symmetry argument as its basis. The normal modes of the octahedron are expanded in terms of the modes of a three-dimensional harmonic-oscillator, whose solutions are the spherical harmonics. Ceulemans pointed out that the asymmetric stretch resembles a dipole so that the leading harmonic in the expansion is of order *one* (P), while the bending modes are predominantly octupolar, the leading harmonic being of order *three* (F). The t_{2u} and t_{1u} bending modes have displacements characterised by the $f_{z(x^2-y^2)}$ and f_{z^3} functions respectively.

The effective operator which brings about the vibronic transition is obtained by coupling a spherical harmonic of order *one*, arising from the dipolar radiation field, with a second spherical harmonic representing the linear vibronic coupling, eqn.(22). Together they must ensure that the matrix element

$$\langle {}^4A_{2g} | O | {}^4T_{2g} \rangle \; ; \quad \text{where } O = \{er.(\delta H/\delta Q)\}$$

is non-zero and so the operator O must have T_{1g} symmetry. We seek spherical harmonics which form a basis for this representation. The first harmonic for which this is true has order *four* (G), the representations spanned by the functions being A_{1g}, E_g, T_{1g} and T_{2g}, and the operator has hexadecapole symmetry. It is then easy to see that the hexadecapolar operator can only be generated by coupling the octupolar bending modes with the electric-dipole operator and that the dipolar stretching modes fail to generate intensity.

While this argument is illuminating it is necessary to perform a full normal coordinate analysis and to make assumptions about the electronic nature of the parity-allowed states from which the intensity is drawn in order to make a quantitative assessment. Calculations of this kind have been done by Acevedo and Flint[29], giving remarkable agreement with observed intensity ratios.

3.3.3. *The Linear Vibronic Coupling.* The presence of the Ham quenching is confirmed by the luminescence spectrum in figure 12. The detail is sufficiently well resolved for the intensity ratios to be found for progressions in both the a_{1g} and e_g modes. The results of section 2.3. give $S(e_g) = 1.1$ and $S(a_{1g}) = 1.6$. the former value agrees well with that from the Ham quenching. The band maximum is expected at $S(e_g)\hbar\,\omega_{eg} + S(a_{1g})\hbar\,\omega_{a1g} + \hbar\,\omega_v$, where ω_v is the mean frequency of the vibronic enabling modes. This expression predicts a Stokes shift (i.e. the separation between the absorption and emission maxima) of 2 x $880cm^{-1}$ which may be compared with an experimental value of $1780cm^{-1}$ All the spectral features of this transition are therefore satisfactorily explained by these two parameters.

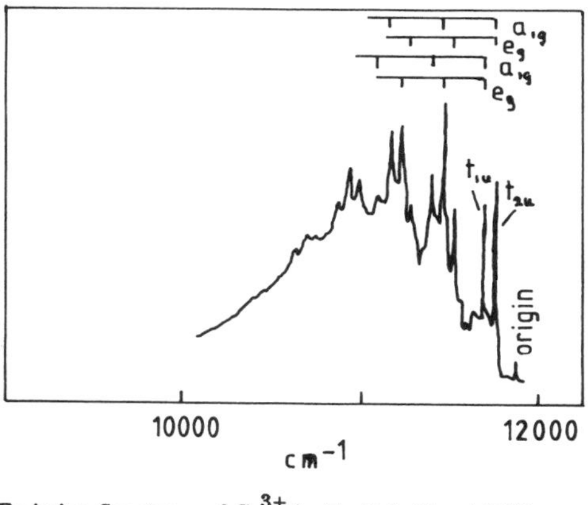

Figure 12. Emission Spectrum of Cr^{3+} in $Cs_2NaInCl_6$ at 10K.

If a force-field analysis is performed on this species we may use (13) and (14) to compute the mean displacements in these two modes. Eqn.(13) does not allow the determination of a sign of the displacement, nevertheless $(\delta V/\delta Q)$ can be estimated by differentiating the crystal field potential with respect to each normal coordinate. For the totally symmetric mode the excitation to an anti-bonding orbital predicts $\Delta Q(a_{1g})$ positive. The Jahn-Teller active e_g mode is more interesting.

Selecting one component of the $^4T_{2g}$ wavefunction in the strong-field basis the transition may be seen as a one-electron excitation from d_{xy} to $d_{x^2-y^2}$. Thus the sign of the deformation in the e_g mode entails an expansion in the xy plane where the anti-bonding interaction is concentrated, and a contraction in the z direction. It is the electron-electron repulsion which determines this excited state configuration, so that we anticipate that a Jahn-Teller distortion will always occur in this sense. A summary of the net atomic displacements, from both modes, is shown in Figure 13. It is to be expected that distortions of this kind are important in stereospecific photochemical reactions. This method of analysis has also been applied to Mn(II) and Co(III) compounds[30].

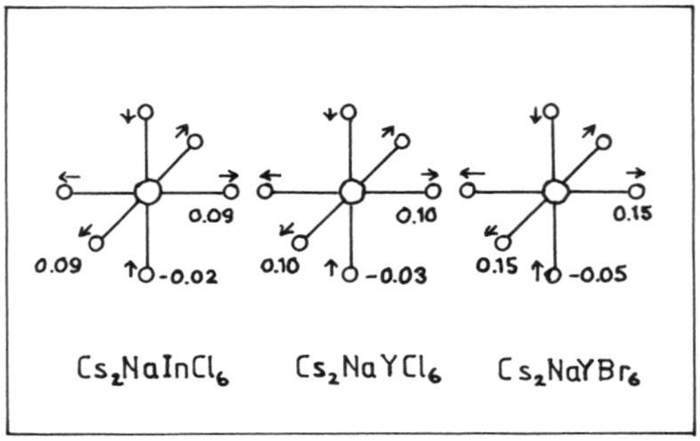

Figure 13. Distortions of the CrX_6^{3-} unit in the $^4T_{2g}$ state in three different lattices.

3.4. Vibronic Coupling in Two-Photon Absorption (TPA)

An extensive study has been made of the TPA spectrum of $Cs_2UO_2Cl_4$ [31]. Here we examine only two of the fourteen electronic excited states, with transitions at 26,197.3 and 26,247.6 cm^{-1}. Zeeman measurements establish them as a pair which would be degenerate in D_{4h}, having B_{2g} and B_{3g} symmetry respectively in D_{2h}. Their splitting, 50 cm^{-1}, is much less than the interval of 1500 cm^{-1} to the next electronic state. In TPA both show sidebands, at 174 cm^{-1}, due to the symmetric two-dimensional rocking mode of the UO_2 group, whose b_{2g} and b_{3g} components are not resolved. It is the mechanism of the vibronic intensity in these modes which is of interest here. The Table shows the expected two-photon polarisations of these modes.

Symmetries in D_{2h} Electronic	Vibrational	Vibronic	Two-Photon Polarisation Tensor
B_{2g}	b_{2g}	A_g	xx, yy, zz
	b_{3g}	B_{1g}	xy
B_{3g}	b_{2g}	B_{1g}	xy
	b_{3g}	A_g	xx, yy, zz

TPA spectra, obtained in the single colour and single beam configuration, are shown in Figure 14 for three polarisations. The vibronic sidebands occur, in conjunction with one quantum of the UO_2 symmetric stretching vibration, at 27,096 and 27,146 cm^{-1} and are most intense in the xx, yy and xy polarisations, as expected from the Table. A careful study of the angular dependence of this intensity [18] finds that in the B_{2g}

electronic state the intensities are such that I(xx) ~ I(xy) >> I(yy), but in the B_{3g} state I(yy) ~ I(xy) >> I(xx), as implied by Figure 14.

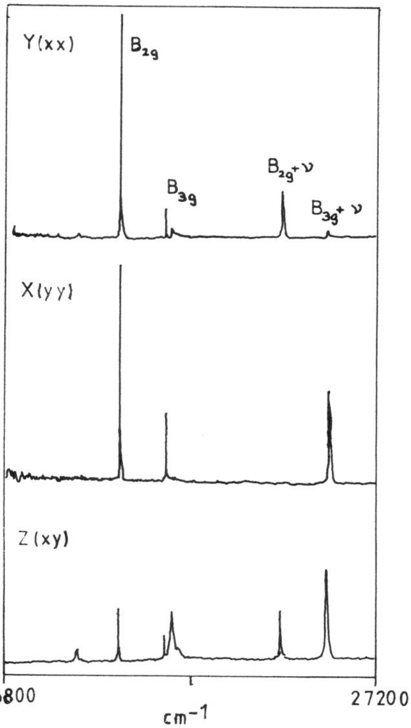

Figure 14. Single crystal polarised TPA Spectra of $Cs_2UO_2Cl_4$ at 4.2K. For notation see[19].

Neither the symmetry arguments in the Table, nor the properties of the initial and final states predict these results. Suppose the final vibronic states of A_g symmetry draw intensity from the nearest electronic state with A_g symmetry and strong TPA intensity, which is found 3000 cm^{-1} to higher energy and is equally strong in the xx and yy polarisations [31]. Because the orthorhombic distortion is so small, the b_{2g} mode on the B_{2g} state and the b_{3g} mode on the B_{3g} state should both induce equal intensity in xx and yy polarisations. They do not. This final state coupling scheme is shown in Figure 15(a). Since both vibronic states draw their intensity from a common electronic state there should be no difference between their polarisation properties.

The difficulty can be resolved by considering linear coupling in the intermediate states. Figure 15(b) and (c) show the elements in the perturbation sequence giving TPA intensity. We can make a distinction between parallel vibronic coupling, V_x and V_y, where the deformation is in the same plane as the intermediate states accessed by the first photon, and orthogonal coupling, V'_x and V'_y, where it is perpendicular to this plane.

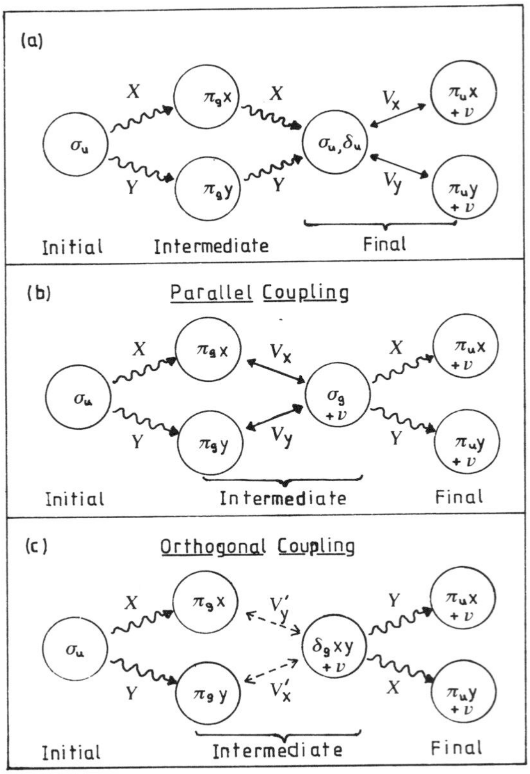

Figure 15. Scheme of the electronic matrix elements providing vibronic intensity in TPA. (a) Vibronic coupling in the final state, (b) parallel coupling in the intermediate state, (c) orthogonal coupling in the intermediate state. X and Y are photon polarisations. V_x and V_y are the parallel, and V'_x and V'_y the orthogonal coupling coefficients. One-electron $D_{\infty h}$ wavefunctions are illustrative.

If the assumption is made that only parallel coupling is significant Figure 15(b) predicts the observed intensity pattern. In this scheme the B_{2g} state is accessible only in the xx and xy polarisations and the B_{3g} state is accessible only in the yy and xy polarisations. The difference in the coupling may be interpreted in terms of the spatial properties of the one-electron orbitals involved in the mechanism [18].

In summary the anisotropy in the TPA intensity arises because the radiation field prepares a polarised intermediate state, from which the directional nature of the vibronic coupling creates states determining the polarisation of radiative access to the final states. The important role of the intermediate state is probably due to a larger linear coupling in the intermediate d-electron states than that applying in the initial and final states, a feature which would be expected to apply to both lanthanide and actinide f-f transitions.

[1] C.J. Ballhausen, *Molecular Electronic Structures of Transition Metal Complexes*. McGraw-Hill, 1979.

[2] C.J. Ballhausen, *Theor. Chim. Acta.*, **1**, 285 (1963)

[3] I.B. Bersuker, *The Jahn-Teller Effect and Vibronic Interactions in Modern Chemistry*. Plenum, 1984.

[4] F.S. Ham, *Phys. Rev.*, **138A**, 1727 (1965)

[5] K. Huang and A. Rhys, *Proc. Roy. Soc.(London)*, **A204**, 406 (1950)
S. Pekar, *Zh. Eksp. Theo. Fiz.*, **20**, 510 (1950)
R.C. O'Rourke, *Phys. Rev.*, **91**, 265 (1953)

[6] D. Curie, 'Absorption and Emission Spectra', in *Optical Properties of Ions in Solids*, ed. B. Di Bartolo, Plenum, 1975.

[7] B. Di Bartolo, *Optical Interaction in Solids*, Wiley, 1968.

[8] R. Orbach, 'Quantum Theory of Lattice Vibrations', in *Optical Properties of Ions in Solids*, ed. B. Di Bartolo, Plenum, 1975.

[9] F.J. Bergin, J.F. Donegan, T.J. Glynn and G.F. Imbusch, *J. Lumin.*, **34**, 307 (1986)

[10] W.M. Yen and P.M. Selzer, 'High Resolution Spectroscopy of Ions in Solids' , in *Laser Spectroscopy of Solids*, eds. W.M. Yen and P.M. Selzer, Springer, 1981.

[11] D.E. McCumber and M.D. Sturge, *J. Appl, Phys.*, **34**, 1682 (1963)

[12] J.L. Skinner and D. Hsu, *Adv. Chem. Phys.*, **65**, 1 (1986)

[13] I.S. Osad'ko, in *Spectroscopy and Excitation Dynamics of Condensed Matter Systems*, eds. V.M. Agranovich and R.M. Hochstrasser, North-Holland, 1983.

[14] G.J. Small, *Chem. Phys. Lett.*, **57** , 501 (1978)

[15] D. Hsu and J.L. Skinner, *J. Chem. Phys.*, **87**, 54 (1987)

[16] D.W. Pack and D.S. McClure, *J. Chem. Phys.*, **87**, 5161 (1987)

[17] A.B. Goldberg, S.A. Payne and D.S. McClure, *J. Chem. Phys.*, **81**, 1523, (1984)

[18] T.J. Barker, R.G. Denning and J.R.G. Thorne, *J. Lumin.*, **38**, 144 (1987)

[19] R.G. Denning, T.R. Snellgrove and D.R. Woodwark, *Mol. Phys.*, **30**, 1819 (1975)
R.G. Denning, T.R. Snellgrove and D.R. Woodwark, *Mol. Phys.*, **32**, 419 (1976)

[20] J.R.G. Thorne, R.G. Denning T.J. Barker and D.I. Grimley, *J. Lumin.*, **34**, 147 (1985)

[21] J.R.G. Thorne, D. Phil. thesis, Oxford, 1982.

[22] W. Moffitt and A. Moscowitz, *J. Chem. Phys.*, **30**, 648 (1959)

[23] R.G. Denning, D.N.P. Foster, T.R. Snellgrove and D.R. Woodwark, *Mol. Phys.*, **37**, 1089 (1979)

[24] O.E. Weigang, *J. Chem. Phys.*, **42**, 2224 (1965)
O.E. Weigang, *J. Chem. Phys.*, **43**, 3609 (1965)

[25] H.U. Gudel and T.R. Snellgrove, *Inorg. Chem.*, **17**, 1617 (1978)
R.W. Schwartz, *Inorg. Chem.*, **15**, 2817 (1976)
P. Shaw, D. Phil. thesis, Oxford, 1975.
R. Knochenmuss, C. Reber, M. Rajasekharan and H.U. Gudel, *Inorg. Chem.*, **85**, 4280 (1986)

[26] P.J. Stephens, *Ann. Rev. Phys. Chem.*, **25**, 201 (1974)

[27] C.D. Flint and P. Greenough, *J. Chem. Soc. Faraday Trans. II*, **68**, 897 (1972)
L. Helmholz and M.E. Russo, *J. Chem. Phys.*, **59**, 5455 (1973)

[28] A. Ceulemans, *Chem. Phys. Lett.*, **97**, 365 (1983)

[29] R. Acevedo and C.D.Flint, *Mol. Phys.* **58**, 1033(1986)

[30] E.I. Solomon and D.S. McClure, *Phys. Rev. B*, **9**, 4690 (1974)
R.B. Wilson and E.I. Solomon, *J. Amer. Chem. Soc.*, **102**, 4085 (1980)

[31] T. J. Barker, R. G. Denning and J. R. G. Thorne, *Inorg. Chem.*, **26**, 1721 (1987)

THE GENERAL THEORY OF RADIATIVE AND NON RADIATIVE TRANSITIONS IN CENTROSYMMETRIC COORDINATION COMPOUNDS OF THE TRANSITION METAL IONS

ROBERTO ACEVEDO
Department of Chemistry
Faculty of Physical and Mathematical Sciences
University of Chile
Tupper 2069. P.O. Box 2777
Santiago - Chile

ABSTRACT. The theory of radiative and non radiative transitions in centrosymmetric inorganic compounds of the transition metal ions, having monoatomic and polyatomic ligand subsystems, is reviewed on the basis of a symmetrized combined crystal field-ligand polarization model. A new set of symmetry adapted equations, applicable to both reducible and non simply reducible groups is derived throughout the course of this work to account for both radiative and non radiative rates of decay from the lowest lying excited electronic states to the ground state for a series of electronic transitions of spectroscopic interest. Some applications of the general formalism developed here, are given to evaluate the non radiative relative rates of decay of the $^4T_{1g} \rightarrow {}^4A_{2g}$ and the $^4T_{2g} \rightarrow {}^4A_{2g}$ radiationless transitions of the $Cr(NH_3)_6^{3+}$ complex ion and the relative vibronic intensity distribution due to the three false origins of the $ReBr_6^{2-}$ ion in the Cs_2ZrBr_6: $ReBr_6$ system.

1. INTRODUCTION

The crystal field theory and its applications to coordination inorganic compounds of the transition metal ions, lanthanides and actinides has been a very active and productive research area for scientists for many decades (1-7).

There are several books (1, 2, 4) dealing with the physics and a wide range of applications of this theory to coordination inorganic compounds. Here we shall focus our attention upon the vibronic crystal field theory and will consider some applications to the actual evaluation of the relative vibronic intensity distribution, due to the false origins, in centrosymmetric octahedral inorganic complex ions of the transition metal ions.

As it is well known, within the independent system model the total non relativistic molecular Hamiltonian may be written as (6-14):

$$H = H_M^0 + \sum_L H_L^0 + H^1 \tag{1}$$

C. D. Flint (ed.), Vibronic Processes in Inorganic Chemistry, 139–193.
© 1989 by Kluwer Academic Publishers.

Here H_M^0 and H_L^0 are the zeroth-order Hamiltonians for the central metal ion and the Lth-ligand subsystem, respectively. The crystal field vibronic operator is written in the standard form as:

$$H_{CF}^1 = \sum_{\Gamma,\gamma,k} \{\delta V_{CF}/\delta Q_\gamma^\Gamma(k)\}_0 Q_\gamma^\Gamma(k) \tag{2}$$

Here k labels the various odd parity normal modes of vibration, Q_γ^Γ stands for the γth-component or a normal coordinate Q which transforms under the Γth-irreducible representation of the relevant point molecular group and V_{CF} represents the crystal field component to the total interaction potencial among the central metal ion (M) and the ligand subsystem (L). Within the framework of this approximation the central metal ion's charge distribution is regarded as a collection of electric multipoles (2^l ; l = 1, 2, 3, 4,...) and the ligand subsystems are considered as effective point charges "$q_L e$", being "e" the electron charge and "q_L" is a parameter to be determined either from experiment or from molecular orbital type-calculations. The crystal field potential V_{CF} may be written in a symmetrized form as given below (15):

$$V_{CF} = \sum_L q_L e \sum_{\Gamma,\gamma,k} G_{\Gamma\gamma}^L(k) M_\gamma^\Gamma(k) \tag{3}$$

where the geometric dependence of the Coulombic interaction between the 2^l-central metal ion's multipoles and the effective ligand charges $q_L e$ is afforded by the symmetrized crystal field geometric factor $G_{\Gamma\gamma}^L(k)$, and the symmetry adapted central metal ion's multipoles are represented by $M_\gamma^\Gamma(k)$.

From eqs |2| and |3|, we write the vibronic crystal field perturbation as:

$$H_{CF}^1 = \sum_L q_L e \sum_{\Gamma_1,\gamma_1,k} Q_{\gamma_1}^{\Gamma_1}(k) \sum_{\Gamma_2,\gamma_2,\tau} M_{\gamma_2}^{\Gamma_2}(\tau) \left| \partial G_{\Gamma_2\gamma_2}^L(\tau)/\partial Q_{\gamma_1}^{\Gamma_1}(k) \right|_0$$

It is seen from the above equation, that this vibronic coupling operator is also a function of the normal coordinates $Q_\gamma^\Gamma(k)$ of the molecular system.

It is now recognized that to solve the vibrational equation of motion for a vibrating N-body system is by no means trivial. This arises from the fact that the interaction vibrational potential, among the N-particles is an unknown function of the displacement coordinates of each nuclei ion. However, it is well known that this interaction potential V is a function of the relative displacement coordinates among the particles.

It is custonary,in the literature, to expand this interaction potential in a Taylor series of the mass wighted nuclear Cartesian displacement coordinates. Thus, let q_i, i = 1, 2,...,s; be a complete set of these Cartesian coordinates, then the vibrational potential function may be written as:

$$\tilde{V} = V - V_0 = (1/2) \sum_{i,j} (f_x)_{ij} q_i q_j + \text{higher order terms} \qquad (5)$$

here $(f_x)_{ij} = (f_x)_{ji}$ are the force constants, expressed in the basis of the mass weighted Cartesian displacement coordinates, of the molecular system.

The second derivatives of the potential function V with respect to the Cartesian displacement coordinates should be evaluated at the nuclear equilibrium configuration of the system. To chemists, it is indeed more suitable to carry out normal coordinates analysis on the basis of the standard internal coordinates, namely stretching coordinates, bending coordinates, torsion coordinates, etc. Let s_1 , i = 1, 2,...,s be a complete set of internal coordinates for the molecular system,then we may write s = B R in matrix notation, where the B matrix relates the set of the standard internal coordinates s to the set of the nuclear Cartesian displacement coordinates R, of the molecular system.

Then, since $q_i = \sqrt{m_i}\, R_{ii}$, where $R_{11} = x_1$, $R_{22} = y_1$, $R_{33} = z_1, \ldots$, etc.; we may write $q = M^{1/2}R$, then it follows that the potential function V, see eq |5| may be written as follows: $2\ V = q' f_x\ q = (M^{1/2}R)' f_x (M^{1/2}R) = (M^{1/2}B^{-1}s)' f_x\ (M^{1/2}B^{-1})s) = s'\ f_{int}\ s$, where $f_{int} = (M^{1/2}B^{-1})' f_x\ (M^{1/2}B^{-1})$. Observe that the higher order terms in the expression for the potential function V have been neglected.

In addition, since the symmetry coordinates S are related to the standard internal coordinates, by means of the well known identity S = U s, where U is symmetry determined, we may rewrite the potential function in the symmetry coordinates basis, as follows:

$$2V = S'Uf_{int}U'S = S'F_{sym}S \qquad (6)$$

Once the equilibrium geometry of the molecule is known, it is a simple matter, to evaluate the B-matrix relating the internal coordinates to the nuclear Cartesian coordinates of the molecular system, in terms of the equilibrium bond angles and the equilibrium bond distances among the particles of the molecular system (16). The B-matrix may be made squared by adding the three translations and the three rotations of the molecule as a whole, and in this way we may write the identity: $S = U\ s = (UB)\ R$ and hence $R = (UB)^{-1}S$. It is seen from eq |6| that the symmetrized F_{ij} - matrix elements may be obtained from the set of the internal coordinates by means of a similarity transformation, that is: $F_{sym} = U\ f_{int}\ U'$. Similary the kinetic energy T may be expressed in the symmetry coordinate basis set as (16):

$$2T = \dot{S}'\ G^{-1}_{sym}\ \dot{S} \qquad (7)$$

where the $(G_{sym})_{ij}$ - matrix elements may be found the transformations: $G_{sym} = Ug\ U'$ and $g = B\ M^{-1}\ B'$, respectively. In terms of the F_{sym} and the G_{sym} matrices, it is straightforward to show that the dynamic of the vibrating system may be described by solving the set of equations of motion:

$$G_{sym}F_{sym}L_{sym} = L_{sym}\Lambda \tag{8}$$

Here Λ is a diagonal matrix whose matrix elements, except a constant factor, correspond to the square of the vibrational frequency associated with each normal mode of vibration. Similary, since $LL' = G$ then $ULL'U' = G_{sym} = L_{sym}L'_{sym}$. The L-matrix depends strongly on the details of the force field employed to solve the vibrational equations of motion and as it will be shown later on, it is of a paramount importance in vibronic coupling theory. The L-matrix relates the symmetry coordinates to the normal coordinates for the molecular systems, that is: $S = L\,Q$, where Q represents the set of the normal coordinates.

For large inorganic systems of spectroscopic interest, to undertake accurate normal coordinate analysis is by no means a trivial task. A fair approximation to describe the dynamics of these molecular systems may be achieved provided we have available a complete set of vibrational frequencies, obtained from Infrared, Raman and Luminiscence spectroscopies. However for many systems, the experimental data is either scare or incomplete so that to obtain a reliable description of the dynamic behaviour of the system may become a formidable task to achieve! Several different force fields have become available during the last decades to perform normal coordinate analysis for polyatomic systems. The most common force fields used so far, are the UBFF, the MUBFF, the GVFF, the MGVFF, the OVFF and the MOVFF (16). We have recently developed a modified general valence force field (MGVFF) for cubic lanthanide hexachloro-elpasolite crystals (18) and have shown that a simple parametrization scheme along with some physical constraints upon the solutions of the vibrational equations of motion may be used as a guide to resolve controversial assignments and may also be useful to carry out normal coordinate analysis in a systematic way for the lanthanide elpasolites of the kind $CsNaLn(III)Cl_6$. The lack of experimental data for large inorganic polyatomic molecules, is indeed, a major source of error in normal coordinate analysis, since we often have more unknown symmetrized force constants $(F_{sym})_{ij}$ than well determined experimental vibrational frequencies. There are some attempts in the literature to quantized the F-matrix and some degree of success has been achieved, mainly for small polyatomic inorganic and organic molecules (16) and References there in. This procedure becomes very complex indeed for large polyatomic molecules and very complicated computing programmes are needed.

Our laboratory is engaged in a project to quantized the F-matrix for coordination compounds of the transition metal ions, and the results will be published somewhere else (19). As it will be shown, the vibronic coupling theory, to account for the observed spectral intensities, in coordination compounds of the transition metal ions, should also include the so called ligand polarization contribution to the total transition dipole moment.

The ligand polarization method has been applied to both centrosymmetric and non centrosymmetric coordination compounds of the transition metal ions and lanthanides (6-14, 17, 20, 21). Both the crystal field and the ligand polarization methods are to a first order correction, comple mentary to one another and therefore to discuss vibronic intensities in coordination compounds, we need to develop this model in some detail. This method was first applied to spectral intensities in inorganic coordination compounds by Mason and Richardson (6, 7, 23, 24). Some specific applications to coordination compounds of the transition metal ions, were carried out by Acevedo and Flint (8-13).

The ligand polarization model assumes that the radiation field induces transient ligand dipoles and then induces electric multipoles on the central metal ion of the molecular system. The interaction between the transient ligand dipoles and the central metal ion's electric multipoles is taken to be the Coulombic potential energy, defined as: $V = \sum_i \sum_j e_i(M)e_j(L)/r_{ij}$.

Then, within the framework of a combined crystal field-ligand polarization model, let us write the interaction potential as given below:

$$V = V_{CF} + V_{LP} \tag{9}$$

where the crystal field potential is given by eq |3| and the symmetrized ligand polarization contribution to the total interaction potential may be expressed as follows (8, 14, 25).

$$V_{LP} = \sum_L \sum_i \sum_{\Gamma\gamma,\alpha} G^L_{\Gamma\gamma,\alpha}(i)\mu^\alpha(L)M^\Gamma_\gamma(i) \tag{10}$$

where the geometric dependence of the Coulombic interaction is afforded by the symmetrized ligand polarization geometric factors $G^L_{\Gamma\gamma,\alpha}(k)$; $\mu^\alpha(L)$ stands for the αth-component of the transient induced ligand dipole and $M^\Gamma_\gamma(i)$ represents the γth-component of a central metal ion's multipole which transforms under the Γth-irreducible representation of the relevant point molecular group.

The vibronic coupling component, derived from the ligand polarization approach is given as follows:

$$H^1_{LP} = \sum_L \sum_{i,k} \sum_{\Gamma_1\gamma_1} Q^{\Gamma_1}_{\gamma_1}(k) \sum_{\Gamma_2\gamma_2,\alpha} \{\partial G^L_{\Gamma_2\gamma_2,\alpha}(i)/\partial Q^{\Gamma_1}_{\gamma_1}(k)\}_0 \mu^\alpha(L)M^{\Gamma_2}_{\gamma_2}(i) \tag{11}$$

For both d-d and f-f type of excitations in centrosymmetric octahedral complexes, a complete tabulation of both the crystal field and the ligand polarization geometric factors as well as the symmetrized central metal ion's multipoles may be found in the References (8, 15, 25). In the next section of the present work, we shall discuss in detail the so called combined vibronic crystal field-ligand polarization method to account for the observed spectral intensities in coordination compounds of the transition metal ions.

2. A COMBINED VIBRONIC CRYSTAL FIELD-LIGAND POLARIZATION MODEL. RADIATIVE TRANSITIONS IN CENTROSYMMETRIC INORGANIC COMPOUNDS OF THE TRANSITION METAL IONS.

Let $|A\alpha\rangle \rightarrow |B\beta\rangle$ be a given electronic transition in a centrosymmetric coordination compounds of a transition metal ion. We may write the total electric dipole moment associated with this particular electronic transition as follows (8, 14).

$$\vec{\mu}_{A\alpha \rightarrow B\beta} = \vec{\mu}^{CF}_{A\alpha \rightarrow B\beta} + \vec{\mu}^{LP}_{A\alpha \rightarrow B\beta} \tag{12}$$

in obvious notation. Then the total dipole strength of the $A\alpha\rangle \rightarrow B\beta\rangle$ excitation, becomes:

$$D_{A\alpha \rightarrow B\beta} = D^{CF}_{A\alpha \rightarrow B\beta} + D^{LP}_{A\alpha \rightarrow B\beta} + D^{DF,LP}_{A\alpha \rightarrow B\beta} \tag{13}$$

where the interference term $D^{CF,LP}_{A\alpha - B\beta}$ (cross term), couples together the vibronic crystal field and the vibronic ligand polarization transition dipole moments, that is we write:

$$D^{CF,LP}_{A\alpha \rightarrow B\beta} = \vec{\mu}^{CF*}_{A\alpha\rightarrow B\beta}\,\vec{\mu}^{LP}_{A\alpha\rightarrow B\beta} + \vec{\mu}^{LP*}_{A\alpha\rightarrow B\beta}\,\vec{\mu}^{CF}_{A\alpha\rightarrow B\beta} \tag{14}$$

The evaluation of the interference term should be handled with care, since it is a signed quantity. Here we follow Griffith's convention of phases (4, 5) to define our wavefunctions, operators and symmetry coordinates for the system. A very systematic work on the choice of phases has been carried out by Butler (32), and his approach seems to be , by far the most appropiate to deal with point group symmetry applications, however in this work we follow Griffith's work since we feel that Chemists are more get used to this formulism rather than to Butler's. Next, having defined the total dipole strength of the $|A\alpha\rangle \rightarrow |B\beta\rangle$ electronic transition, see eq |13| we shall introduce the total dipole strength associated with the overall excitation $|A\rangle \rightarrow |B\rangle$ as given below:

$$D(A - B) = \sum_{\alpha,\beta} \sum_{\nu_i} D_{A\alpha - B\beta}(\mu_i) \tag{15}$$

where the summations are to be carried out over all the α, β-components of the electronic states $|A\rangle$ and $|B\rangle$, the false origins μ_i and the progressions in even parity modes based upon them, respectively.
Next, let us consider each of the contributions to the total transition dipole moment associated with the $|A\alpha\rangle \rightarrow |B\beta\rangle$ electronic transition, individually.
Back in 1957, Liehr and Ballhausen (26, 29, 31) put forward the vibronic crystal field method to evaluate the total spectral intensity in centrosymmetric octahedral coordination compounds of the transition metal ions.

These authors employed a truncated basis set to evaluate the spectral intensity associated with a 3d → 3d electronic transition for an octahedral coordination compound of the first row transition metal ions. Since a typical 3d → 3d electronic transition is to zeroth-order electric dipole forbidden, they suggested that this electronic transition borrows its intensity from a 3d → 4p electric dipole allowed excitation through the cooperation of the odd parity normal modes of vibration of the complex ion. They showed that this mechanism could, in principle explains the observed spectral intensities for a series of electric dipole forbidden electronic transitions in centrosymmetric coordination compounds (vibronic coupling mechanism) of the transition metal ions, lanthanides and actinides.

Following their work, Acevedo and Flint (14, 15, 17) developed general symmetrized equations to evaluate the crystal field vibronic contribution to the total transition dipole moment for a parity forbidden excitation in a centrosymmetric coordination compounds. Thus, let us define $V_{kt} = (\delta V_{CF}/\delta S_{kt})_0$, where V_{CF} stands for the crystal field potential given by eq |3|.

It is shown that for the $|A\alpha\rangle \rightarrow |B\beta\rangle$ electronic transition, the εth-component to the transition dipole moment, may be expressed in general terms as follows:

$$\mu^{CF,\varepsilon}_{A\alpha\rightarrow B\beta} = \frac{1}{\Delta E}\sum_{kt} S_{kt} \sum_{C,\gamma} \{\langle A\alpha|V_{kt}|C\gamma\rangle\langle C\gamma|\mu^{\varepsilon}|B\beta\rangle + + \langle A\alpha|\mu^{\varepsilon}|C\gamma\rangle\langle C\gamma|V_{kt}|B\beta\rangle\} \quad (16)$$

where ΔE represents an average energy gap corresponding to a parity and spin allowed electronic transition and the intermediate electronic states are denoted by the kets $|C\gamma\rangle$. Observe that the intermediate electronic states must have parity other than the terminal electronic states of the electronic transition to have a non vanishing firs order transition dipole moment.

A typical matrix element, involving the initial electronic state and the whole possible set of intermediate electronic states may be written in a symmetry adapted form as given below:

$$\langle A\alpha|V_{kt}|C\gamma\rangle = qe \sum_{D,\delta,j} V\begin{pmatrix} A & C & D \\ \alpha & \gamma & \delta \end{pmatrix} A^{D\delta}_{kt}(j)\langle A||M^{D}(j)||C\rangle \quad (17)$$

where the crystal field vibronic coupling constants $A^{D\delta}_{kt}(j)$ have been evaluated and tabulated by Acevedo et al (15) for both d → d and f → f type of electronic transitions in octahedral symmetry for D = T_{1u}, T_{2u} and j = 1, 3, 5 and 7. Similary, the reduced matrix elements $\langle AM^{D}(j)B\rangle$ are also listed in the Reference (15). The next step in the intensity calculation is to replace the symmetry coordinates by the normal coordinates of the system. This assumes that we have already carried out the normal coordinate analysis for the molecular system and therefore we know the L-matrix, which relates the space of the symmetry coordinates to the space of the normal coordinates, that is the transformation $S = LQ$ is known. Then we may write this transformation as follows:

$$S^D_\delta(k) = \sum_{C,\gamma,i} L_{ki}\delta_{CD}\delta_{\gamma\delta}Q^C_\gamma(i) \qquad (18)$$

From eqs |16|, |17| and |18|, we observe that the transition dipole moment associated with the $|A\alpha\rangle \rightarrow |B\beta\rangle$ electronic transition is a function of the effective energy gap corresponding to a parity and spin allowed electronic transition, a linear combination of the vibronic crystal field coupling constants, and the effective charges carry by the ligands(L). The vibronic crystal field method developed by Liehr and Ballhausen (26) was applied by Acevedo and Flint (9) to the $^2E_g \rightarrow {}^4A_{2g}$ electronic transition of the MnF_6^{2-} ion in octahedral symmetry. It was shown that the relative vibronic intensity distribution due to these three vibronic origins is independent of the radial wavefunctions, but depends on the details of the force field. The calculated values for the relative vibronic intensity distribution agree quite well with experiment. Though the relative vibronic intensity distribution is independent on the choice of the central metal radial wavefunctions, the total oscillator strenght $f = \{2m\Delta E_{A \rightarrow B}/3\hbar^2 e^2\}D_{A \rightarrow B}$ does depend on the choice of the radial wavefunctions and therefore on the central metal ion's charge.

One year later, back in 1958, Koide and Pryce formulated a model to account for the observed spectral intensity in parity forbidden electronic transitions in centrosymmetric coordination compounds (27, 28, 30). The basic assumptions of the Koide and Pryce intensity model for centrosymmetric coordination compounds are: {1} The intermediate electronic states occur at the same energy, say ΔE and {2} They form a complete set of wavefunctions in the Hilbert Space. Indeed, when these sets of assumptions are adopted in the intensity model, the detailed nature of the intermediate electronic states becomes totally irrelevant to the intensity calculation, except on the choice of the effective energy gap $\overline{\Delta E}$. Although this method has been applied to intensity calculations in complex ions of the rare earths, it does seem to us that this so called closure approximation should either be relaxed or abandoned. We have recently shown that for non centrosymmetric complex ions in tetrahedral symmetry, the crystal field component to the total transition dipole moment vanishes identically (21). At first glance, there is no obvious reason we can think of, to explain the failure of the Koide and Pryce's method to account for the observed spectral intensity in perfect tetrahedral complex ions of the kind CoX_4^{2-} ($X = Cl^{1-}$, Br^{1-} and I^{1-}). We have shown that when this procedure is adopted to calculate the spectral intensity, an arithmetic cancellation occur, so that the crystal field transition dipole moment vanishes identically.

Within the framework of the closure approximation for the wavefunctions, the Koide and Pryce's calculation method, the εth-vector component of the electric dipole transition moment associated with the $|A\alpha\rangle \rightarrow |B\beta\rangle$ electronic transition may be written in a symmetry adapted form as follows (8):

$$\mu^{CF,\varepsilon}_{A\alpha \to B\beta} = \left(\frac{2}{\Delta E}\right) \sum_{C,D} \sum_{\gamma,\delta} \sum_{i,k} \lambda^{1/2}(C)\, V\begin{pmatrix} A & B & C \\ \alpha & \beta & \gamma \end{pmatrix} V\begin{pmatrix} D & T_1 & C \\ \delta & \varepsilon & \gamma \end{pmatrix}$$

$$L_{ki} \langle A || O^C(k) || B \rangle Q^D_\delta(i) \tag{19}$$

In our present notation, $\lambda(C)$ stands for the dimension of the Cth-irreducible representation of the relevant point molecular group, ΔE corresponds to the effective energy gap of a parity and spin allowed excitation, $Q^D_\delta(i)$ is the δth-component of a normal coordinate Q which transforms under the Dth-irreducible representation. Furthermore, the reduced matrix elements $\langle A||O^C(k)||B\rangle$, evaluated within the Koide and Pryce's method have been tabulated for both d-d and f-f types of electronic transitions by Acevedo et al (15), and several applications to intensity calculations in centrosymmetric inorganic complex ions of the transition metal ions and lanthanides may be found in References (6, 7, 8-15, 17, 25).
Next, we mentioned earlier on in this section, that both the vibronic crystal field and the vibronic ligand polarization contributions to the total transition dipole moment of the $|A\alpha\rangle \to |B\beta\rangle$ transition are complementary to one another to a first order correction, then we shall discuss in some detail the vibronic ligand polarization transition dipole moment for centrosymmetric coordination compounds of the transition metal ions. As we stated in section I, within the independent system model, the total non relativistic molecular Hamiltonian is given by eq |1|. The zeroth-order Hamiltonian $H^0 = H^0_M + H^0_L$ suggests zeroth-order wavefunctions of the type:

$$|\Psi^0_i\rangle = |M_a\rangle \prod_b |L_b\rangle \tag{20}$$

Let H^1, the vibronic operator, be a small perturbation, upon the eigenvalues and the eigenfunctions of the molecular system, we may then make use of standard perturbation theory for non degenerate electronic states and write our electronic states to a first order correction as given below:

$$|M_a L_0\rangle' = |M_a L_0\rangle + \sum_{k,l} (E_a + E_0 - E_k - E_l)^{-1}$$

$$\langle M_k L_k | H^1 | M_a L_0 \rangle | M_k L_l \rangle \tag{21}$$

Next, we write our total electric dipole operator, in a tensorial notation as follows:

$$D^1_q = D^1_q(M) + D^1_q(L) \ ; \ q = 0, \pm 1. \tag{22}$$

observe that the total electric dipole tensor operator as written above, excludes from our formulism effects such charge transfer, etc. For the sake of simplicity, we shall choose our zero energy levels, so that $E_0(M) = E_0(L) = 0$, and therefore to a first order approximation, we may write the total transition dipole moment associated with the $|0\rangle \rightarrow |a\rangle$ excitation as:

$$\langle D_q^1 \rangle_{0a} = \langle M_0 L_0 | D_q^1 | M_a L_0 \rangle + \tag{23.0}$$

$$+ \sum_{k \neq 0} (-E_k)^{-1} \langle M_0 L_0 | H^1 | M_k L_0 \rangle \langle M_k | D_q^1(M) | M_0 \rangle + \tag{23.1}$$

$$+ \sum_{k \neq a} (-E_a - E_k)^{-1} \langle M_k L_0 | H^1 | M_a L_0 \rangle \langle M_0 | D_q^1(M) | M_k \rangle + \tag{23.2}$$

$$+ \sum_{l \neq 0} (-E_a - E_l)^{-1} \langle M_0 L_0 | H^1 | M_a L_l \rangle \langle L_l | D_q^1(L) | L_0 \rangle + \tag{23.3}$$

$$+ \sum_{l \neq 0} (E_a - E_l)^{-1} \langle M_0 L_l | H^1 | M_a L_0 \rangle \langle L_0 | D_q^1(L) | L_l \rangle \tag{23.4}$$

From the above set of equations, it is seen that for a centrosymmetric complex ion, the zeroth-order electric dipole transition dipole moment vanishes identically (Lapporte forbidden electronic transition). The equations (23.1) and (23.2) represent the basis of the vibronic crystal field method, whereas the second pair of identities, namely equations (23.3) and (23.4) are the basis of the vibronic ligand polarization method to evaluate vibronic intensities in centrosymmetric coordination compounds (6, 22, 23). This latter pair of equations, in addition to the central metal ion's wavefunctions involve ligand ground state and excited states wavefunctions.

We have shown that the εth-vector component of the transition dipole moment associated with the $|A\alpha\rangle \rightarrow |B\beta\rangle$ excitation may be written in a symmetry adapted form as given below (8, 13, 14):

$$\mu^{LP,\varepsilon}_{A\alpha \rightarrow B\beta} = \sum_{C,D} \sum_{\gamma,\delta} \sum_{k,\tau} V\begin{pmatrix} A & B & C \\ \alpha & \beta & \gamma \end{pmatrix} B^{C\gamma,\varepsilon}_{D\delta}(k,\tau) \langle A || M^C(\tau) || B \rangle S^D_\delta(k) \tag{24}$$

where for isotropic ligands, the ligand polarization vibronic coupling constants are defined as:

$$B^{C\gamma,\varepsilon}_{D\delta}(k,\tau) = - \bar{\alpha}_L \sum_L \{\partial G^{L,\tau}_{C\gamma,\varepsilon} / \partial S^D_\delta(k)\}_0 \tag{25}$$

It is worth mentioning that for simply reducible groups and anisotropic ligands, such as H_2O, NH_3 and CN^{1-}, equations |24| and |25| should be modified in order to allow for the anisotropic components of the second rank ligand polarizability tensor operator (23, 24). It is known that for isotropic ligand subsystems, the polarizability tensor operator is diagonal and therefore we may introduce the so called mean ligand polarizability, $\bar{\alpha}_L$, defined in the standard form as:

$$\bar{\alpha}_L = \alpha_{xx} = \alpha_{yy} = \alpha_{zz} = \sum_{l\neq o} \frac{2E_l}{E_l^2 - E_k^2} |\mu_{ol}^{\varepsilon}|^2 \tag{26}$$

where $\mu_{ol}^{\varepsilon} = \langle L_o|\mu^{\varepsilon}|L_l\rangle$, being $|L_o\rangle$ and $|L_l\rangle$ eigenfunctions of the zero-th-order ligand Hamiltonian H_L^0.
For centrosymmetric octahedral complex ions, the ligand polarization vibronic coupling constants $B_{D\delta}^{C\gamma,\varepsilon}(\kappa,\tau)$ and the reduced matrix elements $\langle A||M^C(\tau)||B\rangle$ have been evaluated and tabulated by Acevedo and Flint (8, 25). From equations |18| and |24| we may rewrite the εth-vector component of the transition dipole moment associated with the $|A\alpha\rangle \rightarrow |B\beta\rangle$ excitation, within the framework of the vibronic ligand polarization method as given below:

$$\mu_{A\alpha \rightarrow B\beta}^{LP,\varepsilon} = \sum_{C,D} \sum_{\gamma,\delta} \sum_{i,k,\tau} V\begin{pmatrix} A & B & C \\ \alpha & \beta & \gamma \end{pmatrix} L_{ki} B_{D\delta}^{C\gamma,\varepsilon}(k,\tau) \langle A||M^C(\tau)||B\rangle Q_{\delta}^{D}(i) \tag{27}$$

Having already defined the vibronic crystal field and the vibronic ligand polarization components to the total transition dipole moment of the $|A\alpha\rangle \rightarrow |B\beta\rangle$ excitation, we may work out an explicit equation for the interference term, which couples together the crystal field and the ligand polarization transition dipole moments. As for the crystal field component, we shall make use of the so called closure approximation, see eq |19| and as for the ligand polarization component to the total transition dipole moment we will employ the identity given by eq |27|, respectively. The same procedure may be utilized, when the Liehr and Ballhausen method is adopted.
To evaluate the total dipole strength associated with the $|A\alpha\rangle \rightarrow |B\beta\rangle$ electronic transition, we shall assume that the potential energy-surfaces associated with the terminal electronic states of this transition have the same shape and are only vertically displaced to one another. When this approximation is adopted, the total dipole strength $D_{A\alpha \rightarrow B\beta}$ becomes:

$$D_{A\alpha \rightarrow B\beta} = \sum_{\varepsilon=x,y,z} |\langle 0|\mu_{A\alpha \rightarrow B\beta}^{\varepsilon}|1\rangle|^2 \tag{28}$$

and therefore, the total dipole strength associated with the overall electronic transition $|A\rangle \rightarrow |B\rangle$ may be written as follows:

$$D_{A\rightarrow B} \quad \sum_{\nu_i} \sum_{\alpha,\beta} D_{A\alpha \rightarrow B\beta}(\nu_i) \tag{29}$$

Thus, it is straightforward to show that the various contributions to the total dipole strength of the overall electronic transition $|A\alpha\rangle \rightarrow B\beta\rangle$ may be written in a symmetry adapted form, for centrosymmetric inorganic complex ions and for simply reducible groups as given below:

THE VIBRONIC CRYSTAL FIELD COMPONENT TO THE TOTAL TRANSITION DIPOLE MOMENT

$$D^{CF}_{A\to B}(\nu_i) = \left(\frac{2}{\Delta E}\right)^2 \sum_{\alpha}\sum_{C,D}\sum_{\gamma,\delta} V\begin{pmatrix} D & T_1 & C \\ \delta & \alpha & \gamma \end{pmatrix}^2 \sum_{k,m} L_{ki}L_{mi} \times$$

$$\times \langle A||O^C(k)||B\rangle\langle A||O^C(m)||B\rangle |\langle 0|Q^D_\delta(i)|1\rangle|^2 \qquad (30)$$

THE VIBRONIC LIGAND POLARIZATION COMPONENT TO THE TOTAL TRANSITION DIPOLE MOMENT

$$D^{LP}_{A\to B}(\nu_i) = \sum_{\alpha}\sum_{C,D}\sum_{\gamma,\delta} \lambda^{-1}(C) \sum_{k,m} L_{ki}L_{km} \sum_{\tau,\tau'} B^{C\gamma,\alpha}_{D\delta}(k,\tau) \times$$

$$\times B^{C\gamma,\alpha}_{D\delta}(m,\tau')\langle A||M^C(\tau)||B\rangle\langle A||M^C(\tau')||B\rangle$$

$$|\langle 0|Q^D_\delta(i)|1\rangle|^2 \qquad (31)$$

THE VIBRONIC INTERFERENCE TERM TO THE TOTAL TRANSITION DIPOLE MOMENT

$$D^{CF,LP}_{A\to B}(\nu_i) = 2\left(\frac{2}{\Delta E}\right) \sum_{\alpha}\sum_{C,D}\sum_{\gamma,\delta} \lambda^{-1/2}(C)\; V\begin{pmatrix} D & T_1 & C \\ \delta & \alpha & \gamma \end{pmatrix} \sum_{k,m} L_{ki}L_{mi} \times$$

$$\times \langle A||O^C(k)||B\rangle\{\sum_{\tau} B^{C\gamma,\alpha}_{D\delta}(m,\tau)\langle A||M^C(\tau)||B\rangle\}$$

$$|\langle 0|Q^D_\delta(i)|1\rangle|^2 \qquad (32)$$

The above set of equations hold for centrosymmetric coordination compounds and for simply reducible groups. Equations |30|, |31| and |32| should be scaled by a factor equals to $\lambda^{-1}(A)$, where $\lambda(A)$ corresponds to the orbital degeneracy of the initial electronic state, represented by the ket $|A\rangle$.

It is also worth mentioning that no correction for the refractivity index of the medium has been included in the above set of symmetrized equation.

In the coming section IV, we shall correct our set of equations to evaluate the relative vibronic intensity distribution due to the three false origins for the $ReBr_6^{2-}$ complex ion in the octahedral double group.

3. THE THEORY OF NON RADIATIVE TRANSITIONS IN COORDINATION COMPOUNDS. THE CRYSTAL FIELD AND THE LIGAND POLARIZATION COMPONENTS TO THE TOTAL NON RADIATIVE RATE OF DECAY.

In this section, we shall discuss the theory of radiationless transi -

tions from a formal point of view, by employing both the crystal field and the ligand polarization models. To a first order correction these two contributions are complementary to one another, as it will shown later on this section.

Here we shall develop a symmetry adapted approach to account for the electronic factor which together with the vibrational factor determines the rate of non radiative decays from the lowest lying electronic states to the ground state of the complex ion. The relative importance of both promoting and accepting modes to induce radiationless transitions will be discussed in some detail, throughout the course of the present work. We shall focus our attention upon centrosymmetric inorganic complex ions of the first row transition metal ions.

As a matter of fact, the theory of non radiative transitions in both organic and inorganic molecular systems has received the attention of many scientists for many decades (33-44). As for coordination compounds of the transition metal ions, Robbins and Thomson (33,38) devoted their work to the crystal field component to the total non radiative rate of decay from the lowest lying excited electronic states to the ground state. Based upon the transformation properties of both the adiabatic Born-Oppenheimer states and the non radiative operator, these authors derived a set of static selection rules, which hold for both internal conversion and intersystem crossing processes. Furthermore, the normal modes of vibrations of a polyatomic molecular systems were classified into three categories, namely: A) Vibrations of the cluster, B) Pure ligand vibrations and C) Framework-Ligand coupling vibrations. Robbins and Thomson argue that when the promoting modes belong to the framework's vibrations then, their efficiency to induce radiationless transitions should be greater, whereas when the promoting modes belong to the second category then, their efficiency to promote radiationless transitions should be much smaller, indeed!

In the work done by Robbins and Thomson (33,38), we feel that the framework-ligand coupling vibrations have been rather overlooked. As we shall see in the coming section V, these latter vibrations play an important role to account for the relative non radiative rates of decay in systems such as the $Cr(NH_3)_6^{3+}$ ion, in cubic environments. Similary, Ballhausen and Strek (37) considered the ligand polarization component to the total non radiative rate of decay, and showed that within this scheme, the number of active Hidrogen atoms in the ligand subsystems should be correlated with the electronic factor rather than with the vibrational factor as it was pointed out by Robbins and Thomson (33,38).

A) NON RADIATIVE TRANSITIONS. THE CRYSTAL FIELD COMPONENT. A SYMMETRY ADAPTED CRYSTAL FIELD APPROACH.

Here we shall focus our attention upon the crystal field component to the non radiative transition rate, of a given radiationless transition in a centrosymmetric inorganic compound of the transition metal ions. The non radiative rate of decay from the lowest lying excited electronic states to the ground state of a transition metal complex ion, is given by the Fermi-Golden Rule Number Two (2).

$$k = \frac{2\pi}{\hbar}|V_{nm}|^2 \rho_m \tag{33}$$

where $V_{nm} = \langle n|V|m\rangle$ represent an off diagonal matrix element connecting the Born Oppenheimer vibronic states $|n\rangle$ and $|m\rangle$, through a tensorial operator, which we will define, as given below, see eq |34|. As a matter of fact, for internal conversions, that is for radiationless transitions between vibronic states of the same spin multiplicity, $\mu = (2S+1)$, we shall follow Bryan and Siebrand (44) and define:

$$V_{nm} = \langle {}^{\mu}\Psi_m(q,Q)|T_N|{}^{\mu}\Psi_n(q,Q)\rangle \tag{34}$$

Here, T_N is the nuclear kinetic energy operator and the set (q,Q) stands for the electronic and nuclear coordinates of the system, respectively. Within the framework of the Adiabatic Born Oppenheimer approximation, the vibronic states are given in the standard form, as follows:

$${}^{\mu}\Psi_i(q,Q) = {}^{\mu}\Phi_i(q,Q)\Lambda_i(Q) \tag{35}$$

The nuclear wavefunctions $\Lambda_i(Q)$, in the harmonic approximation, are currently written as follows:

$$\Lambda_i(Q) = \prod_k \chi_k^{(i)} \ (\nu_i) = \prod_k \nu_k^{(i)} \tag{36}$$

where $\nu_k^{(i)}$ represent the frequency of the kth-normal mode of vibration in the ith-electronic state. From equations |34|, |35| and |36|, the off diagonal matrix elements V_{nm} may be written as (39):

$$V_{nm} \simeq - \sum_k \frac{\hbar^2}{\mu_k} \langle\Lambda_m'|\Lambda_n'\rangle\langle\Phi_m|\frac{\partial}{\partial Q_k}|\Phi_n\rangle\langle\nu_k^{(m)}|\frac{\partial}{\partial Q_k}|\nu_k^{(n)}\rangle \tag{37}$$

being μ_k the reduced mass corresponding to the kth-normal mode of vibration.

It is worth mentioning that the above equation holds, under the basic assumption that the vibronic coupling integral, defined as given below:

$$\langle\Phi_m(q,Q)|\delta/\delta Q|\Phi_n(q,Q)\rangle = \frac{\langle\Phi_m(q,Q|\partial V(q,Q)/\partial Q|\Phi_n(q,Q)\rangle}{E_n(Q) \text{ --} E_m(Q)} \tag{38}$$

should be independent of the nuclear coordinates Q. Thus the vibronic integral given by eq |38| should be taken over all the electronic coordinates, exclusively. Then, should $E_n(Q) - E_m(Q)$ be independent of Q, then it follows that the V_{nm} matrix elements should vanish identically, unless $E_n(Q) - E_m(Q) = \hbar w_k$, since parallel potential energy surfaces associated with the terminal vibronic states do have the same set of nuclear wavefunctions and therefore, the vibrational overlap integral

$<\Lambda_n|\Lambda_m>$ becomes diagonal in the vibrational quantum numbers. Indeed, it is not straightforward to show that the matrix element $<\Phi_m(q,Q)|\delta V(q,Q)/\delta Q|\Phi_n(q,Q)>$ is Q-independent, however it one makes reasonable assumptions about the potential energy surfaces of the terminal electronic states, we find that eq |38| should be multiplied by a correction factor p defined as:

$$p = 1 + \sum_{i=1}^{\nu} \{\frac{i!}{i}\} \tag{39}$$

When the p correction factor is included explicitly in the above set of equations, we can then evaluate the derivatives $\{\delta V(q,Q)/\delta Q\}_0$ at the nuclear equilibrium configuration of the molecular system. Next, let us define the matrix elements P_{mn} as:

$$P_{nm} = p \frac{<\Phi_m(q,Q)|\{\partial V(q,Q)/\partial Q\}_0|\Phi_n(q,Q)>}{E_n(Q) - E_m(Q)} \tag{40}$$

and hence we may-write the identities:

$$<\Lambda_m|<\Phi_m|\{\delta/\delta Q\}|\Phi_n>|\Lambda_n> = P_{mn}<\Lambda_m|\Lambda_n> \tag{41}$$

and also, as for the vibrational overlap integral, we write:

$$<\Lambda_m|\Lambda_n> = \prod_{k,1} <\nu_k^{(m)}|\nu_{1=}^{(n)}> \tag{42}$$

The identity given by equation |42| may still be simplified further on, by assuming that the potential energy surfaces associated with the terminal Born-Oppenheimer vibronic states have the same shape and are only vertically displaced to one another. When this further assumption is adopted, the above vibrational overlap integral may be re-written as follows:

$$<\Lambda_m|\Lambda_n> = \prod_k <\nu_k^{(m)}|\nu_k^{(n)}> \tag{43}$$

Furthermore, for $\nu = 0$ and $\nu' = 1$ (low temperature, limit case), the vibrational momentum integral may be written in a simple form as:

$$<1_k^{(m)}|\{\delta/\delta Q_k\}|0_k^{(n)}> = -\sqrt{\frac{\mu_k w_k^{(n)}}{2h}} \tag{44}$$

and by collecting all of these results together, we write the off diagonal matrix elements as:

$$V_{nm} \simeq \sum_k \sqrt{\frac{\hbar^3 w_k^{(n)}}{2\mu_k}} \; \frac{p\langle\Phi_m|\{\partial V/\partial Q_k\}_0|\Phi_n\rangle}{E_n(Q) - E_m(Q)} \langle\Lambda_m'|\Lambda_n'\rangle \tag{45}$$

In the above set of general equations, the enabling modes are written as Q_k and the accepting modes are those normal modes of vibrations such as, the vibrational overlap integrals $\langle\Lambda_m'|\Lambda_n'\rangle$ are other than zero. Next, we shall make use of a symmetrized approach (24) to deal with the actual evaluation of the non radiative rates of decay for centrosym - metric inorganic compounds of the transition metal ions. To achieve this purpose, we shall utilize the irreducible tensor algebra for point molecular groups putforward by Griffith (5) and extended to the octahedral double group O^* by Dobosh (46). We shall deal mainly with the electronic factor P_{mn}, given by eq |40|.
Let $V^{(1)}$ be an irreducible tensor operator defined in a symmetry adapted form as follows:

$$V^{(1)} = \sum_{D,\delta,k} O_\delta^D(k) X_\delta^D(k) \tag{46}$$

where: $O_\delta^D(k) = \{\delta V_{CF}/\delta S_\delta^D(k)\}_0$ and $X_\delta^D(k) = \{\delta/\delta S_\delta^D(k)\}$, respectively.

Here V_{CF} stands for the crystal field contribution to the total Coulombic potential energy and S_δ^D stands for the δth-component of an even parity symmetry coordinate S which transforms under the Dth-irreducible representation of the relevant point molecular group and k is a repeated representation label. A typical matrix element involving the irreducible tensor operator $V^{(1)}$ and the terminal vibronic states $|A\alpha\rangle$ and $|B\beta\rangle$ may be written as:

$$\langle \alpha|V^{(1)}|B\beta\rangle = \sum_{D,\delta,k} (-1)^{A+\alpha^+} \sum_m V_m\begin{pmatrix} A & B & D \\ \alpha^+ & \beta & \delta\end{pmatrix} \langle A||O^D(k)||B\rangle_m X_\delta^D(k) \tag{47}$$

For the octahedral double groups, the V_m-coefficients have been tabulated by Dobosh (45, 46) and the reduced matrix elements $\langle A||O^D(k)||B\rangle_m$ may be evaluated as follows:
Define the irreducible tensor operator as given below:

$$O_\delta^D(k) = \sum_L \left\{\frac{\partial V_{CF}}{\partial X_L}\frac{\partial X_L}{\partial S_\delta^D(k)} + \frac{\partial V_{CF}}{\partial Y_L}\frac{\partial Y_L}{\partial S_\delta^D(k)} + \frac{\partial V_{CF}}{\partial Z_L}\frac{\partial Z_L}{\partial S_\delta^D(k)}\right\}_0 \tag{48}$$

where the coefficients $\{\delta X_L/\delta S_\delta^D(k)\}$, $\{\delta Y_L/\delta S_\delta^D(k)\}$ and $\{\delta Z_L/\delta S_\delta^D(k)\}$ can be computed directly from the well known identity $R = (UB)^{-1}S$, see the Introduction. Similary, since $S = L\,Q$, then we may write the set of identities

$$W_\delta^D(k) = \{\delta V_{CF}/\delta Q_\delta^D(k)\}_0 = \sum_{C,\gamma,i} L_{ki} \delta_{CD} \delta_{\gamma\delta} O_\gamma^C(i) \tag{49}$$

and also:

$$\delta/\delta S_\delta^D(k) = X_\delta^D(k) = \sum_{C,\gamma,i} (L^{-1})_{ik} \delta_{CD} \delta_{\gamma\delta} \{\delta/SQ_\gamma^C(i)\} \tag{50}$$

From equation $|50|$ it is seen that the actual evaluation of the electronic factor $P_{mn}(k)$ depends also on the details of the force field employed to describe the dynamic behaviour of the molecular system. As for the vibrational factor $\langle \Lambda_m | \Lambda_n \rangle$, see haarhoff (47).

B) NON RADIATIVE TRANSITIONS. THE LIGAND POLARIZATION COMPONENT.

Here we shall follow Ballhausen and Strek (37) to discuss the ligand polarization component to the total non radiative rate of decay k. The non radiative vibronic operator is written, within the independent system model, as follows (48):

$$H_{NR} = H_{NR}(M) + H_{NR}(L) \tag{51}$$

where $H_{NR}(M) = \sum_i (\delta V/\delta Q_i)_0 Q_i$ and $H_{NR}(L) = \sum_j (\delta V/\delta Q_j)_0 Q_j$, respectively.

In our present notation, $\{Q_i\}$ stands for the set of normal coordinates corresponding to the chromophore (M) and $\{Q_j\}$ are the normal coordinates of the ligand subsystems (L), respectively. The framework-ligand coupling vibrations are not included explicity, in the Hamiltonian defined by the equation $|51|$.

Thus to a first order correction, a typical matrix element involving the terminal states $|M_o L_o\rangle$ and $|M_a L_o\rangle$ and the vibronic perturbation given by equation $|51|$, may be expressed as:

$$\langle M_o L_o | H_{NR} | M_a L_o \rangle = \tag{52}$$

$$\langle M_o L_o | H_{NR}(M) | M_a L_o \rangle + \tag{52.0}$$

$$\sum_{k \neq 0} (-E_k)^{-1} \langle M_o L_o | V_{LP} | M_k L_o \rangle \langle M_k | H_{NR}(M) | M_a \rangle + \tag{52.1}$$

$$\sum_{k \neq a} (E_a - E_k)^{-1} \langle M_a L_o | V_{LP} | M_k L_o \rangle \langle M_k | H_{NR}(M) | M_o \rangle + \tag{52.2}$$

$$\sum_{1\neq 0}(-E_a-E_1)^{-1}<M_oL_o|V_{LP}|M_aL_1><L_1|H_{NR}(L)|L_o> + \quad (52.3)$$

$$\sum_{1\neq 0}(E_a-E_1)^{-1}<M_aL_o|V_{LP}|M_oL_1><L_1|H_{NR}(L)|L_o> \quad (52.4)$$

To zeroth-order approximation, the integral given by eq |52.0| is other than zero, when and only when, there is at least one normal mode of the framework (enabling mode) able to couple the terminal states $|M_oL_o>$ and $|M_aL_o>$, through the non radiative operator $H_{NR}(M)$, otherwise this integral vanishes identically. When the static selection rules worked out by Robbins and Thomson (33, 38) exclude radiationless transitions, then we consider the first order correction to account for the observed non radiative rate of decay.

It is seen from the above set of equations, that the identities given by equations |52.1| and |52.2| are the basis of the crystal field contribution to the total non radiative rate of decay, whereas the second pair of identities, |52.3| and |52.4|, form the basis of the so called ligand polarization component to the total non radiative rate of decay k.

Here we shall be mainly concerned with the set of equations |52.3| and |52.4| to study on a semi-quantitative basis the dynamic contribution to the total rate of decay from the lowest lying electronic excited states to the ground state of the complex ion. These two equations, may be written as follows:

$$<M_oL_o|H_{NR}|M_aL_o>^{LP} = - \sum_{1\neq 0}\frac{2E_1}{E_1^2-E_a^2}<M_oL_o|V_{LP}|M_aL_1><L_o|H_{NR}(L)|L_1> \quad (53)$$

Next, let us write the vibronic states as given below:

$$|M_a> = |M_a'>|av> \quad (54.1)$$

for the central metal ion and for the ligands we write:

$$|L_1> = |L_1'>|1w> \quad (54.2)$$

where the labels v and w stand for the vibrational quantum numbers for the metal and the ligand subsystems, respectively.

Then from equations |53|, |54.1| and |54.2|, the ligand polarization matrix element may be re-written as:

$$<M_oL_o|H_{NR}|M_aL_o>^{LP} = - \sum_{1\neq 0}\frac{2E_1}{E_1^2-E_a^2}<M_o'L_o'|V_{LM}|M_a'L_o'><ov|av'> \times$$
$$<ow|1w'> \times \sum_i <L_o'|\{\partial V/\partial Q_i\}_0|L_1'><ow|Q_i|1w'> \quad (55)$$

where the vibrational integral $<ow|Q_i|1w'>$ may be evaluated using the harmonic oscillator approximation to give:

$$<1w|Q_i|ow'> = \sqrt{\frac{\hbar}{2\mu_i w_i}} \{\sqrt{n_i+1}\ \delta(n_i', n_i+1) + \sqrt{n_i}\ \delta(n_i', n_i-1)\}$$

$$<1w|ow'> \tag{56}$$

In this notation, n_i represents the vibrational quantum number associated with the ith-normal mode Q_i, which has the angular frequency w_i in the state $|L_1>$, whereas n_i' plays the same rol as n_i, but it is linked to the state $|L_o>$, respectively.
It is then straightforward to show that the non radiative transition probability, say W_{oa} may be expressed as follows:

$$W_{oa} = \frac{2\pi}{\hbar} (Z_n^M)^{-1} \sum_{av} \exp(-\frac{E_{av}^M}{kT}) \delta(E_{av}^M - E_{ov'}^M) \times$$
$$\times \left|\sum_i \sum_{1w} \sum_{1\neq 0} \frac{2E_1}{E_1^2 - E_a^2} <M_o'L_o'|V_{LP}|M_a'L_o'><L_o'|(\frac{\partial V}{\partial Q_i})_0|L_1'>\right.$$
$$<ov'|av><ow'|1w> \times |<ow|Q_i|1w'>|^2 \tag{57}$$

Therefore, and from the above equation, it is easy to see that the electronic factor associated with the ith-normal mode of vibration may be written as follows:

$$I_i = \left|\sum_{1\neq 0} \frac{2E_1}{E_1^2 - E_a^2} \sum_{C\gamma,\delta} \sum_k G_{C\gamma,\delta}^L(k) <M_o'|M_\gamma^C(k)|M_a'><L_o'|\mu^\delta|L_1'> \times\right.$$
$$\times <L_1'|\partial V/\partial Q_i)_0|L_o'>|^2 \tag{58}$$

and the non radiative ligand polarizability tensor operator, corresponding to the ith-normal mode of vibration takes the form:

$$\alpha_{NR}^{-L} = \sum_{1\neq 0} \frac{4E_1^2}{|E_1^2 - E_a^2|^2} |<L_o'|\mu^\delta|L_1'>|^2 |<L_1'|(\partial V/\partial Q_i)_0|L_o'>|^2 \tag{59}$$

and therefore, the electronic factor may be expressed as given below:

$$I_i = \alpha_{NR}^{-L} \sum_{C\gamma,\delta} \sum_{F\phi,\delta'} \sum_{k,k'} G_{C\gamma,\delta}^{L}(k)\, G_{F\phi,\delta'}^{L}(k') \times$$

$$\times \; <M_o'|M_\gamma^C(k)|M_a'><M_o'|M_\phi^F(k')|M_a'><L_o'|\mu^\delta|L_1'><L_o'|\mu^{\delta'}|L_1'> \quad (60)$$

Finally, for isotropic ligand subsystems and $C \equiv F$, the electronic factor associated with the ith-normal mode of vibration becomes:

$$I_i = \sum_L \alpha_{NR}^{-L} \sum_{C\gamma,\delta} \sum_k |G_{C\gamma,\delta}^{L}(k)|^2 |<M_o'|M_\gamma^C(k)|M_a'>|^2 |<L_o'|\mu^\delta|L_1'>|^2 \quad (61)$$

A fairly general discussion of the vibrational factor, which together with the electronic factor determines the non radiative rate of decay, may be found in References (37, 39, 40, 44, 47-49).

APPLICATIONS

4. THE $\Gamma_8(^2T_{2g}) \rightarrow \Gamma_8(^4A_{2g})$ ELECTRONIC TRANSITION OF THE $ReBr_6^{2-}$ ION IN THE Cs_2ZrBr_6: $ReBr_6^{2-}$ SYSTEM.

Here we shall carried out, a vibronic intensity calculation for the $ReBr_6^{2-}$ complex ion in the octahedral double group. The calculation will be undertaken by employing a symmetrized combined crystal field-closure-ligand polarization model, to account for the observed vibronic intensity due to the three false origins, corresponding to the $\Gamma_8(^2T_{2g}) \rightarrow \Gamma_8(^4A_{2g})$ electronic transition of the $ReBr_6^{2-}$ complex ion. For the Cs_2ZrBr_6: $ReBr_6^{2-}$ system, at least nine electronic transitions have been observed and assigned by Flint and Paulusz (50, 51), and we have chosen this particular electronic transition for illustrative purposes. Ideally an intensity calculation for this system would require a complete lattice dynamic calculation, however at this first stage, we show that the vibronic intensity calculation based upon a molecular model seems to be quite satisfactory to describe on a quantitative basis the observed vibronic intensity distribution for this electronic transition.

Several spectroscopic and magnetic studies have been carried out for systems such as ReX_6^{2-}, where $X^- = Cl^-$ and Br^-, during the last years (50-58).

In particular, Flint and Paulusz (50, 51) have undertaken accurate spectroscopic studies for these systems and have obtained high quality spectra for both the $ReBr_6^{2-}$ and the $ReCl_6^{2-}$ complex ions. In fact, for the $ReBr_6^{2-}$ ion, they observed and assigned at least nine luminiscence electronic transitions. In this system, the corresponding vibronic intensity

distribution vary from transition to transition significantly. To study those factor upon which the vibronic intensity depends on, we have chosen for illustrative puropose the $\Gamma_8(^2T_{2g}) \rightarrow \Gamma_8(^4A_{2g})$ electronic transition of the $ReBr_6^{2-}$ complex ion in the Cs_2ZrBr_6: $ReBr_6^{2-}$ system, by assuming a molecular model, although we recognize that the vibrational frequencies are both host and temperature dependent.
Re^{4+} is a heavy metal, then it is bound to have a large spin orbit coupling constant, say about 2,000 cm^{1-}, which will produce an important splitting of the multiplet electronic states of this cation. Both Eisenstein (52) and Dorian (54) carried out full crystal field calculations for the $ReCl_6^{2-}$ in the octahedral double group O^*, using a total molecular Hamiltonian of the standard form: $H = H^0 + \sum_{i,j}'\{e^2/r_{ij}\} + H_{SO} + V_{CF}$.

The calculations performed by these authors are not very accurate, because the energy-levels derived from the strong field t_{2g}^3 configuration are weakly dependent on the Dq-values. It is indeed feasible to produce a better fitting among the observed and the experimental energies, however this procedure leads us to rather low calculated values for the Dq-parameter.
On the basis of the recent experimental data available (50, 51), we have re-studied the energy levels for the $ReBr_6^{2-}$ in the octahedral double group, and the best fit has been obtained for the following set of parameters: $B = 327$ cm^{1-}, $C = 1{,}818$ cm^{1-}, $Dq = 3{,}034.7$ cm^{1-} $\lambda_{SO} = 2{,}392 cm^{1-}$ respectively. We list our results in the Table I.

MOLECULAR MODEL AND NORMAL COORDINATE ANALYSIS:

We have undertaken a normal coordinate analysis, for the seven atom system $ReBr_6^{2-}$, in octahedral symmetry, assuming a weak coupling among the internal and external vibrations of Cs_2ZrBr_6: $ReBr_6^{2-}$ system. In several previous works in centrosymmetric inorganic complex ions of the first row transition metal ions, we have shown that the relative vibronic intensity distribution of a series of electronic transitions depends on several factors, such as: The details of the vibrational force field, the quality of the central metal ion's radial wavefunctions, the charge on the central metal ion, the effective energy gap (corresponding to a parity and a spin allowed electronic transition), and the effective charge on the ligand subsystems (8-15, 17-21, 24) and references there in). In this section, we shall investigate the vibronic intensity dependence on the details of the interaction vibrational force field.
Here, we shall develop a general valence force field (GVFF) for this complex ion based upon a simple parametrization scheme along with some physical constraints on the solutions of the vibrational equations of motion (18), for the vibrating seyen atom system. As it is known, the normal modes of vibration of $ReBr_6^{2-}$ complex ion are distributed among the irreducible representations of the octahedral point molecular group as follows:

$$\Gamma_{vib} = \alpha_{1g}(R) + \varepsilon_g(R) + 2\ \tau_{1u}(IR,\ vibronic) + \tau_{2g}(R) + \tau_{2u}$$

Furthermore, in most cases of interest to inorganic chemistry, the interaction between two opposite bonds (f_{dd}) is bigger than the interaction among two bonds at right angles ($f_{dd'}$), so that we may introduce a factor, say k defined as $k = \{f_{dd}/f_{dd'}\}$, where $k > 1$. In principle, the k-value is unknown and could be determined by using an additional physical criterion, The k value is fixed so that to minimize both the mean error deviation (among the calculated and observed vibrational frequencies) and the cross term which arises from the definition of the so called potential energy distribution, PED, defined as (24, 59, 60):

$$CT = 100 \sum_{i,j}{}' L_{ji} L_{ki} F_{jk}$$

where the L_{mn}-matrix elements relate the space of the symmetry coordinates S to the space of the normal coordinates. Q, by means of the well known identity $S = L\,Q$, and the F_{mn}-matrix elements are the symmetrized force constants, expressed as symmetry adapted linear combinations of the internal force constants for the molecular system.

To investigate the influence of the details of the vibrational force field on the relative vibronic intensity distribution associated with the $\Gamma_8(^2T_{2g}) \rightarrow \Gamma_8(^4A_{2g})$ electronic transition of the $ReBr_6^{2-}$ complex ion, we have employed six different force fields, namely: GVFF[a] (k = 1.17), GVFF[a] (k = 2.25), GVFF[b], OVFF, UBFF, MOVFF and MUBFF, respectively (61-63). In the Table II, we list the symmetrized force constants (in mdyne/A°) and the observed and calculated vibrational frequencies are displayed in the Table III. Finally, in the Tables IV and V, we list the values for the internal force constants by using different force fields and also the corresponding L_{mn}-matrix elements, taking into account the various approximations to the real interaction vibrational force field.

SPECTRAL INTENSITIES.

In the octahedral double group, the total transition dipole moment associated with the $|\Gamma_1\gamma_1 j\rangle \rightarrow |\Gamma_2\gamma_2 l\rangle$ excitation may be expressed as (24):

$$\vec{\mu}_{\Gamma_1\gamma_1 j \rightarrow \Gamma_2\gamma_2 l} = \vec{\mu}^{CF}_{\Gamma_1\gamma_1 j \rightarrow \Gamma_2\gamma_2 l} + \vec{\mu}^{LP}_{\Gamma_1\gamma_1 j \rightarrow \Gamma_2\gamma_2 l} \qquad (62)$$

where both j and l are repeated representation labels, and therefore the total dipole strength $D(\Gamma_1\gamma_1 j \rightarrow \Gamma_2\gamma_2 l)$ becomes:

$$D(\Gamma_1\gamma_1 j \to \Gamma_2\gamma_2 l) = D^{CF}_{\Gamma_1\gamma_1 j \to \Gamma_2\gamma_2 l} + D^{LP}_{\Gamma_1\gamma_1 j \to \Gamma_2\gamma_2 l} + $$

$$+ D^{CF,LP}_{\Gamma_1\gamma_1 j \to \Gamma_2\gamma_2 l} \qquad (63)$$

When the closure approximation is adopted, and within the framework of crystal field method, the αth-vector component of the transition dipole moment associated with this excitation, may be written in the space of the symmetry coordinates S as given below:

$$\mu^{CF,\alpha}_{\Gamma_1\gamma_1 j \to \Gamma_2\gamma_2 l} = \left(\frac{2}{\Delta E}\right) \sum_{\bar{\Gamma},\bar{\gamma},k} \sum_{\Gamma,\gamma,i} \lambda^{1/2}(\Gamma)(-1)^{\Gamma+\gamma^+}(-1)^{\Gamma_1+\gamma_1^+}$$

$$V\begin{pmatrix} \bar{\Gamma} & T_1 & \Gamma \\ \bar{\gamma} & \alpha & \gamma^+ \end{pmatrix} \times V_i\begin{pmatrix} \Gamma_1 & \Gamma_2 & \Gamma \\ \gamma_1^+ & \gamma_2 & \gamma \end{pmatrix} \langle\Gamma_1 j || O^{\Gamma}(k) || \Gamma_2 l\rangle_i S^{\Gamma}_{\gamma}(k)$$

$$(64)$$

where, the γth-component of the irreducible tensor operator $O^{\Gamma}(k)$ becomes:

$$O^{\Gamma}_{\gamma}(k) = \lambda^{1/2}(\Gamma)(-1)^{\Gamma+\gamma^+} \sum_{\bar{\gamma},\alpha} V\begin{pmatrix} \bar{\Gamma} & T_1 & \Gamma \\ \bar{\gamma} & \alpha & \gamma^+ \end{pmatrix} W^{\bar{\Gamma}}_{\bar{\gamma}}(k)\, \mu^{T_1}_{\alpha} \qquad (65)$$

In the above identity, the vibronic coupling operator $W^{\bar{\Gamma}}_{\bar{\gamma}}(k)$, is written in the space of the symmetry coordinates as follows:

$$W^{\bar{\Gamma}}_{\bar{\gamma}}(k) = \{\partial V_{CF}/\partial S^{\bar{\Gamma}}_{\bar{\gamma}}(k)\}_0 \qquad (66)$$

For the octahedral double group, the V_m-coefficients and the phase factors $(-1)^{\Gamma+\gamma^+}$ can be found in References (45, 46). Equation |64| may be written in the space of the normal coordinates Q, by employing the identity given by the equation |18|, which relates the symmetry coordinates S to the normal coordinates Q for the molecular system. The L_{mn}-matrix elements for the various force fields are listed in the Table V, as well as the calculated potential energy distribution (PED). Similary, in the octahedral double group O^*, the αth-vector component of transition dipole moment, associated with the $|\Gamma_1\gamma_1 l\rangle \to |\Gamma_2\gamma_2 l\rangle$ excitation, may be written, within the framework of the ligand polarization model (8, 14, 15, 24) as given below:

$$\mu^{LP,\alpha}_{\Gamma_1\gamma_1 j\to\Gamma_2\gamma_2 1} = (-1)^{\Gamma_1+\gamma_1^+} \sum_{\overline{\Gamma},\overline{\gamma}} \sum_{\Gamma,\gamma} \sum_{k,i} \sum_{\tau} B^{\Gamma\gamma'\alpha(k,i)}_{\overline{\Gamma}\overline{\gamma}} \times$$

$$\times V_\tau \begin{pmatrix} \Gamma_1 & \Gamma_2 & \Gamma \\ \gamma_1^+ & \gamma_2 & \gamma \end{pmatrix} <\Gamma_1 j||M^{\Gamma}(i)||\Gamma_2 1>_\tau \qquad (67)$$

In the above expression the index "i" labels the rank of the relevant central metal ion's electric multipoles, the index "k" labels the different odd parity symmetry coordinates and finally the index τ is a repeated representation label. Similary, $\overline{\alpha}_L$ is the mean ligand polarizability measured at the frequency of the electronic transition $|\Gamma_1\gamma_1 j> \to |\Gamma_2\gamma_2 1>$. The symmetrized ligand polarization geometric factors have been tabulated in the Reference (25) and the real expressions for the ligand polarization vibronic coupling constants, in the case of an octahedral system such as ML_6 are given in the Reference (8).
Since $\Gamma_8 \times \Gamma_8 = A_{1g} + A_{2g} + E_g + 2T_{1g} + 2T_{2g}$, it follows that the $\Gamma_8(^2T_{2g}) \to \Gamma_8(^4A_{2g})$ electronic transition is electric quadrupole allowed, magnetic dipole allowed, electric hexadecapole allowed and electric dipole vibronically allowed. For this particular electronic transition, a typical reduced matrix element to be evaluated is of the form:

$$<^2T_2U'j||O^{\Gamma}||^4A_2U'1>_i = p_1<t_2^3\ ^4A_2U'||O^{\Gamma}||t_2^3\ ^4A_2U'1>_i +$$

$$+ p_2<t_2^3\ ^4A_2U'j||O^{\Gamma}||t_2^2\ e\ ^4T_2U'1>_i +$$

$$+ p_3<t_2^3\ ^2EU'j||O^{\Gamma}||t_2^3\ ^2T_2U'1>_i +$$

$$+ p_4<t_2^3\ ^2T_1U'j||O^{\Gamma}||t_2^3\ ^2T_2U'1>_i +$$

$$+ p_5<t_2^3\ ^2T_2U'j||O^{\Gamma}||t_2^3\ ^2T_2U'1>_i \qquad (68)$$

where the expansion coefficients p_i are as follows: $p_1 = -0.1643$; $p_2 = -0.017$; $p_3 = +0.0519$; $p_4 = -0.0657$ and $p_5 = +0.1773$. For the purposes of the vibronic intensity calculation, we have neglected expansion coefficients smaller than 0.01. In the above expression, the relevant eigenvectors for the $ReBr_6^{2-}$ complex ion have been taken from the Table I.
As we shall see, the algebra in the octahedral double group 0* leads us to use complex components of the symmetry coordinates $S^{\Gamma}_{\gamma}(k)$, where $\gamma = 0, \pm 1$.
The complex ligand polarization vibronic coupling constants are related to the real ligand polarization vibronic coupling constants, by means of the relationships given in the Appendix I, see References (8).

The space of the complex symmetry coordinates is related to the space of the real symmetry coordinates by means of the relation:

$$S_0^{\Gamma}(k) = S_{ka}^{\Gamma} \tag{69}$$

$$S_{\pm 1}^{\Gamma}(k) = \mp 1/\sqrt{2}\,(S_{kc}^{\Gamma} \pm i\, S_{kb}^{\Gamma})$$

where, for k = 3, 4; $\Gamma = T_{1u}$ and k = 6, we have $\Gamma = T_{2u}$, respectively. Next, let us introduce the quantities:

$$q_{jl}^{\Gamma,i}(k) = \langle {}^2T_2U'j||C^{\Gamma}(k)||{}^4A_2U'l\rangle_i \tag{70}$$

These polyelectronic reduced matrix elements may be expressed as linear combinations of monoelectronic reduced matrix elements, by using the standard method of the irreducible tensor algebra (5, 45, 46). The final expressions are given in the Appendix II. The relevant monoelectronic reduced matrix elements evaluated within the framework of the vibronic crystal field-closure method are listed in the Appendix III. Next and for the purposes of the vibronic intensity calculation, it is convenient to introduce the quantities defined below:

$$\beta_{\gamma_1\gamma_2}^{i,3} = C_{\gamma_1\gamma_2}^{i}(3)L_{33} + C_{\gamma_1\gamma_2}^{i}(4)L_{43}$$

$$\beta_{\gamma_1\gamma_2}^{i,4} = C_{\gamma_1\gamma_2}^{i}(3)L_{34} + C_{\gamma_1\gamma_2}^{i}(4)L_{44}$$

$$\beta_{\gamma_1\gamma_2}^{i,6} = C_{\gamma_1\gamma_2}^{i}(6)L_{66} \tag{71}$$

where the L_{mn}-matrix elements for the six different force fields are given in the Table V and the quantities $C_{\gamma_1\gamma_2}^{i}(k)$ for k = 3, 4, 6 and γ_1, γ_2 = κ, λ, μ and ν are listed in terms of the quantities $q_{jl}^{\Gamma,i}(k)$, in the Appendix IV.

Next, in the Table VI, we inform the zth-vector components of transition dipole moment associated with the $\Gamma_8(^2T_{2g}) \rightarrow \Gamma_8(^4A_{2g})$ electronic transition for the $ReBr_6^{2-}$ complex ion, evaluated within the framework of the vibronic crystal field-closure method.

As for the vibronic ligand polarization contribution to the total transition dipole moment of the $\Gamma_8(^2T_{2g}) \rightarrow \Gamma_8(^4A_{2g})$ electronic transition, we observe from the equation |67|, that the relevant polyelectronic reduced matrix elements to be evaluated are of the form:

$$R_{jl}^{\Gamma,\tau}(i) = \langle {}^2T_{2g}U'j||M^\Gamma(i)||{}^4A_{2g}U'l\rangle_\tau =$$

$$p_1\langle t_2^3\ {}^4A_2U'j||M^\Gamma(i)||t_2^3\ {}^4A_2U'l\rangle_\tau +$$

$$+ p_2\langle t_2^3\ {}^4A_2U'j||M^\Gamma(i)||t_2^2\ e\ {}^4T_2U'l\rangle_\tau +$$

$$+ p_3\langle t_2^3\ {}^2EU'j||M^\Gamma(i)||t_2^3\ {}^2T_2U'l\rangle_\tau +$$

$$+ p_4\langle t_2^3\ {}^2T_1U'j||M^\Gamma(i)||t_2^3\ {}^2T_2U'l\rangle_\tau +$$

$$+ p_5\langle t_2^3\ {}^2T_2U'j||M^\Gamma(i)||t_2^3\ {}^2T_2U'l\rangle_\tau \qquad (72)$$

where the expansion coefficients p_i have been defined in the text, previously.
The explicit form of these polyelectronic reduced matrix elements in terms of the monoelectronic reduced matrix elements $\langle t_2||m^\Gamma(i)||t_2\rangle$ and $\langle t_2||m^\Gamma(i)||e\rangle$ is given by a similar tabulation as presented in the Appendix II for the $q_{jl}^{\Gamma,i}(k)$ polyelectronic reduced matrix elements.
Next, let us introduce the quantities $n_{\gamma_1\gamma_2}^{i,3}$, $n_{\gamma_1\gamma_2}^{i,4}$ and $n_{\gamma_1\gamma_2}^{i,6}$ as given

$$n_{\gamma_1\gamma_2}^{i,3} = m_{\gamma_1\gamma_2}^{i}(3)L_{33} + m_{\gamma_1\gamma_2}^{i}(4)L_{43}$$

$$n_{\gamma_1\gamma_2}^{i,4} = m_{\gamma_1\gamma_2}^{i}(3)L_{34} + m_{\gamma_1\gamma_2}^{i}(4)L_{44}$$

$$n_{\gamma_1\gamma_2}^{i,6} = m_{\gamma_1\gamma_2}^{i}(6)L_{66} \qquad (73)$$

where the quantities $m_{\gamma_1\gamma_2}^{i}(k)$ for $\gamma_1 = \kappa$ and $\gamma_2 = \kappa, \lambda, \mu, \nu$ and $k = 3,4$ and 6 are listed in the Appendix V.
In terms of the above defined quantities, we can find the zth-vector components to the total transition dipole moment associated with the $\Gamma_8({}^2T_{2g}) \rightarrow \Gamma_8({}^4A_{2g})$ electronic transition, within the framework of the vibronic ligand polarization model. We list our results in the Table VII.
From the information, listed in the Tables VI and VII, it is straightforward to obtain explicit equations for the total dipole strength $D({}^2T_{2g} \rightarrow {}^4A_{2g})$ associated with the overall electronic transition.
For the sake of completeness, we list in the Table VIII, the analytical expressions to evaluate the total dipole strength associated with each of the vibronic origins of the total electronic transition, so that the

overall dipole strength associated with the $\Gamma_8(^2T_{2g}) \rightarrow \Gamma_8(^4A_{2g})$ transition, can be evaluated from the identity given below:

$$D(^2T_{2g}U' \rightarrow {}^4A_{2g}U') = \sum_{\nu_i} \{D^{CF}(\nu_i) + D^{LP}(\nu_i) + D^{CF,LP}(\nu_i)\} \quad (74)$$

As it can be seen from the above set of equations, the total dipole strength associated with the overall electronic transition depends on several factors: The effective energy gap ΔE corresponding to a parity and spin allowed electronic transition, the effective nuclear charge on each ligand subsystem "$q_L e$", the charge on the central metal ion, the quality of central metal ion's radial wavefunctions, the equilibrium molecular parameter (bond distance), the mean ligand polarizability $\bar{\alpha}_L$ and the details of the vibrational force field.

In addition to these factors, the model employed suffers from intrinsic limitations. Firstly, our model calculation is based upon an independent system model, that is no overlapping between the central metal ion and the ligand subsystems charge distributions is taken into account, therefore effects such as charge transfer, etc. may not be explained on the basis of our model calculation and secondly our model calculation is based upon a molecular model, so that the coupling among internal and external vibrations is neglected.

Throughout the course of this work, we have derived a new set of general equations-symmetry determined to evaluate, the total dipole strength associated with the $\Gamma_8(^2T_{2g}) \rightarrow \Gamma_8(^4A_{2g})$ radiative electronic transition for the $ReBr_6^{2-}$.

As we said in the text previously, the total dipole strength associated with this electronic transition also depends on the expectation values of r^k, that is:

$$\langle r^k \rangle = \langle R_{5d} | r^k | R_{5d} \rangle \quad (75)$$

These expectation values of r^k depend on the quality of the radial wave functions R_{5d} and indeed on the central metal ion's charge. For Re^+, Bash and Gray (64) worked out an analytic expression for the $R_{5d}(+1)$ radial function which is given below:

$$R_{5d} = 0.1230\, R_3(\mu_3) - 0.3342\, R_4(\mu_4) + 0.6662\, R_5(\mu_5) + + 0.5910\, R_5(\mu'_5) \quad (76)$$

where $\mu_3 = 20.255$, $\mu_4 = 10.409$, $\mu_5 = 5.343$ and $\mu'_5 = 2.277$ respectively. The radial wavefunctions $R_k(\mu_k)$ are analytic Slater-type wavefunctions. An alternative set of radial wavefunctions for neutral Re has been worked out by Desclaux (65). In his Tables, we find values for $\langle r^k \rangle_{5d,5d}$ (k = 2, 4 and 6) corresponding to Re(0) for j = 1 + 1/2 and to Re(0*) for j = 1 - 1/2, respectively. The calculated non relativistic values of r^k (k = 2, 4 and 6) for Re^+ (64) and the calculated relativistic values of r^k (k = 2, 4 and 6) for Re(0) and Re(0*) are listed below:

$$
\begin{aligned}
Re^+ &: \langle r^2\rangle = 0.923103 \quad \langle r^4\rangle = 1.750281 \quad \langle r^6\rangle = 5.175324 \\
Re(0) &: \langle r^2\rangle = 1.219653 \quad \langle r^4\rangle = 2.978715 \quad \langle r^6\rangle = 13.582437 \\
Re(0^*) &: \langle r^2\rangle = 1.123240 \quad \langle r^4\rangle = 2.502811 \quad \langle r^6\rangle = 10.372674
\end{aligned}
\tag{77}
$$

where the $\langle r^k\rangle$-values are given in $(\mathring{A})^k$.
Similary, for the $ReBr_6^{2-}$ complex ion, the equilibrium bond distance is unknown, however has been estimated for the $ReCl_6^{2-}$ complex ion to be about 2.37 A° (52). Then we shall use for the $ReBr_6^{2-}$ ion, an equilibrium bond distance of 2.40 A°. From the work of Le febre (66), the mean ligand polarizability value for the Br^- ion is equals to 4.85 $\mathring{A}^3$, and then we shall adopt this value to estimate the relative vibronic intensity distribution.
Furthermore, the effective energy gap ΔE corresponding to a parity and spin allowed electronic transition, from which the $\Gamma_8(^2T_{2g}) \rightarrow \Gamma_8(^4A_{2g})$ electronic transition borrows its intensity from, has been estimated to be about 80.000 cm^{-1}. (For the complex ions of the first row transition metal ions this value is taken to be about 100.000 cm^{1-}).
An additional complication arises in the vibronic intensity distribution associated with the $\Gamma_8(^2T_{2g}) \rightarrow \Gamma_8(^4A_{2g})$ excitation. In fact, it is customary in the literature (26, 27, 30, 31) to estimate the q-value from the well known identity:

$$D_q = \{qe^2\langle r^4\rangle/6R_0^5\} \tag{78}$$

where "qe" is the effective ligand charge and R_0 is taken to be the equilibrium metal-ligand bond distance. Since D_q = 3.035 cm^{1-} (50.51), then we obtain:
For Re^+: q = 7.1379, for Re(0): q = 4.1942 and finally for Re(0*); q = 4.9900, respectively. Within the framework of the pure electrostatic crystal field method, it is clear than these calculated q-values are not very meaningful and therefore we rather estimate the ionic charges on both the central metal ion and the ligand subsystems from the identity: δ(Br) + 6δ(Br) = -2, since the total charge on the $ReBr_6^{2-}$ complex ion is equals to -2.
We have carried ou the vibronic intensity calculation for a wide range of central metal ion charges and also for different value of the effective nuclear charge on each ligand position. The best agreement with the experimental data, on the relative vibronic intensity distribution due to the three false origins of the $\Gamma_8(^2T_{2g}) \rightarrow \Gamma_8(^4A_{2g})$ electronic transition of the $ReBr_6^{2-}$ complex ions, is achieved for Re(0) and q = -(1/3). We list our results in the Table IX.
When the q values used in the vibronic intensity calculation,

obtained from the equation |78| by using the experimental D_q=3.035 cm^{1-}, it follows that the vibronic intensity due to the $\nu_6(\tau_{2u})$-vibronic origin becomes very small indeed, when compared with the vibronic intensities induced by the $\nu_3(\tau_{1u})$ and the $\nu_4(\tau_{1u})$ false origins, although the total oscillator strength associated with the overall electronic transition moves in the range of 10^{-4} to 10^{-5}. In addition, for these large values of q, except for the $\nu_6(\tau_{2u})$ vibronic origin, the crystal field component to the total dipole strength is at least two order of magnitude bigger than the corresponding ligand polarization component to the total dipole strength of the overall electronic transition. This latter result is some how unexpected, since the Br^{1-} anion has an appreciable value for the mean ligand polarizability, and one would expect that these two, namely the crystal field and the ligand polarization components to the total dipole strength of the electronic transition should roughly induced comparable values for the vibronic intensities. We believe that these arguments support our vibronic intensity calculation, by choosing δ(Br) = 0 and δ(Br) = - (1/3), as the ionic partial charges on the central metal and ligand subsystems, respectively.
For the sake of comparision, we give below the calculated values for both the relative vibronic intensity distribution and the total oscillator strength for the overall $\Gamma_8(^2T_{2g}) \rightarrow \Gamma_8(^4A_{2g})$ electronic transition of this complex ion when a GVFF[a] (k = 1.17) is employed and the following set of charges are used δ(Re) = +0.1 and δ(Br) = 0.35, respectively.

TABLE X. Relative vibronic intensity distribution for the $\Gamma_8(^2T_{2g}) \rightarrow \Gamma_8(^4A_{2g})$ electronic transition of the $ReBr_6^{2-}$ complex ion in the Cs_2ZrBr_6: $ReBr_6^{2-}$ system. A GVFF[a] (k = 1.17), δ(Re) = + 0.10 and δ(Br) = -0.35.

Vibronic Origins	D^{CF}	D^{LP}	$D^{CF,LP}$
ν_3	0.512391E-3 (0.155605E-3)	0.768872E-4 (0.317754E-4)	-0.303587E-3 (-0.10829 E-3)
ν_4	0.105136E-2 (0.370127E-3)	0.663987E-5 (0.229596E-5)	0.806641E-6 (0.417570E-4)
ν_6	0.158815E-6 (0.151242E-6)	0.114857E-3 (0.500137E-4)	-0.796935E-6 (-0.568137E-6)

Vibronic intensity distribution:
D_3: D_4: D_6 = 2.50126: 9.26999: 1.00000
(1.74509: 8.35093: 1.00000)*

Total oscillator strength: $f\{\Gamma_8(^2T_{2g}) \rightarrow \Gamma_8(^4A_{2g})\}$ = 1.57870E-5
(5.95593E-6)

These results should be compared with those reported in the Table IX, for $\delta(Re) = 0$ and $\delta(Br) = -(1/3)$, for the same general valence force field, GVFF[a] (k = 1.17). These latter values are given in brackets in the above Table X.
As we have seen, throughout the course of this section, the model calculation employed to account for both the relative vibronic intensity distribution and the total oscillator strength for the $\Gamma_8(^2T_{2g}) \rightarrow \Gamma_8(^4A_{2g})$ electronic transition of the $ReBr_6^{2-}$ complex ion, in the Cs_2ZrBr_6: $ReBr_6^{2-}$ system seems to be satisfactory, although the intrinsic limitations of the combined crystal field-closure-ligand polarization model. Ideally, a vibronic intensity calculation for this electronic transition would require a complete lattice dynamic calculation, however at this first stage, the model calculation employed may explain on a quantitative basis the observed experimental results for this system (50, 51).

5. THE $^4T_{1g} \rightarrow {}^4A_{2g}$ AND THE $^4T_{2g} \rightarrow {}^4A_{2g}$ RADIATIONLESS TRANSITIONS FOR THE HEXAMMINOCHROMIUM (III) ION, IN OCTAHEDRAL SYMMETRY.

For illustrative purposes, in this final section, we shall consider on a quantitative basis, the actual evaluation of the crystal field electronic factor $P_{mn}(k)$ given by the equation |40| for two non radiative transition of the Hexamminochromium (III) ion in octahedral symmetry. Flint et al (70-73) studied the $^2E_g \rightarrow {}^4A_{2g}$ luminiscence of the $Cr(NH_3)_6^{3+}$ complex ion in cubic environments. On the basis of this experimental data, full normal coordinate analysis have been undertaken, by employing a 7-atom, a 13-atom and a 25-atom system models (67-69), and we have shown that the 13-atom system model is appropiate to describe the dynamic of this complex ion.
For the $^2E_g \rightarrow {}^4A_{2g}$ electronic transition of this complex ion, Flint et al (70-73, 74), observed that the vibronic intensity induced by the false origins $\nu_7(\tau_{1u})$: δ(Cr-N-H), $\nu_9(\tau_{1u})$: δ(N-Cr-N), $\nu_{12}(\tau_{2u})$: δ(Cr-N-H) and $\nu_{13}(\tau_{2u})$: δ(N-Cr-N); are roughly comparable in magnitude. This is indeed, a rather important experimental result, since the normal modes (ν_9, ν_{13}) and the normal modes (ν_7, ν_{12}) correspond to vibrations of the cluster $|CrN_6|$ and to framework-ligand coupling vibrations (in this case, rocking vibrations), respectively. In a way, this experimental result could be rather unexpected, since one may argue that the efficiency of the rocking modes to induce vibronic intensity should be smaller than those corresponding to the normal modes of the cluster. A similar experimental result has been found for the $Cr(CN)_6^{3-}$ complex in octahedral symmetry, where it has been found that the framework-ligand coupling normal modes of vibrations ν_7 and ν_{12} are associated to the strongest vibronic features in the spectrum (72).
Based upon, qualitative arguments, Flint (73) has argued that it is the lone pair motion, associated with both the $-NH_3$ and the -CN ligand subsystems, out of the internuclear axis $Cr-NH_3$ and Cr-CN, respectively which is responsible for the observed vibronic intensities of the $Cr(NH_3)_6^{3+}$ and the $Cr(CN)_6^{3-}$ complex ions in cubic environments.

Acevedo and Flint (75), considered on a quantitative basis, the electronic factors associated with both the $^4T_{1g} \rightarrow {}^4A_{2g}$ and the $^4T_{2g} \rightarrow {}^4A_{2g}$ non radiative transitions for the hexamminochromium (III) ion in octahedral symmetry, and proved that when the lone pair motion is included, in the calculation of these electronic factors, the relative rate of decay k defined as given below:

$$k = \{k'(^4T_{1g} \rightarrow {}^4A_{2g})/k''(^4T_{2g} \rightarrow {}^4A_{2g}\} \qquad (79)$$

comes closer to 100. This result shows that the $\nu_5(\tau_{1g})$: δ(Cr-N-H) rocking mode is rather inefficient to promote radiationless transitions in comparison to the $\nu_{11}(\tau_{2g})$: δ(N-Cr-N) and $\nu_{10}(\tau_{2g})$: δ(Cr-N-H) normal modes of vibrations (68, 69, 75). The calculation performed by Acevedo and Flint (75) for the non radiative rates of decay of the hexammino - chromium (II) ion in octahedral symmetry, has shown that the inclussion of the lone pair motion of the $-NH_3$ group, is essential to account for the expected non radiative rates of decay and also for the observed radiative rates of decay for this complex ion. Preliminary results, obtained in our laboratory suggest that the relative vibronic intensity distribution for both the $Cr(NH_3)_6^{3+}$ and the $Cr(CN)_6^{3-}$ complex ions in octahedral symmetry can be rationalized in terms of the inclussion of the lone pair motion (76).

Recently, Acevedo and Díaz (77) has carried out a non radiative calculation for the $^4T_{1g} \rightarrow {}^4A_{2g}$ and the $^4T_{2g} \rightarrow {}^4A_{2g}$ radiationless transitions of the hexamminochromium (III) ion in the octahedral double group O*, without including the lone pair motion. We used the strong field configurations t_{2g}^3, $t_{2g}^2e_g$, $t_{2g}e_g^2$ and e_g^3 and carried out a full crystal field calculation for the system. The total non relativistic Hamilto - nian employed in the calculation is of the form: $H = H^0 + \sum_{i,j} \{e^2/r_{ij}\} + \sum_i (\xi_i) l_i s_i + V_{CF}$.

The full interaction matrices for a d^3 electronic configuration, in the strong field coupling scheme, were taken from Eisenstein (52) and the observed experimental frequencies from Flint et al (70-73, 74). In our model calculation we used the following set of parameters: B = 672 cm^{-1}, C = 3,024 cm^{-1} Dq = 2,150 cm^{-1} and $\lambda_{SO} \simeq 200$ cm^{-1}, respectively.

5.a) THE $\Gamma_8(^4T_{1g}) \rightarrow \Gamma_8(^4A_{2g})$ NON RADIATIVE TRANSITION.

As it was discussed in Section III-A, a typical polyelectronic reduced matrix element to be evaluated is of the form $\langle\Gamma_8(^4T_{1g})||O^C||\Gamma_8(^4A_{2g})\rangle_m$ where O^C represents a polyelectronic irreducible tensor operator, which transforms under the Cth-irreducible representation of the octahedral double group O* and "m" is a repeated representation label. From the Table XI, this polyelectronic reduced matrix element, may be expressed as given below:

$$<\Gamma_8(^4T_{1g})||O^C||\Gamma_8(^4A_{2g})>_m =$$
$$0.868250974<t_2^2e\,^4T_1\Gamma_8 1||O^C||t_2^3\,^4A_2\Gamma_8 1>_m -$$
$$- 0.219320195<t_2e^2\,^4T_1\Gamma_8 1||O^C||t_2^3\,^4A_2\Gamma_8 1>_m -$$
$$- 0.429697406<t_2^2e\,^4T_1\Gamma_8 2||O^C||t_2^3\,^4A_2\Gamma_8 1>_m +$$
$$+ 0.108531371<t_2e^2\,^4T_1\Gamma_8 2||O^C||t_2^3\,^4A_2\Gamma_8 1>_m -$$
$$- 0.022939190<t_2^2e\,^4T_2\Gamma_8 2||O^C||t_2^3\,^4A_2\Gamma_8 1>_m +$$
$$+ 0.009611538<t_2^2e\,^4T_2\Gamma_8 1||O^C||t_2^3\,^4A_2\Gamma_8 1>_m \quad (80)$$

where interaction coefficients smaller than 10^{-3} have been neglected. These polyelectronic reduced matrix elements can be expressed in terms of monoelectronic reduced matrix elements, by using the algebra put forward by Griffith (5) for point symmetry groups and extended by Piepho (45) and Dobosh (46) for octahedral double group O*. It is straightforward to show that the following identities hold:

$$<\Gamma_8(^4T_1)||O^{T_2}||\Gamma_8(^4A_2)>_1 = - 0.6748<t_2||O^{T_2}||e>$$
$$<\Gamma_8(^4T_1)||O^{T_2}||\Gamma_8(^4A_2)>_2 = - 0.8922<t_2||O^{T_2}||e>$$
$$<\Gamma_8(^4T_1)||O^{T_1}||\Gamma_8(^4A_2)>_1 = + 0.0200<t_2||O^{T_1}||e>$$
$$<\Gamma_8(^4T_1)||O^{T_1}||\Gamma_8(^4A_2)>_2 = - 0.0287<t_2||O^{T_1}||e> \quad (81)$$

By-employing the V_m-coupling coefficients tabulated by Dobosh (46), it is easy to find the relevant matrix elements of the form $<A\alpha|V^{(1)}|B\beta>$, see the equation (47). Section III-A, which are relevant to the evaluation of the electronic factor P_{nm} defined by the equation |40|. We list these matrix elements in the Table XII. In our present notation, we write: $g_5 = <t_2||O^{T_1}(5)||e>$, $g_{10} = <t_2||O^{T_2}(10)||e>$ and $g_{11} = <t_2||O^{T_2}(11)||e>$, respectively. We also use the identities, given below:

$$X_{+1}^{\Gamma}(k) = - 1/\sqrt{2}(|\Gamma_x> + i|\Gamma_y>)_k$$
$$X_0^{\Gamma}(k) = |\Gamma_z>_k$$
$$X_{-1}^{\Gamma}(k) = + 1/\sqrt{2}(|\Gamma_x> - i|\Gamma_y>)_k \quad (82)$$

where $\Gamma = \tau_{1g}, \tau_{2g}$ and k = 5,10 and 11.

The monoelectronic reduced matrix elements $<t_2||O^{\Gamma}(k)||e>$ for k = 5,10 and 11, refer to the $\nu_5(\tau_{1g})$: δ(Cr-N-H) rocking vibration, the $\nu_{10}(\tau_{2g})$: δ(Cr-N-H) rocking vibration and to the $\nu_{11}(\tau_{2g})$: δ(N-Cr-N) bending vibration, respectively.
The relevant monoelectronic reduced matrix elements have been evaluated, within the framework of the vibronic crystal field method, using the standard procedure, that is: Firstly, within the 13-atom model system, a full normal coordinate analysis for the $Cr(NX)_6^{3+}$ system was underta - ken (67, 69). We then introduced the three translations and the three rotations of the molecule as a whole to obtain the transformation: $R = (UB)^{-1}$ S, relating the nuclear Cartesian displacement coordinates to the standard symmetry coordinates for the molecular system. Secondly: The ammino group was considered as an extended electric dipole, and the ionic partial charges on the central metal and the ligand subsystems were evaluated by employing a MO-LCAO type calculation. According to this procedure the effective charges are: δ(Cr) = + 0.6900; δ(N) = - 0.1100 and δ(H) = + 0.1650, respectively. Finally, we evaluated the electronic integrals $<\Gamma_1\gamma_1|\{\delta V_{CF}/\delta S_{\gamma}^{\Gamma}(k)\}_0|\Gamma_2\gamma_2>$.

Thus, the relevant monoelectronic reduced matrix elements are given by:

$$<t_2||O^{T_1}(5)||e> = (5\sqrt{3}/6)\{e^2<r^4>\}\{0.175268(\delta_H/r^6) - 0.290728(\delta_N/R^6)\}$$

$$<t_2||O^{T_2}(10)||e> = - 0.283333\{e^2<r^4>\}\{\delta_H/r^4\}$$

$$<t_2||O^{T_2}(11)||e> = 1.022000\{e^2<r^2>\}\{\delta_H/r^4\} \quad (83)$$

where r = 2.38 Å and R = 2.00 Å. The expectation values of r^k have been evaluated by using the radial wavefunctions put forward by Richardson et al (78). Thus for Chromium (+ 0.6900), we obtain by linear interpolation, the following expectation values of r^k, between two 3d-radial wavefunctions: $<r^2> = 0.7626$ A°^{2} and $<r^4> = 1.4070$ A°^{4}. Then, by collecting all this information together, we obtain for the $\Gamma_8(^4T_{1g}) \rightarrow \Gamma_8(^4A_{2g})$, the non radiative rate of decay k as given below:

$$k \propto 4.4597732E\text{-}32|<\Lambda^4T_{1g}|\Lambda^4A_{2g}>|^2 \quad (84)$$

in cgs units. The above result for the non radiative rate of decay k holds, except an unknoen parameter p^2, see Section III-A.

5.b) THE $\Gamma_8(^4T_{2g}) \rightarrow \Gamma_8(^4A_{2g})$ NON RADIATIVE TRANSITION.

By using the eigenvectors tabulated in the Table XI, a typical polyelectronic reduced matrix element to be evaluated is of the form:

$$\begin{aligned}&<\Gamma_8(^4T_{2g})||O^C||\Gamma_8(^4A_{2g})> = \\ &+ 0.43178177<t_2^2e\,^4T_2\Gamma_8 1||O^C||t_2^3\,^4A_2\Gamma_8 1>_m + \\ &+ 0.842008134<t_2^2e\,^4T_2\Gamma_8 2||O^C||t_2^3\,^4A_2\Gamma_8 1>_m\end{aligned} \tag{85}$$

and since $T_{2g} \times A_{2g} = T_{1g}$, then $C = T_1$ and the relevant polyelectronic reduced matrix elements are expressed in terms of the monoelectronic reduced matrix elements $<t_2||O^{T_1}(5)||e>$ by means of the identities given below:

$$\begin{aligned}&<\Gamma_8(^4T_{2g})||O^{T_1}||\Gamma_8(^4A_{2g})>_1 = -\ 0.880753425<t_2||O^{T_1}||e> \\ &<\Gamma_8(^4T_{2g})||O^{T_1}||\Gamma_8(^4A_{2g})>_2 = 0.646651115<t_2||O^{T_1}||e>\end{aligned} \tag{86}$$

By following the same procedure as we did for the $\Gamma_8(^4T_{1g}) \rightarrow \Gamma_8(^4A_{2g})$ non radiative transition, it is straightforward to show that the following identities hold:

$$\begin{aligned}&<\Gamma_8 k|V^{(1)}|\Gamma_8 \kappa> = -\ 0.4246\ g_5\ X_0^{T_1}(5) \\ &<\Gamma_8 k|V^{(1)}|\Gamma_8 \lambda> = \ \ 0.1763\ g_5\ X_{+1}^{T_1}(5) \\ &<\Gamma_8 k|V^{(1)}|\Gamma_8 \mu> = \ \ 0.0000 \\ &<\Gamma_8 k|V^{(1)}|\Gamma_8 \nu> = -\ 0.2952\ g_5\ X_{-1}^{T_1}(5)\end{aligned} \tag{87}$$

As for the λ, μ and ν components of the Γ_8-irreducible representation, we can obtain the relevant matrix elements from equation |47|. After some algebra, we can easily found that the non radiative rate of decay, associated with this electronic transition may be expressed as follows:

$$k \ \alpha\ 1.3545198E\text{-}33|<\Lambda\,^4T_{2g}|\Lambda\,^4A_{2g}>|^2 \tag{88}$$

except an unknown p^2 parameter, see Section III-A.

Next, we shall assume that the square of the vibrational overlap integrals, that is: $|<\Lambda^4T_{1g}|\Lambda^4A_{2g}>|^2$ and $|<\Lambda^4T_{2g}|\Lambda^4A_{2g}>|^2$ are roughly the same in magnitude, as well as the p^2 factors, associated with the $\Gamma_8(^4T_{1g}) \rightarrow \Gamma_8(^4A_{2g})$ and the $\Gamma_8(^4T_{2g}) \rightarrow \Gamma_8(^4A_{2g})$ radiationless trans - tions, see equations |84| and |88|.
Under these assumptions, we observe that the relative rate of decay k associated with these two radiationless transitions, is given by:

$$k = \{\Gamma_8(^4T_{1g}) \rightarrow \Gamma_8(^4A_{2g})/\Gamma_8(^4T_{2g}) \rightarrow \Gamma_8(^4A_{2g})\} = 33 \qquad (89)$$

It is clear that the introduction of the lone pair motion associated with the ammino group, in hhe non radiative calculation, represents a sensible way to interpret the experimental data for the hexammichromium (III) complex ion (75), supporting Flint's claim that for the $Cr(NH_3)_6^{3+}$ and the $Cr(CN)_6^{3-}$ complex ions, it is the lone pair motion out of the internuclear axis which is responsable for the observed vibronic intensities in the luminiscence spectra of these complex ions (73). Vibronic intensity calculations for these complex ions are in progress in our laboratory, and initial results suggest that the inclusion of the lone pair motion is essential to describe on a quantitative basis the experimental vibronic features in the corresponding spectra.

ACKNOWLEDGEMENTS

R.A. would like to express his gratitude to Fondecyt. Grant No0623 (1988) for partial finantial support, as well as to the DTI (University of Chile). Grant No Q-2494-8833. The British Council is also gratefully acknowledged for a travel award to cover local expenses in London (Auguest-September, 1988).
Professor C.D. Flint, Director of NATO-ASI is also thanked for his kind invitation to participate and deliver these lectures in this Advanced Study Institute, held in Italy. September 7-18, 1988.

REFERENCES.

1. Ballhausen, C.J. Introduction to Ligand Field Theory. Mc-Graw Hill, N.Y., 1962.
2. Ballhausen, C.J. Molecular Electronic Structures of Transition Metal Complexes. Mc-Graw Hill, INC, 1979.
3. Ballhausen, C.J., Hansen, A.E. Ann. Rev. Phys. Chem., 23, 15 (1972).
4. Griffith, J.S. The Theory of Transition Metal Ions. Cambridge, University Press, 1964.
5. Griffith, J.S. The irreducible Tensor Method for Molecular Symmetry Groups. Prentice-Hall. Englewood Cliffs, N.J., 1962.
6. Faulkner, T.R., Richardson, F.S. Mol. Phys., 35, 1141 (1978).
7. Richardson, F.S. Chem. Phys. Letters, 86(1), 47 (1982).
8. Acevedo, R., Flint, C.D. Theor. Chim. Acta (Berl.), 69(3), 225 (1986).
9. Acevedo, R., Flint, C.D. Mol. Phys., 49(5), 1065 (1983).
10. Acevedo, R., Flint, C.D. Mol. Phys., 56(3), 683 (1985).
11. Acevedo, R., Flint, C.D. Mol. Phys., 53(1), 129 (1984).
12. Acevedo, R., Flint, C.D. Mol. Phys., 58(6), 1033 (1986).
13. Flint, C.D., Acevedo, R. In "Understanding Molecular Properties", pages 195-203. Editors: Avery. J., Dahl, P., Hansen, A.E. D. Reidel Publishing Company, 1987.
14. Acevedo, R., Díaz, G. Annals. Chem. (Spain), 83, 135 (1987).
15. Acevedo, R., Meruane, T., Letelier, J.R. Theor. Chim. Acta(Berl.), 64, 339 (1984).
16. Califano, S. Vibrational States. John Wiley and Sons. Ltd., 1976.
17. Acevedo, R. PhD. Thesis. Univ. of London, 1981.
18. Acevedo, R., Díaz, G. Spectroscopy Letters, 21(1), 19 (1988).
19. Acevedo, R., Letelier, J.R. To be published.
20. Acevedo, R., Díaz, G., Flint, C.D. J.C.S. Faraday Trans II, 81, 1861 (1985).
21. Acevedo, R., Díaz, G., Meruane, T. Annals. Chem. (Spain), 1988. In press.
22. Mason, S.F., Peacock, R.D., Stewart, B. Mol. Phys., 30 1829 (1975).
23. Mason, S.F. Unpublished Lectures Notes, 1980.
24. Díaz, G. PhD. Thesis. Univ. of Chile, 1988.
25. Meruane, T., Acevedo, R. Theor. Chim. Acta (Berl.), 62, 301 (1983).
26. Liehr, A.D., Ballhausen, C.J. J. Phys. Rev., 106, 1161 (1957).
27. Koide, S., Pryce, M.H.L. Phil. Mag., 3, 607 (1958).
28. Hammer, N.K. Mol. Phys., 5, 455 (1961).
29. Chakravarty, A.S. J. Phys. Chem., 74, 4347 (1970).
30. Koide, S. Phil. Mag., 4, 243 (1959).
31. Liehr, A.D. Adv. Chem. Phys., 37, 511 (1962).
32. Butler, P.H. Point Group Symmetry Applications. Plenum Press, N.Y. 1981.
33. Robbins, D.J., Thomson, A.J. Mol. Phys., 25(5), 1103 (1973).
34. Bison, M., M., Jortner, J. J. Chem. Phys., 50(8), 3284 (1969).
35. Bison, M., Joprther, J. J. Chem. Phys., 48, 715 (1968).
36. Bison, M., Jortner, J. J. Chem. Phys., 50(9), 4061 (1969).
37. Ballhausen, C.J., Strek, W. Mol. Phys., 36(5), 1321 (1978).
38. Robbins, D.J., Thomson, A.J. Phil. Mag., 36(4), 999 (1977).

39. Englman, R., Jortner, J. Mol. Phys., 18, 145 (1970).
40. Witkowski, A., Moffit, W. J. Chem. Phys., 33, 872 (1961).
41. Chatterjee, K.K., Forster, L.S. Spectrochim. Acta, 20, 1603 (1964).
42. De Armond, K., Forster, L.S. Spectrochim. Acta, 19, 1787 (1963).
43. Jortner, J., Rice, A.S., Hoschtrasser, R.M. In "Advances in Photochemistry". Vol. VII, 149. John Wiley and Sons, 1962.
44. Henry, B.R., Siebrand, W. In "Organic Molecular Photophysics"., Vol. I, 1531. John Wiley and Sons, 1973.
45. Piepho, S.B. In "Recent Advances in Group Theory and their Applica tion to Spectroscopy". Editor. Donini, J. Plenum Press, N.Y. 1979.
46. Dobosh, P.A. Phys. Rev., A-5, 3276 (1972).
47. Haarhoff, P.C. Mol. Phys., 7, 101 (1963).
48. Witkowski, A., Moffit, W. J. Chem. Phys., 33, 872 (1969).
49. Lim, E.C. Excited States. Academic Press, INC., Vol. VI, 1982.
50. Flint, C.D., Paulusz, A.G. Chem. Phys. Letts, 62(2), 259 (1979).
51. Flint, C.D., Paulusz, A.G. Mol. Phys., 43, 321 (1981).
52. Einsenstein, J.C. J. Chem. Phys., 34(5), 1628 (1981).
53. Black, A.M., Flint, C.D. J.C.S. Faraday. Trans. II, 73, 877 (1977).
54. Dorian, P.B. Transition Metal Chemistry, 4, 1, 1968.
55. Chodos, S.L., Satten, R.A. J. Chem. Phys., 67(2), 2411 (1975).
56. Chodos, S.L., Black, A.M., Flint, C.D. J. Chem. Phys. 65(11), 4816 (1976).
57. Johnston, P., Satten, R.A., Wong, E.Y. J. Chem. Phys., 44, 687 (1966).
58. Satten, R.A., Wong, E.Y. J. Chem. Phys., 43, 3025 (1965).
59. Cortés, E. MSC. Thesis. Univ. of Chile, 1988.
60. Cortés, E., Acevedo, R. Annals. Chem. (Spain), 1988. In press.
61. Pandey, S.L. Singh, B.P. Indian Journal of Pure and Applied Physics, 14, 815 (1976).
62. Labonville, P., Ferraro, L.J., Wall, M.C., Basile, L.J. Coordina - tion Chem. Rev., 7, 257 (1972).
63. Pistorius, C.W.F.T. J. Chem. Phys., 29, 1328 (1958).
64. Bash, H., Gray, H.B. Theoretica Chimica Acta (Berl.), 4, 367 (1966).
65. Desclaux, J.P. Atomic Data and Nuclear Data Tables, 12(4), 312 (1973).
66. Le Febre, R.J.W. Adv. Org. Chem., 3, 1 (1965).
67. Acevedo, R., Díaz, G. Spectroscopy Letters, 16(3), 199 (1984).
68. Acevedo, R., Díaz, G., Flint, C.D. Spectrochimica Acta A, 41(12), 1397 (1985).
69. Acevedo, R., Díaz, G. Annals. Chem. (Spain) B, 83, 135 (1988).
70. Flint, C.D., Greenough, P. J.C.S. Faraday Trans. II, 68, 897 (1972).
71. Greenough, P. PhD. Thesis Univ. of London, 1972.
72. Flint, C.D., Greenough, P. J.C.S. Faraday Trans. II, 70, 815 (1974).
73. Flint, C.D. J.C.S. Faraday Trans. II, 72, 721 (1976).
74. Hempleman Unpublished Results.
75. Acevedo, R., Flint, C.D. Mol. Phys. 54(3), 619 (1985).
76. Acevedo, R., Díaz, G., Flint, C.D. To be published.
77. Acevedo, R., Díaz, G. Annals. Chem. (Spain), 1988. In press.
78. Richardson, J.W., Nieuwpoort, W.C., Powell, R.R., Edgell, W.F. J. Chem. Phys., 36, 1057 (1962).

APPENDIX I. THE COMPLEX LIGAND POLARIZATION VIBRONIC COUPLING CONSTANTS IN TERMS OF THE REAL LIGAND POLARIZATION VIBRONIC COUPLING CONSTANTS*

$$B^{\Gamma 1,0}_{\bar{\Gamma}+1} = \frac{i}{2}\{-B^{\Gamma X,Z}_{\bar{\Gamma}X} + i\ B^{\Gamma Y,Z}_{\bar{\Gamma}X} + i\ B^{\Gamma X,Z}_{\bar{\Gamma}Y} + B^{\Gamma Y,Z}_{\bar{\Gamma}Y}\}$$

$$B^{\Gamma 1,0}_{\bar{\Gamma}0} = i/\sqrt{2}\{B^{\Gamma X,Z}_{\bar{\Gamma}Z} - i\ B^{\Gamma Y,Z}_{\bar{\Gamma}Z}\}$$

$$B^{\Gamma 1,0}_{\bar{\Gamma}-1} = \frac{i}{2}\{B^{\Gamma X,Z}_{\bar{\Gamma}X} - i\ B^{\Gamma Y,Z}_{\bar{\Gamma}X} + i\ B^{\Gamma X,Z}_{\bar{\Gamma}Y} + B^{\Gamma Y,Z}_{\bar{\Gamma}Y}\}$$

$$B^{\Gamma -1,0}_{\bar{\Gamma}+1} = \frac{i}{2}\{B^{\Gamma X,Z}_{\bar{\Gamma}X} + i\ B^{\Gamma Y,Z}_{\bar{\Gamma}X} - i\ B^{\Gamma X,Z}_{\bar{\Gamma}Y} + B^{\Gamma Y,Z}_{\bar{\Gamma}Y}\}$$

$$B^{\Gamma -1,0}_{\bar{\Gamma}0} = i/\sqrt{2}\{B^{\Gamma X,Z}_{\bar{\Gamma}Z} + i\ B^{\Gamma Y,Z}_{\bar{\Gamma}Z}\}$$

$$B^{\Gamma -1,0}_{\bar{\Gamma}-1} = \frac{i}{2}\{-B^{\Gamma X,Z}_{\bar{\Gamma}X} - i\ B^{\Gamma Y,Z}_{\bar{\Gamma}X} - i\ B^{\Gamma X,Z}_{\bar{\Gamma}Y} + B^{\Gamma Y,Z}_{\bar{\Gamma}Y}\}$$

$$B^{\Gamma 0,0}_{\bar{\Gamma}+1} = -i/\sqrt{2}\{-B^{\Gamma Z,Z}_{\bar{\Gamma}X} + i\ B^{\Gamma Z,Z}_{\bar{\Gamma}Y}\}$$

$$B^{\Gamma 0,0}_{\bar{\Gamma}0} = -i\ B^{\Gamma Z,Z}_{\bar{\Gamma}Z}$$

$$B^{\Gamma 0,0}_{\bar{\Gamma}-1} = -i/\sqrt{2}\{B^{\Gamma Z,Z}_{\bar{\Gamma}X} + i\ B^{\Gamma Z,Z}_{\bar{\Gamma}Y}\}$$

* $\Gamma, \bar{\Gamma} = T_1$ and T_2 and $B^{\Gamma\gamma,\alpha}_{\bar{\Gamma}\gamma}(k,\tau) = -\bar{\alpha}_L \sum \left|\frac{\partial G^{L,\tau}_{\Gamma\gamma,\alpha}}{\partial S^{\bar{\Gamma}}_{\gamma}(k)}\right|_0$

APPENDIX II. THE $q_{j1}^{\Gamma,i}(k)$ POLYELECTRONIC REDUCED MATRIX ELEMENTS IN TERMS OF THE MONOELECTRONIC REDUCED MATRIX ELEMENTS.

$$q_{11}^{A_1,1} = \sqrt{3}\, p_5 \, \langle t_2 || O^{A_1} || t_2 \rangle$$

$$q_{11}^{E,1} = - p_4/\sqrt{2} \, \langle t_2 || O^{E} || t_2 \rangle$$

$$q_{11}^{T_1,1} = 2p_2/\sqrt{15} \, \langle t_2 || O^{T_1} || e \rangle - p_5/\sqrt{10} \, \langle t_2 || O^{T_1} || t_2 \rangle$$

$$q_{11}^{T_1,2} = p_2/\sqrt{15} \, \langle t_2 || O^{T_1} || e \rangle - 4p_5/3\sqrt{10} \, \langle t_2 || O^{T_1} || t_2 \rangle$$

$$q_{12}^{T_1,1} = p_2/\sqrt{15} \, \langle t_2 || O^{T_1} || e \rangle$$

$$q_{12}^{T_1,2} = - 2p_2/\sqrt{15} \, \langle t_2 || O^{T_1} || e \rangle$$

$$q_{11}^{T_2,1} = (- p_3/\sqrt{3} + \sqrt{2}\, p_4/3) \, \langle t_2 || O^{T_2} || t_2 \rangle$$

$$q_{11}^{T_2,2} = (p_3/\sqrt{3} + \sqrt{2}\, p_4/6) \, \langle t_2 || O^{T_2} || t_2 \rangle$$

APPENDIX III. THE VIBRONIC CRYSTAL FIELD-CLOSURE METHOD. RELEVANT MONOELECTRONIC REDUCED MATRIX ELEMENTS*,**

$$\langle t_2||O^{A_1}(3)||t_2\rangle = \sqrt{2}\{2\langle\gamma_2\rangle - \frac{16}{21}\langle\gamma_4\rangle - \frac{10}{11}\langle\gamma_6\rangle\}$$

$$\langle t_2||O^{A_1}(4)||t_2\rangle = \{-2\langle\gamma_2\rangle - \frac{4}{7}\langle\gamma_4\rangle + \frac{10}{11}\langle\gamma_6\rangle\}$$

$$\langle t_2||O^{A_1}(6)||t_2\rangle = 0$$

$$\langle t_2||O^{E}(3)||t_2\rangle = \{-\frac{8}{7}\langle\gamma_2\rangle - \frac{8}{21}\langle\gamma_4\rangle + \frac{100}{77}\langle\gamma_6\rangle\}$$

$$\langle t_2||O^{E}(4)||t_2\rangle = \sqrt{2}\{\frac{4}{7}\langle\gamma_2\rangle - \frac{1}{7}\langle\gamma_4\rangle - \frac{5}{77}\langle\gamma_6\rangle\}$$

$$\langle t_2||O^{E}(6)||t_2\rangle = \{\frac{5\sqrt{6}}{7}\langle\gamma_4\rangle - \frac{5\sqrt{6}}{11}\langle\gamma_6\rangle\}$$

$$\langle t_2||O^{T_1}(3)||e\rangle = -\sqrt{3}\{\frac{20}{21}\langle\gamma_4\rangle - \frac{10}{11}\langle\gamma_6\rangle\}$$

$$\langle t_2||O^{T_1}(4)||e\rangle = -\sqrt{6}\{\frac{5}{14}\langle\gamma_4\rangle + \frac{5}{11}\langle\gamma_6\rangle\}$$

$$\langle t_2||O^{T_1}(6)||e\rangle = -\sqrt{6}\{\frac{5}{14}\langle\gamma_4\rangle + \frac{5}{11}\langle\gamma_6\rangle\}$$

$$\langle t_2||O^{T_1}(k)||t_2\rangle = 0, \quad \text{for } k = 3, 4, 6$$

$$\langle t_2||O^{T_2}(3)||t_2\rangle = \sqrt{6}\{\frac{4}{7}\langle\gamma_2\rangle - \frac{16}{21}\langle\gamma_4\rangle + \frac{20}{77}\langle\gamma_6\rangle\}$$

$$\langle t_2||O^{T_2}(4)||t_2\rangle = \sqrt{3}\{-\frac{4}{7}\langle\gamma_2\rangle - \frac{4}{7}\langle\gamma_4\rangle + \frac{40}{77}\langle\gamma_6\rangle\}$$

$$\langle t_2||O^{T_2}(6)||t_2\rangle = 0$$

* The reduced matrix elements listed in this Appendix are given in units of "e", being -e, the electron charge.

** $\langle\gamma_k\rangle = qe^2\langle r^k\rangle/R_0^{k+1}$, where R_0 is the equilibrium bond distance. "q" stands for the effective ligand charge.

APPENDIX IV. THE CRYSTAL FIELD $C^{i}_{\gamma_1\gamma_2}(k)$ QUANTITIES RELEVANT TO THE INTENSITY CALCULATION FOR THE $\Gamma_8(^2T_{2g}) \rightarrow \Gamma_8(^4A_{2g})$ ELECTRONIC TRANSITION.

$$C^1_{kk}(3) = -\frac{1}{2\sqrt{3}}\, q^{A_1}_{11}(3) + \frac{1}{2\sqrt{3}}\, q^{E}_{11}(3)$$

$$C^1_{kk}(4) = -\frac{1}{2\sqrt{3}}\, q^{A_1}_{11}(4) + \frac{1}{2\sqrt{3}}\, q^{E}_{11}(4)$$

$$C^1_{kk}(6) = 0$$

$$C^2_{k\lambda}(3) = -\frac{1}{2\sqrt{5}}\, q^{T_1,1}_{11}(3) - \frac{1}{4\sqrt{5}}\, q^{T_1,2}_{11}(3) + \frac{1}{4\sqrt{3}}\, q^{T_2,1}_{11}(3) - \frac{1}{2\sqrt{3}}\, q^{T_2,2}_{11}(3)$$

$$C^2_{k\lambda}(4) = -\frac{1}{2\sqrt{5}}\, q^{T_1,1}_{11}(4) - \frac{1}{4\sqrt{5}}\, q^{T_1,2}_{11}(4) + \frac{1}{4\sqrt{3}}\, q^{T_2,1}_{11}(4) - \frac{1}{2\sqrt{3}}\, q^{T_2,2}_{11}(4)$$

$$C^2_{k\lambda}(6) = -\frac{1}{2\sqrt{5}}\, q^{T_1,1}_{11}(6) - \frac{1}{4\sqrt{5}}\, q^{T_1,2}_{11}(6)$$

$$C^3_{k\mu}(3) = C^3_{k\mu}(4) = 0 \;;\; C^3_{k\mu}(6) = \frac{1}{2\sqrt{3}}\, q^{E}_{11}(6)$$

$$C^4_{k\nu}(3) = \frac{\sqrt{5}}{4\sqrt{3}}\, q^{T_1,2}_{11}(3) + \frac{1}{4}\, q^{T_2,1}_{11}(3)$$

$$C^4_{k\nu}(4) = \frac{\sqrt{5}}{4\sqrt{3}}\, q^{T_1,2}_{11}(4) + \frac{1}{4}\, q^{T_2,1}_{11}(4)$$

$$C^4_{k\nu}(6) = \frac{\sqrt{5}}{4\sqrt{3}}\, q^{T_1,2}_{11}(6)$$

$$C^5_{kk}(3) = C^5_{kk}(4) = C^5_{kk}(6) = 0$$

$$C^6_{k\lambda}(3) = -\frac{1}{2\sqrt{5}}\{q^{T_1,1}_{12}(3) + \frac{1}{2}\, q^{T_1,2}_{12}(3)\}$$

$$C^6_{k\lambda}(4) = -\frac{1}{2\sqrt{5}}\{q^{T_1,1}_{12}(4) + \frac{1}{2}\, q^{T_1,2}_{12}(4)\}$$

$$C^6_{k\lambda}(6) = -\frac{1}{2\sqrt{5}}\{q^{T_1,1}_{12}(6) + \frac{1}{2}\, q^{T_1,2}_{12}(6)\}$$

$$C^7_{k\mu}(3) = C^7_{k\mu}(4) = C^7_{k\mu}(6) = 0$$

Cont. Appendix IV.

$$c^{8}_{k\nu}(3) = \frac{\sqrt{15}}{12}\, q^{T_1,2}_{12}(3)$$

$$c^{8}_{k\nu}(4) = \frac{\sqrt{15}}{12}\, q^{T_1,2}_{12}(4)$$

$$c^{8}_{k\nu}(6) = \frac{\sqrt{15}}{12}\, q^{T_1,2}_{12}(6)$$

APPENDIX V. THE LIGAND POLARIZATION $m^{i}_{\gamma_1\gamma_2}(k)$ QUANTITIES RELEVANT TO THE INTENSITY CALCULATION FOR THE $\Gamma_8(^2T_{2g}) \rightarrow \Gamma_8(^4A_{2g})$ ELECTRONIC TRANSITION.

$$m^1_{kk}(3) = -\frac{6\sqrt{6}}{7}\,p_4\ e\langle r^2\rangle\bar{\alpha}_L R_0^{-5}$$

$$m^1_{kk}(4) = -\frac{9\sqrt{3}}{14}\,p_4\ e\langle r^2\rangle\bar{\alpha}_L R_0^{-5}$$

$$m^1_{kk}(6) = 0$$

$$m^2_{k\lambda}(3) = \frac{1}{\sqrt{2}}\,\{\frac{6\sqrt{6}}{7}\,p_3\,\langle r^2\rangle R_0^{-5} + (-5p_2/6 - \frac{15\sqrt{6}}{21}\,p_3)\langle r^4\rangle R_0^{-7}\}e\,\bar{\alpha}_L$$

$$m^2_{k\lambda}(4) = \frac{1}{\sqrt{2}}\,\{\frac{9\sqrt{3}}{14}\,p_3\langle r^2\rangle R_0^{-5} + (\frac{5\sqrt{2}}{4}\,p_2 - \frac{5\sqrt{3}}{7}\,p_3)\langle r^4\rangle R_0^{-7}\}e\,\bar{\alpha}_L$$

$$m^2_{k\mu}(3) = m^3_{k\mu}(4) = 0\ ;\ m^3_{k\mu}(6) = (3p_4/14)e\,\langle r^2\rangle\bar{\alpha}_L R_0^{-5}$$

$$m^4_{k\nu}(3) = \frac{1}{\sqrt{2}} - \frac{2\sqrt{6}}{7}(-\sqrt{3}\,p_3+\sqrt{2}\,p_4)\langle r^2\rangle R_0^{-5} + (\frac{5\sqrt{3}}{6}\,p_2 + \frac{5\sqrt{2}}{7}\,p_3 + \frac{5\sqrt{3}}{21}\,p_4)\times$$
$$\times\ \langle r^4\rangle R_0^{-7}\}e\,\bar{\alpha}_L$$

$$m^4_{k\nu}(4) = \frac{1}{\sqrt{2}}\{-\frac{3\sqrt{3}}{14}(-\sqrt{3}\,p_3+\sqrt{2}\,p_4)\langle r^2\rangle R_0^{-5} + (-\frac{5\sqrt{6}}{12}\,p_2 - \frac{10}{7}\,p_3 + \frac{10\sqrt{6}}{21}\,p_4)\times$$
$$\times\ \langle r^4\rangle R_0^{-7}\}e\,\bar{\alpha}_L$$

$$m^4_{k\nu}(6) = \frac{1}{\sqrt{2}}\{-\frac{5\sqrt{3}}{14}(-\sqrt{3}\,p_3+\sqrt{2}\,p_4)\langle r^2\rangle R_0^{-5} - \frac{5\sqrt{6}}{12}\,p_2\langle r^4\rangle R_0^{-7}\}e\,\bar{\alpha}_L$$

$$m^5_{kk}(3) = m^5_{kk}(4) = m^5_{kk}(6) = 0$$

$$m^6_{k\lambda}(3) = m^6_{k\lambda}(4) = m^6_{k\lambda}(6) = 0$$

$$m^7_{k\mu}(3) = m^7_{k\mu}(4) = m^7_{k\mu}(6) = 0$$

$$m^8_{k\nu}(3) = \frac{1}{\sqrt{2}}\,\{-\frac{\sqrt{3}}{3}\,p_2\langle r^4\rangle R_0^{-7}\}e\,\bar{\alpha}_L$$

$$m^2_{k\lambda}(6) = \frac{1}{\sqrt{2}}\,\{\frac{15\sqrt{3}}{14}\,p_3\langle r^2\rangle R_0^{-5}\}e\,\bar{\alpha}_L$$

Cont. Appendix V.

$$m_{k\nu}^{8}(4) = \frac{1}{\sqrt{2}}\{\frac{5\sqrt{6}}{6} p_2 <r^4> R_0^{-7}\} e \, \overline{\alpha}_L$$

$$m_{k\nu}^{8}(6) = \frac{1}{\sqrt{2}}\{\frac{5\sqrt{6}}{6} p_2 <r^4> R_0^{-7}\} e \, \overline{\alpha}_L$$

TABLE I. THE SET OF EIGENVECTORS AND EIGENVALUES FOR THE $ReBr_6^{2-}$ COMPLEX ION.

STATE	ENERGY*(cm^{-1})	EIGENVECTORS
$\Gamma_8(^4A_{2g})$	0(0)	$0.9670\|1>+0.2026\|15>+$ $0.1002\|21>+...$
$\Gamma_8(^2T_{1g})$	7,446(7,507.75)	$0.1525\|1>+0.5978\|2>-$ $-0.6752\|6>-0.3773\|15>+...$
$\Gamma_8(^2E_g)$	8,773(8,531.72)	$0.7496\|2>+0.6430\|6>+...$
$\Gamma_6(^2T_{1g})$	9,146(8,927.63)	$0.9977\|2>+...$
$\Gamma_7(^2T_{2g})$	13,150(14,122.93)	$0.1300\|2>+0.9847\|4>+...$
$\Gamma_8(^2T_{2g})$	14,922(15,341.87)	$-0.1699\|1>+0.2560\|2>-$ $-0.3242\|6>+0.8750\|15>+...$

where: Γ_8-block: $|1> = |t_{2g}^3\ ^4A_{2g}>$, $|2> = |t_{2g}^3\ ^2E_g>$,
$|6> = |t_{2g}^3\ ^2T_{1g}>$, $|15> = |t_{2g}^3\ ^2T_{2g}>$,
$|21> = |t_{2g}^2\ eg\ ^4T_{2g}>$

Γ_7-block: $|2> = |t_{2g}^2\ eg\ ^4T_{1g}>$, $|4> = |t_{2g}^3\ ^2T_{2g}>$

Γ_6-block: $|2> = |t_{2g}^3\ ^2T_{1g}>$

and expansion coefficients smaller than 10^{-1} have not been included in this Table.

* As for the energies, the values given in brackets are the calculated ones, using the set of parameters: $B = 327\ cm^{-1}$, $C = 1,818\ cm^{-1}$, $Dq = 3,034,7\ cm^{-1}$ and $\xi_{SO} = 2,392\ cm^{-1}$, respectively.

TABLE II: SIMMETRIZED FORCE CONSTANTS Fij (in mdyne/A°) FOR THE $ReBr_6^{2-}$ COMPLEX ION.

SPECIES		GVFF[a*] (k=1.17)	GVFF[b]	OVFF	UBFF	MOVFF	MUBFF
α_{1g}	F_{11}	2.0756	2.0756	2.0719	2.0800	2.2719	2.3180
εg	F_{22}	1.5412	1.5412	1.5448	1.3000	1.7447	1.7985
τ_{1u}	F_{33}	1.5123	1.1817	1.7228	1.5600	1.5228	1.4715
	$F_{34}=F_{43}$		0.0734	0.1589	0.2200	0.1589	0.1418
	F_{44}	0.1755	0.1590	0.1672	0.1600	0.1621	0.1688
τ_{2g}	F_{55}	0.1556	0.1556	0.1588	0.1400	0.1588	0.1629
τ_{2u}	F_{66}	o.1544	0.1544	0.1505	0.1200	0.1505	0.1571

* The F_{ij}-values for the GVFF[a] (k=2.25) are:
F_{11}= 2.0756 , F_{22}= 1.5412 , F_{33}= 1.3196 , F_{34}= 0.1481 , F_{44}= 0.1590 , F_{55}= 0.1556 and F_{66}= 0.1544, respectovely.

b. See Reference (61)

TABLE III: OBSERVED AND CALCULATED VIBRATIONAL FREQUENCIES (cm^{-1}) FOR THE $ReBr_6^{2-}$ COMPEX ION.

SPECIES	(cm^{-1})[50,51]	GVFF[a*] (k=1.17)	GVFF[b]	OVFF	UBFF	MOVFF	MUBFF
α_{1g}	ν_1=210	210	210	210	211	220	221
εg	ν_2=181	181	181	181	166	193	195
τ_{1u}	ν_3=216	216	217	249	227	232	231
	ν_4=114	114	112	117	113	117	117
τ_{2g}	ν_5=115	115	115	116	110	116	118
τ_{2u}	ν_6= 81	81	81	80	71.4	80	82

* For k=2.25 and a GVFF[a], the calculated vibrational frequencies are: ν_1= 210 , ν_2= 181 , ν_3= 216 , ν_4= 114 , ν_5= 115 and ν_6= 81 (in cm^{-1}) respectively.

TABLE IV: SIMMETRIZED INTERNAL FORCE CONSTANTS (in mdyne/A°) FOR THE $ReBr_6^{2-}$ COMPLEX ION

	GVFF[a] (k=1.17)	GVFF[a] (k=2.25)	GVFF[b]	OVFF	MOVFF	UBFF	MUBFF
f_d	1.6154	1.5195	1.4503	K: 1.4025	1.4025	K: 1.1200	1.4380
f_{dd}	0.1031	0.2004	0.2687	F: 0.1674	0.1674	F: 0.2400	0.1575
$f_{dd'}$	0.0892	0.0890	0.0891	F': -0.0084	-0.0084	F': -0.0200	-0.0157
$f_{d\alpha}$	0.1393	0.0740	0.367	D/2: 0.0710	0.0710	H: 0.0100	0.0762
f_{α}	0.1650	0.1567	0.1566	K: 0.0000	0.2000	h: 0.0000	-0.0100
$f_{\alpha\alpha}$	0.0053	0.0012	-0.0023	—	—	K: 0.0000	0.2500
$f_{\alpha\alpha'}$	0.0047	0.0005	0.0010	—	—	—	—

TABLE V: THE L-MATRICES AND THE CALCULATED POTENTIAL ENERGY DISTRIBUTION FOR THE τ_{1u} AND THE τ_{2u} SYMMETRY SPECIES FORTHE $ReBr_6^{2-}$ COMPLEX ION.

SPECIES		GVFF[a] (k=1.17)	GVFF[a] (k=2.25)	GVFF[b]	OUFF	MOUFF	UBFF	MUBFF
THE L_{ij}-MATRIX ELEMENTS								
τ_{1u}	L_{33}	0.1490	0.1520	0.1510	0.1520	0.1520	0.1510	0.1520
	L_{34}	-0.0310	-0.0010	-0.0190	-0.0010	-0.0010	-0.0170	-0.0030
	L_{43}	-0.0940	-0.1410	-0.1670	-0.1400	-0.1410	-0.1170	-0.1460
	L_{44}	0.2430	0.2190	-0.2010	0.2200	0.2230	0.2330	0.2160
τ_{2u}	L_{66}	0.1582	0.1582	0.1582	0.1582	0.1582	0.1582	0.1582
THE PED VALUES*								
τ_{1u}	$(PED)_{33}$	1.0850	1.0000	0.9080	1.0030	1.0040	1.0560	0.9860
	$(PED)_{34}$	-0.0860	0.0000	0.0910	-0.0040	-0.0050	-0.0570	0.0130
	$(PED)_{43}$	-0.0860	0.0000	0.0910	-0.0040	-0.0050	-0.0570	0.0130
	$(PED)_{44}$	1.0850	1.0000	0.9080	1.0030	1.0040	1.0560	0.9860
τ_{2u}	$(PED)_{66}$	1.0000	1.0000	1.0000	1.0000	1.0000	1.0000	1.0000

* Unnornalized PED-Values.

TABLE VI. THE $\Gamma_8(^2T_{2g}) \rightarrow \Gamma_8(^4A_{2g})$ ELECTRONIC TRANSITION OF $ReBr_6^{2-}$ COMPLEX ION, IN THE OCTAHEDRAL DOUBLE GROUP O*. THE CRYSTAL FIELD CONTRIBUTIONS TO THE TOTAL TRANSITION DIPOLE MOMENT.

A) THE $|^2T_{2g}U'k1\rangle \rightarrow |^4A_{2g}U'\gamma_2 1\rangle$ EXCITATION.

(1) $|^2T_{2g}U'k1\rangle \rightarrow |^4A_2U'k1\rangle$

$$\mu^Z = \left(\frac{2}{\Delta E}\right)\{\beta_{kk}^{1,3}Q_{3a} + \beta_{kk}^{1,4}Q_{4a}\}$$

(2) $|^2T_2U'k1\rangle \rightarrow |^4A_2U'\lambda 1\rangle$

$$\mu^Z = \frac{1}{\sqrt{2}}\left(\frac{2}{\Delta E}\right)\{-\beta_{k\lambda}^{2,3}Q_{3c} - \beta_{k\lambda}^{2,4}Q_{4c} + \beta_{k\lambda}^{2,6}Q_{6c}\}$$

(3) $|^2T_2U'k1\rangle \rightarrow |^4A_2U'\mu 1\rangle$

$$\mu^Z = \left(\frac{2}{\Delta E}\right)\{\beta_{k\mu}^{3,6}Q_{6a}\}$$

(4) $|^2T_2U'k1\rangle \rightarrow |^4A_2U'\nu 1\rangle$

$$\mu^Z = \frac{1}{\sqrt{2}}\left(\frac{2}{\Delta E}\right)\{\beta_{k\nu}^{4,3}Q_{3c} + \beta_{k\nu}^{4,4}Q_{4c} - \beta_{k\nu}^{4,6}Q_{6c}\}$$

B) THE $|^2T_{2g}U'k1\rangle \rightarrow |^4A_{2g}U'\gamma_2 2\rangle$ EXCITATION

(5) $|^2T_2U'k1\rangle \rightarrow |^4A_2U'k2\rangle$

$$\mu^Z = 0$$

(6) $|^2T_2U'k1\rangle \rightarrow |^4A_2U'\lambda 2\rangle$

$$\mu^Z = \frac{1}{\sqrt{2}}\left(\frac{2}{\Delta E}\right)\{-\beta_{k\lambda}^{6,3}Q_{3c} - \beta_{k\lambda}^{6,4}Q_{4c} + \beta_{k\lambda}^{6,6}Q_{6c}\}$$

(7) $|^2T_2U'k1\rangle \rightarrow |^4A_2U'\mu 2\rangle$

$$\mu^Z = 0$$

(8) $|^2T_2U'k1\rangle \rightarrow |^4A_2U'\nu 2\rangle$

$$\mu^Z = \frac{1}{\sqrt{2}}\left(\frac{2}{\Delta E}\right)\{\beta_{k\nu}^{8,3}Q_{3c} + \beta_{k\nu}^{8,4}Q_{4c} - \beta_{k\nu}^{8,6}Q_{6c}\}$$

TABLE VII. THE $\Gamma_8(^2T_{2g}) \rightarrow \Gamma_8(^4A_{2g})$ ELECTRONIC TRANSITION OF THE $ReBr_6^{2-}$ COMPLEX ION, IN THE OCTAHEDRAL DOUBLE GROUP O*. THE LIGAND POLARIZATION CONTRIBUTIONS TO THE TOTAL TRANSITION DIPOLE MOMENT.

A) THE $|^2T_{2g}U'\kappa 1\rangle \rightarrow |^4A_{2g}U'\gamma_2 1\rangle$ EXCITATION.

(1) $|^2T_2U'\kappa 1\rangle \rightarrow |^4A_2U'\kappa 1\rangle$

$$\mu^Z = \{n_{\kappa\kappa}^{1,3}Q_{3a} + n_{\kappa\kappa}^{1,4}Q_{4a}\}$$

(2) $|^2T_2U'\kappa 1\rangle \rightarrow |^4A_2U'\lambda 1\rangle$

$$\mu^Z = \{n_{\kappa\lambda}^{2,3}Q_{3c} + n_{\kappa\lambda}^{2,4}Q_{4c} - n_{\kappa\lambda}^{2,6}Q_{6c}\}$$

(3) $|^2T_2U'\kappa 1\rangle \rightarrow |^4A_2U'\mu 1\rangle$

$$\mu^Z = \{n_{\kappa\mu}^{3,6}Q_{6a}\}$$

(4) $|^2T_2U'\kappa 1\rangle \rightarrow |^4A_2U'\nu 1\rangle$

$$\mu^Z = \{-n_{\kappa\nu}^{4,3}Q_{3c} - n_{\kappa\nu}^{4,4}Q_{4c} + n_{\kappa\nu}^{4,6}Q_{6c}\}$$

B) THE $|^2T_2U'\kappa 1\rangle \rightarrow |^4A_2U'\kappa 2\rangle$

(5) $|^2T_2U'\kappa 1\rangle \rightarrow |^4A_2U'\kappa 2\rangle$

$$\mu^Z = 0$$

(6) $|^2T_2U'\kappa 1\rangle \rightarrow |^4A_2U'\lambda 2\rangle$

$$\mu^Z = 0$$

(7) $|^2T_2U'\kappa 1\rangle \rightarrow |^4A_2U'\mu 2\rangle$

$$\mu^Z = 0$$

(8) $|^2T_2U'\kappa 1\rangle \rightarrow |^4A_2U'\nu 2\rangle$

$$\mu^Z = \{-n_{\kappa\nu}^{8,3}Q_{3c} - n_{\kappa\nu}^{8,4}Q_{4c} + n_{\kappa\nu}^{8,6}Q_{6c}\}$$

TABLE VIII. THE $\Gamma_8(^2T_{2g}) \rightarrow \Gamma_8(^4A_{2g})$ ELECTRONIC TRANSITION. THE $ReBr_6^{2-}$ COMPLEX ION, IN THE OCTAHEDRAL DOUBLE GROUP O*. THE DIPOLE STRENGTHS FOR THE THREE VIBRONIC ORIGINS, ν_3, ν_4 AND ν_6, RESPECTIVELY.

I.- THE ν_3-VIBRONIC ORIGIN.

$$D^{CF}(\nu_3) = 3\left(\frac{2}{\Delta E}\right)^2\{|\beta_{kk}^{1,3}|^2 + \frac{1}{2}|\beta_{k\lambda}^{2,3}|^2 + \frac{1}{2}|\beta_{k\nu}^{4,3}|^2 + \frac{1}{2}|\beta_{k\lambda}^{6,3}|^2 + \frac{1}{2}|\beta_{k\nu}^{8,3}|^2\} \, |\langle 0|Q_3|1\rangle|^2$$

$$D^{LP}(\nu_3) = 3\{|n_{kk}^{1,3}|^2 + |n_{k\lambda}^{2,3}|^2 + |n_{k\nu}^{4,3}|^2 + |n_{k\nu}^{8,3}|^2\} \, |\langle 0|Q_3|1\rangle|^2$$

$$D^{CF,LP}(\nu_3) = 3\left(\frac{2}{\Delta E}\right)\{2|\beta_{kk}^{1,3}||n_{kk}^{1,3}| - \sqrt{2}|\beta_{k\lambda}^{2,3}||n_{k\lambda}^{2,3}| - \sqrt{2}|\beta_{k\nu}^{4,3}||n_{k\nu}^{4,3}| - \sqrt{2}|\beta_{k\nu}^{8,3}||n_{k\nu}^{8,3}|\} \times |\langle 0|Q_3|1\rangle|^2$$

II.- THE ν_4-VIBRONIC ORIGIN.

$$D^{CF}(\nu_4) = 3\left(\frac{2}{\Delta E}\right)^2\{|\beta_{kk}^{1,4}|^2 + \frac{1}{2}|\beta_{k\lambda}^{2,4}|^2 + \frac{1}{2}|\beta_{k\nu}^{4,4}|^2 + \frac{1}{2}|\beta_{k\lambda}^{6,4}|^2 + \frac{1}{2}|\beta_{k\nu}^{8,4}|^2\} \, |\langle 0|Q_4|1\rangle|^2$$

$$D^{LP}(\nu_4) = 3\{|n_{kk}^{1,4}|^2 + |n_{k\lambda}^{2,4}|^2 + |n_{k\nu}^{4,4}|^2 + |n_{k\nu}^{8,4}|^2\} \, |\langle 0|Q_4|1\rangle|^2$$

$$D^{CF,LP}(\nu_4) = 3\left(\frac{2}{\Delta E}\right)\{2|\beta_{kk}^{1,4}||n_{kk}^{1,4}| - \sqrt{2}|\beta_{k\lambda}^{2,4}||n_{k\lambda}^{2,4}| - \sqrt{2}|\beta_{k\nu}^{4,4}||n_{k\nu}^{4,4}| - \sqrt{2}|\beta_{k\nu}^{8,4}||n_{k\nu}^{8,4}|\} \, |\langle 0|Q_4|1\rangle|^2$$

III.- THE ν_6-VIBRONIC ORIGIN.

$$D^{CF}(\nu_6) = 3\left(\frac{2}{\Delta E}\right)^2\{\frac{1}{2}|\beta_{k\lambda}^{2,6}|^2 + |\beta_{k\mu}^{3,6}|^2 + \frac{1}{2}|\beta_{k\nu}^{4,6}|^2 + \frac{1}{2}|\beta_{k\lambda}^{6,6}|^2 + \frac{1}{2}|\beta_{k\nu}^{8,6}|^2\} \, |\langle 0|Q_6|1\rangle|^2$$

$$D^{LP}(\nu_6) = 3\{|n_{k\lambda}^{2,6}|^2 + |n_{k\mu}^{3,6}|^2 + |n_{k\nu}^{4,6}|^2 + |n_{k\nu}^{8,6}|\} \, |\langle 0|Q_6|1\rangle|^2$$

$$D^{CF,LP}(\nu_6) = 3\left(\frac{2}{\Delta E}\right)\{-\sqrt{2}|\beta_{k\lambda}^{2,6}||n_{k\lambda}^{2,6}| + 2|\beta_{k\mu}^{3,6}||n_{k\mu}^{3,6}| - \sqrt{2}|\beta_{k\nu}^{4,6}|n_{k\nu}^{4,6}| - \sqrt{2}|\beta_{k\nu}^{8,6}||n_{k\nu}^{8,6}|\} \, |\langle 0|Q_6|1\rangle|^2$$

TABLE IX

THE $\Gamma_8(^2T_{2g}) \rightarrow \Gamma_8(^4A_{2g})$ ELECTRONIC TRANSITION OF THE $ReBr_6^{2-}$ COMPLEX ION.

$\delta(Re)=+0.00$ AND $\delta(Br)=-1/3$. (*), (**), (***).

FORCE FIELD	$D_3/10^{-4}$	$D_4/10^{-4}$	$D_6/10^{-4}$	$D(\nu_3):D(\nu_4):D(\nu_6)$	$f/10^{-5}$
GVFF[a] (k=1.17)	0.865509	4.14179	0.495968	1.74509:8.35093:1.0000	0.595593
GVFF(k=2.25)	1.411940	3.106930	0.495968	2.84685:6.26438:1.0000	0.542733
GVFF[b]	1.75434	2.50672	0.495968	3.53721:5.05419:1.0000	0.514830
OVFF	1.21368	3.05493	0.502167	2.41689:6.08350:1.0000	0.516319
MOVFF	1.31457	3.13870	0.502167	2.61779:6.25031:1.0000	0.536303
UBFF	1.06245	3.70664	0.562653	1.88828:6.58780:1.0000	0.577029
MUBFF	1.38123	2.91100	0.489919	2.81931:5.94181:1.0000	0.517550

(*) The expectation values of $\langle r^2 \rangle$, $\langle r^4 \rangle$, and $\langle r^6 \rangle$ have been taken from Reference (65)

(**) We define: $D_i = D^{CF}(\nu_i) + D^{CP}(\nu_i) + D^{CF,CP}(\nu_i)$

(***) The Depole strength values are given in D^2, where $1 D = 1$ Debye$=10^{-18}$ ues.cm. (cg5: units).

TABLE XI. THE $Cr(NH_3)_6^{3+}$ ION IN OCTAHEDRAL DOUBLE GROUP O* THE EIGENVECTORS FOR THE $|\Gamma_8(^4T_{1g})>$, $|\Gamma_8(^4T_{2g})>$ $|\Gamma_8(^4A_{2g})>$ ELECTRONIC STATES.

$$|\Gamma_8(^4A_{2g})> = 0.99985636|t_2^3\ {}^4A_2\Gamma_8 1>+0.01000000|t_2^3\ {}^2T_2\Gamma_8 1>- - 0.007086982|t_2^2e\ {}^2T_2\Gamma_8 1>+0.0053932252|t_2^2e\ {}^4T_2\Gamma_8 1>+ + 0.01079845|t_2^2e\ {}^4T_2\Gamma_8 2>$$

$$|\Gamma_8(^4T_{2g})> = -0.014737259|t_2^3\ {}^4A_2\Gamma_8 1>+0.01605098|t_2^{3\,2}E\Gamma_8 1>+ + 0.022468004|t_2^2e\,{}^4T_1\Gamma_8 1>-0.011629802|t_2e^{2\,4}T_1\Gamma_8 1>+ + 0.3043454556|t_2^{3\,2}T_2\Gamma_8 1>-0.04565494|t_2^2e\,{}^2T_2\Gamma_8 1>+ + 0.080549648|t_2^2e\,{}^2T_2\Gamma_8 2>+0.043799134|t_2e^{2\,2}T_2\Gamma_8 1>+ + 0.431843801|t_2^2e\,{}^4T_2\Gamma_8 1>+0.842129098|t_2^2e\,{}^4T_2\Gamma_8 2>$$

$$|\Gamma_8(^4T_{1g})> = -0.012878011|t_2^2e\,{}^2E\Gamma_8 2>+0.868375708|t_2^2e\,{}^4T_1\Gamma_8 1>- - 0.219351703|t_2e^{2\,4}T_1\Gamma_8 1>-0.429759137|t_2^2e\,{}^4T_1\Gamma_8 2>+ + 0.108546963|t_2e^{2\,4}T_1\Gamma_8 2>-0.020988640|t_2^{3\,2}T_2\Gamma_8 1>+ + 0.01000000|t_2^2e\,{}^4T_2\Gamma_8 1>-0.022942486|t_2^2e\,{}^4T_2\Gamma_8 2>$$

TABLE XII. REDUCED MATRIX ELEMENTS $\langle\Gamma_1\gamma_1|V^{(1)}|\Gamma_2\gamma_2\rangle$ FOR THE $\Gamma_8({}^4T_{1g}) \rightarrow \Gamma_8({}^4A_{2g})$ NON RADIATIVE TRANSITION.

$\langle\Gamma_8 k|V^{(1)}|\Gamma_8 k\rangle = 0.0045\ g_5\ X_0^{T_1}(5)$

$\langle\Gamma_8 k|V^{(1)}|\Gamma_8\lambda\rangle = 0.0039\ g_5\ X_{+1}^{T_1} - 0.2265|g_{10}X_{-1}^{T_2}(10) + g_{11}X_{-1}^{T_2}(11)|$

$\langle\Gamma_8 k|V^{(1)}|\Gamma_8\mu\rangle = -0.4524|g_{10}X_0^{T_2}(10)+g_{11}X_0^{T_2}(11)|$

$\langle\Gamma_8 k|V^{(1)}|\Gamma_8\nu\rangle = 0.0131\ g_5 X_{-1}^{T_1}(5)-0.2386|g_{10}X_{+1}^{T_2} + g_{11}X_{-1}^{T_2}(11)|$

$\langle\Gamma_8\lambda|V^{(1)}|\Gamma_8 k\rangle = 0.0039\ g_5 X_{-1}^{T_1}(5) - 0.5020|g_{10}X_{+1}^{T_2}(10) + g_{11}X_{+1}^{T_2}(11)|$

$\langle\Gamma_8\lambda|V^{(1)}|\Gamma_8\lambda\rangle = -0.0109\ g_5 X_0^{T_1}(5)$

$\langle\Gamma_8\lambda|V^{(1)}|\Gamma_8\mu\rangle = -0.0086\ g_5 X_{+1}^{T_1}(5) - 0.2386|g_{10}X_{-1}^{T_2}(10) + g_{11}X_{-1}^{T_2}(11)|$

$\langle\Gamma_8\lambda|V^{(1)}|\Gamma_8\nu\rangle = -0.0627|g_{10}X_0^{T_2}(10) + g_{11}X_0^{T_2}(11)|$

$\langle\Gamma_8\mu|V^{(1)}|\Gamma_8 k\rangle = 0.0627|g_{10}X_0^{T_2}(10) + g_{11}X_0^{T_2}(11)|$

$\langle\Gamma_8\mu|V^{(1)}|\Gamma_8\lambda\rangle = 0.0086\ g_5 X_{-1}^{T_1}(5) + 0.2386|g_{10}X_{+1}^{T_2}(10) + g_{11}X_{+1}^{T_2}(11)|$

$\langle\Gamma_8\mu|V^{(1)}|\Gamma_8\mu\rangle = 0.109\ g_5 X_0^{T_1}(5)$

$\langle\Gamma_8\mu|V^{(1)}|\Gamma_8\nu\rangle = 0.0039\ g_5 X_{+1}^{T_1}(5) + 0.5020|g_{10}X_{-1}^{T_2}(10) + g_{11}X_{-1}^{T_2}(11)|$

$\langle\Gamma_8\nu|V^{(1)}|\Gamma_8 k\rangle = -0.0131\ g_5 X_{+1}^{T_1}(5) + 0.2386|g_{10}X_{-1}^{T_2}(10) + g_{11}X_{-1}^{T_2}(11)|$

$\langle\Gamma_8\nu|V^{(1)}|\Gamma_8\lambda\rangle = 0.4524|g_{10}X_0^{T_2}(10) + g_{11}X_0^{T_2}(11)|$

$\langle\Gamma_8\nu|V^{(1)}|\Gamma_8\mu\rangle = -0.0039\ g_5 X_{-1}^{T_1}(5) + 0.2265|g_{10}X_{+1}^{T_2}(10) + g_{11}X_{+1}^{T_2}(11)|$

$\langle\Gamma_8\nu|V^{(1)}|\Gamma_8\nu\rangle = -0.0045\ g_5 X_0^{T_1}(5)$

where: $g_5 = \langle t_2||O^{T_1}(5)||e\rangle$, $g_{10} = \langle t_2||O^{T_2}(10)||e\rangle$, $g_{11} = \langle t_2||O^{T_2}(11)||e\rangle$ and the irreducible tensor operators $X_\gamma^\Gamma(k)$ are defined in terms of the real components by means of the identities.
$X_{\pm1}^\Gamma(k) = \mp 1/\sqrt{2}\ \{|\Gamma_X\rangle \pm i|\Gamma_Y\rangle\}_k$ and $X_0^\Gamma = |\Gamma_Z\rangle_k$
where $k = 5,10,11$ and $\Gamma = \tau_{1g}$ and τ_{2g}, respectively.

ELECTRONIC AND VIBRONIC STRUCTURES OF SINGLE-CRYSTAL $[Ru(bpy)_3](PF_6)_2$ AND DOPED $[Os(bpy)_3]^{2+}$

H. YERSIN, D. BRAUN, G. HENSLER[a], E. GALLHUBER[b]
Institut für Physikalische und Theoretische Chemie
Universität Regensburg D-8400 Regensburg, Fed. Rep. Ger.

ABSTRACT: Investigations of the title compounds with high resolution spectroscopy (e.g. polarized emission, polarized absorption, excitation, site-selective emission, ODMR) yields lines with halfwidths in the order of wavenumbers or less. Pronounced differences in the physical properties of the various excited and emitting states are observed. It is possible to present a more comprehensive description of these compounds. An interesting feature is that the vibronic coupling properties of the lowest excited states are found to be tunable by magnetic fields.

1. Introduction

$[Ru(bpy)_3]^{2+}$ (with bpy = 2,2'-bipyridine) has been studied most extensively during the last fifteen years in many research laboratories. (See reviews [1-10], Fig. 1 shows the structure of the complex.) The stimulation for this engagement results from a unique combination of properties. Chemically, the complex is very stable in its ground state and rather inert in its electronically (lowest) excited states. It absorbs visible light and emits bright red-orange light with a relatively high quantum efficiency even in solution at room temperature. The emitting excited states are long-lived also in fluid solutions. Thus, they can take part in bimolecular radiationless energy or electron transfer processes even under condition of low concentrations of the components.

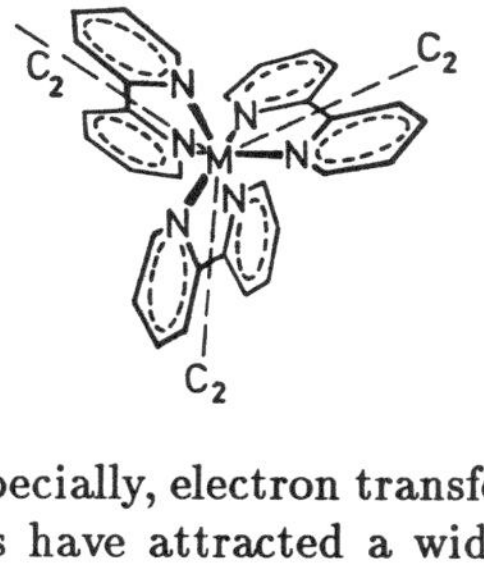

Figure 1. Schematic structure of $[Ru(bpy)_3]^{2+}$ and $[Os(bpy)_3]^{2+}$, reproduced for the D_3 point group symmetry. The $\vec{C}_3$ axis lies perpendicular to the drawing plane.

Especially, electron transfer processes involving the complex in the ground and/or excited states have attracted a widespread attention. Due to the fact that the excited complex represents a good electron donor and at the same time a good electron acceptor the molecule has been used to study a series of photocatalytic processes, of which the most famous example is the decomposition of water into hydrogen and oxygen by solar energy. But also

C. D. Flint (ed.), Vibronic Processes in Inorganic Chemistry, 195–219.
© 1989 by Kluwer Academic Publishers.

several other interesting chemical processes or physical effects have been observed to occur under participation of the excited $[Ru(bpy)_3]^{2+}$ complex. It is used as a reactant for a variety of chemiluminescent processes. Further, light absorbed may be converted into electric energy at suitable electrodes (photogalvanic effect) or vice versa electric energy can be transformed into light (electro-luminescence). Moreover, it is possible to synthesize a large amount of other *Ru(II)* complexes with bidentate polypyridine-type ligands whereby three or two of the ligands can be the same or the ligands can all be different. Thus, in principle more than 10^6 different *Ru(II)*-polypyridine complexes may be constructed. Since the individual ligands strongly influence the electronic structures of the complexes the availability of such a large number of different compounds results in the possibility of *tuning* the electronic properties *chemically*. This idea has also stimulated several research groups, who successfully applied it in physical and chemical investigations.[1,2]

The photochemical and many photophysical properties of these complexes are determined by the properties of the lowest excited states. But although these have been investigated by a large variety of experimental techniques (mainly applied to dissolved complexes) and several theoretical calculations the literature discussions are still very controversial and sometimes became heated. However, one well-established experimental method, the spectroscopy with single crystals, has only very recently been applied to these compounds, despite the fact that especially measurements of single crystals can supply more defined data than measurements of solutions.

Thus, one subject of this contribution is to demonstrate, how powerful the spectroscopy of single crystals can be, when applied to $[Ru(bpy)_3]^{2+}$ and the related $[Os(bpy)_3]^{2+}$ compounds. It will be outlined that by use of these techniques it is even possible to resolve sharp zero-phonon lines and detailed vibronic finestructures (with half-widths in the order of cm^{-1}). This should be compared to the several hundredths of cm^{-1} broad bands observed in solution. It will further be shown that the different electronically excited (and emitting) states exhibit very different behavior (not only with respect to their vibronic coupling properties) and that through application of *physical tuning methods* (like application of pressure, magnetic field, low temperature) more specific information is available. Thus, a deeper understanding of the photophysical properties of these compounds will be deduced.

2. Spectroscopic background

The optical absorption and emission spectra of $[Ru(bpy)_3]^{2+}$ and $[Os(bpy)_3]^{2+}$ (Fig. 2) are easily classified on the basis of transitions between one-electron MOs. The following

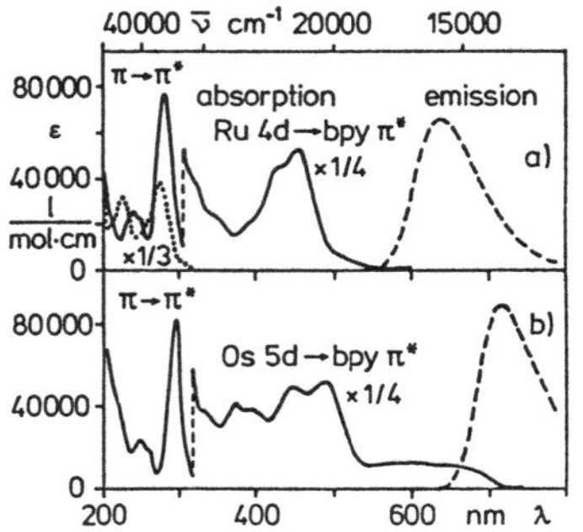

Figure 2. Absorption and emission spectra of $[Ru(bpy)_3]^{2+}$ (a) and $[Os(bpy)_3]^{2+}$ (b) in aqueous solution at $T \approx 300$ K. The dotted spectrum in (a) represents the absorption of trans-*bpy* (in H_2O). The spectra have to be multiplied with the given factors to be in scale.

assignments are accepted throughout the literature. The very intense absorption between about 200 and 300 nm is assigned to be mainly centered on the *bpy*–ligands and thus to be dominantly of $\pi - \pi^*$ origin, which is near at hand when comparing this part of the spectrum with that of the free ligand. The transitions occurring between about 350 and 500 nm are neither present in the free ligand nor in *Ru/Os*–complexes with ligands which have no π^* orbitals (like NH_3). Therefore, these transitions are specified as allowed charge transfer (CT) transitions. Arguments [1,3] on the basis of a correlation of redox potentials and absorption or emission energies allow to ascribe the corresponding absorption bands to metal-to-ligand CT transitions (MLCT) of $Ru4d \rightarrow bpy\pi^*$ or $Os5d \rightarrow bpy\pi^*$ character. The weak bands between about 500 and 600 nm of $[Ru(bpy)_3]^{2+}$ and about 530 nm and 730 nm of $[Os(bpy)_3]^{2+}$, respectively, are assigned to triplet components of MLCT transitions due to their relatively low molar extinction coefficients and due to an overlapping with the long-lived emission. (The ground state is a singlet.)

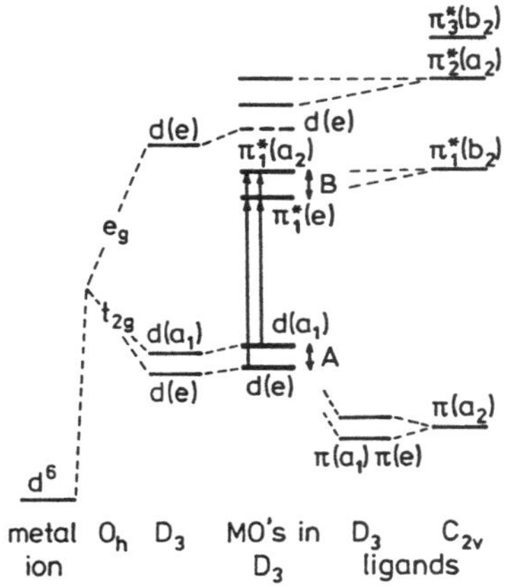

Figure 3. Simplified MO-diagram for $[Ru(bpy)_3]^{2+}$ and $[Os(bpy)_3]^{2+}$. A and B are the trigonal splitting parameters. They have been estimated in [12] and [13], respectively (however, with different signs for B). The four transitions between the one-electron orbitals represent the basis for the many-electron states given in Tab. I.

With the described characterization of involved transitions and under assumption of a D_3 symmetry of the complex a one-electron MO scheme is readily developed (Fig. 3). (For more detailed information see for example [11-13].) The diagram shows the splitting of the metal *d*-orbitals by symmetry reduction from O_h to D_3. Thus, the highest occupied MOs of the complex are of $d(a_1)$ and $d(e)$ character with some π admixture. The lowest unoccupied MOs result from the π_1^*-orbitals of the three ligands, which in the complex split into $\pi_1^*(a_2)$ and $\pi_1^*(e)$. Especially, these four MOs are considered to be relevant with respect to the number and symmetries of the lower excited electronic states. Higher lying MOs generally are not taken into account, since the π_2^* orbital is expected to lie at least $6 \cdot 10^3$ cm^{-1} to higher energy.[11-13] The $d(e)$-orbital, although probably situated only $2-3 \cdot 10^3$ cm^{-1} above the lowest excited one (in $[Ru(bpy)_3]^{2+}$; in $[Os(bpy)_3]^{2+}$ the $d-d$-splitting is about 30 % larger[3]) is spectroscopically not very important, due to the weakness of the $d-d$-transitions compared to those of superimposing ones with allowed MLCT character. On the other hand the metal centered excited states seem to be responsible for the photodecomposition of the complexes which occurs for $[Ru(bpy)_3]^{2+}$ in aqueous solution with a photochemical quantum yield of the order of 10^{-5}.[6]

A comparison of the optical spectra with the one-electron MO scheme shows immediately that this approach cannot lead to a more detailed understanding of the electronic structures. Especially, the approach does not give a satisfying description of the lowest excited states of triplet characters. But one is predominantly interested in these triplets, e.g. for the description of electron transfer processes. A better understanding may result from a quantum mechanical model which takes into account the electron-electron interaction and the spin-orbit coupling leading to a description in the scope of many-electron states.

3. Group theory and many-electron states.

Several quantum mechanical model calculations are found in the literature.[12-15] But no real agreement with experiment has been obtained, neither about the symmetry labelling nor on the number and energies of the lowest excited states. However, the results based on symmetry arguments are valid and very useful if the symmetry of the complex is known. Therefore, group-theoretical considerations allow to structurize the large number of expected many-electron states. Further, selection rules can immediately be written down. A combination of these outcomes with experimental results from single-crystal spectroscopy may indeed deliver a reasonable model for the lowest excited states.

The ground state of the complexes (in D_3) is determined by filling up the d-orbitals with six electrons, giving the ground state configuration $[d(e)]^4[d(a_1)]^2$. Due to the filled shell case the ground state is of ${}^1A_1$ representation. The lowest excited states result from promoting one electron out of the ground state configuration to one of the lowest unoccupied π^* orbitals (Fig. 3). Thus, we have to take into account four types of transitions or four excited state configurations (Tab. I), of which those transitions involving the $d(a_1)$ MOs do not carry intensity in a first order approximation[12].

Table I. Group-theoretical derivation of excited state representations resulting from $d \rightarrow \pi_1^*$ excitations (in D_3 symmetry, s.o.c. = spin-orbit coupling).

Configuration	States without s.o.c.	States$^{a)}$ with s.o.c.
$[d(e)]^4[d(a_1)]^2$	${}^1A_1$	A_1 (ground state)
$d(e) \rightarrow \pi^*(a_2)^{b)}$	${}^1E$	$1E^{c)}$
	${}^3E$	$2E, 3E, 1A_1, 1A_2$
$d(a_1) \rightarrow \pi^*(a_2)$	${}^1A_2$	$2A_2$
	${}^3A_2$	$2A_1, 4E$
$d(e) \rightarrow \pi^*(e)$	${}^1E$	$5E$
	${}^1A_1$	$3A_1$
	${}^1A_2$	$3A_2$
	${}^3E$	$6E, 7E, 4A_1, 4A_2$
	${}^3A_1$	$8E, 5A_2$
	${}^3A_2$	$9E, 5A_1$
$d(a_1) \rightarrow \pi^*(e)$	${}^1E$	$10E$
	${}^3E$	$11E, 12E, 6A_1, 6A_2$

a) We use unprimed symbols to distinguish these group-theoretically determined states from the primed experimentally determined ones (4.). b) Abbreviation for $[d(e)]^3[d(a)_1]^2[\pi_1^*(a_2)]^1$. c) The numbering of states of equal representation is chosen to correspond to ref. [13].

The representations of the resulting many-electron states are determined group-theoretically by multiplying the representations of that MO from which the electron is taken off with that to which it is promoted. (For the simple D_3 point group the Pauli principle is fulfilled when applying this procedure.) Thus, as an example, one finds – using Tab. II – for the second configuration in Tab. I an excited state of E orbital (spatial) character.

Through electron-electron interaction the spins in the considered orbitals couple and one obtains triplets (which are stabilized) and singlets (which are labilized). Further, the states of same symmetry and multiplicity can mix by configuration interaction[16] and may be shifted energetically. Tab. I summarizes in the second column the irreducible representations of the resultant triplet and singlet terms.

For $[Ru(bpy)_3]^{2+}$ and $[Os(bpy)_3]^{2+}$ the spin-orbit coupling (s.o.c.) is large ($\zeta[Ru] \approx 1200$ cm^{-1}, $\zeta[Os] \approx 3000$ cm^{-1}[13]). Therefore, in most cases the states can no longer be classified according to their spin quantum numbers or their spatial representations. The relevant symmetry labels result from products of spin and spatial symmetries. In general, this product gives for the relevant double group reducible representations, that decompose into irreducible ones. Hence, a parent state possibly splits through spin-orbit coupling. Two examples may illustrate how to proceed: In D_3 a spin singlet with $S = 0$ transforms as A_1.[17] Thus, the total symmetries of e.g. $^1A_2$ and ^1E are $A_1 \times A_2 = A_2$ and $A_1 \times E = E$, respectively (use Tab. II). A spin triplet with $S = 1$ transforms as A_2 (for $m_S = 0$) and as E (for $m_S = \pm 1$). Therefore, the total symmetry labels of e.g. $^3A_2$ and ^3E are $(A_2 + E) \times A_2 = A_1 + E$ and $(A_2 + E) \times E = E + A_1 + A_2 + E$, respectively.

With this procedure one finds 24 double group states resulting from the four configurations. Without spin-orbit coupling, however, all subterms to a given triplet parent term will still be degenerate. Under the action of spin-orbit coupling states of the same *total* symmetry may mix. (The spin-orbit coupling operator transforms as A_1.) Therefore, new wavefunctions (states) will result from a mixing of singlet and triplet parents. In general the triplet parents will split and all states will shift to new energies. These double group states (summarized in Tab. I), which contain the spin and spatial (orbital) symmetry components, are considered to be an adequate starting point for a description of the spectroscopic properties of the title compounds.

It is interesting to note that a computational estimate of the energies of these 24 states and their mixing coefficients[13] indicate that the very lowest excited state contains no admixture of singlet character. Consequently, in the limit of this approach[13] this state can still be regarded as triplet.

Table II: Multiplication table for the D_3 point group

D_3	A_1	A_2	E
A_1	A_1	A_2	E
A_2		A_1	E
E			$A_1 + A_2 + E$

Finally, the selection rules for electric dipole transitions are given for the D_3 point group. Due to the transformation properties of z (lying parallel to $\vec{C}_3$) as A_2 and of x, y (perpendicular to $\vec{C}_3$) as E, respectively, one finds (using Tab. II) that the transitions between the ground state A_1 and the excited states are polarized as follows ($\vec{E}$ = electric field vector): $A_1 \leftrightarrow E : \vec{E} \perp \vec{C}_3$; $A_1 \leftrightarrow A_2 : \vec{E} \parallel \vec{C}_3$; $A_1 \leftrightarrow A_1$: forbidden.

4. Spectroscopy with single-crystal $[Ru(bpy)_3](PF_6)_2$

In many cases the spectroscopy of neat single crystals is very advantageous compared to that of dissolved chromophores. This is particularly valid for $[Ru(bpy)_3]^{2+}$, since up to now no suitable matrices, neither crystalline nor glassy ones, have been found, which allow to resolve sharp zero-phonon or vibronic components as could be accomplished in neat single crystals with appropriate counter ions.[18-22] The drastic improvement of resolution with these systems results mainly from reduced inhomogeneities. Moreover, in crystals with appropriate structures the complexes may be arranged in such a way that by recording polarized spectra one can directly apply group-theoretical selection rules to assign the involved states. This method often supplies enormous advantages compared to photoselection measurements which only very indirectly lead to the required assignments and thus are more uncertain.[23,24]

However, a non-critical use of single-crystal spectra may also lead to misinterpretations, which might occur, when intermolecular interactions lead to distinct excitonic energy bands and/or to splittings of these (e.g. Davydov effect). Especially, emission data can be misunderstood if impurities are effectively excited by energy transfer and subsequently mask the emission of the investigated material. (Very instructive examples represent single crystals of tetracyano-palladates and -platinates: Doping of more than 10 % $[Pd(CN)_4]^{2-}$ complexes into $M_2[Pt(CN)_4]$ salts has no drastic influence while less than 10^{-4} % of $[Pt(CN)_4]^{2-}$ in $M_2[Pd(CN)_4]$ totally quenches the host emission. These effects result from the energy order of the lowest excited states of the dopant relative to those of the pure material.[25]) Therefore, the situation concerning neat single-crystal $[Ru(bpy)_3]X_2$ was very carefully checked and it could be demonstrated that the emission properties of dissolved and thus isolated complexes – with respect to intermolecular interactions – are displayed in neat crystalline compounds. This behavior seems to result from the fact that the exciton band widths corresponding to the lowest excited states of $[Ru(bpy)_3]^{2+}$ are smaller than the inhomogeneous broadenings. (See [22], 4.1.1., and 5.1.)

4.1. Lowest excited states

Crosby and coworkers [26-30] have intensively investigated the lowest excited states of dissolved $[Ru(bpy)_3]^{2+}$ in glassy solutions and other matrices. Under the assumption of a thermal equilibration between these states they could rationalize the results of their quantum efficiency and emission lifetime measurements on the basis of an energy level diagram of three close-lying states (of A_1, E, and A_2 representation). On the other hand, Ferguson and Krausz [31-33] proposed a model, which is largely in conflict with that of Crosby, especially with regard to the group-theoretical classification of the three lowest excited states (assigned to three E-terms [31]) and to the assumption of a non-thermal equilibration [31,33] within these. Both groups discussed their results on the basis of a D_3 site symmetry of the dissolved chromophores. Independently, other research groups came to the conclusion that *the* excited *state* (without specifying which state) has to be described in a C_2 point group due to a localization of the excited electron on a single *bpy*-ligand.[34-38] In this situation it is near at hand to focus the investigations on $[Ru(bpy)_3](PF_6)_2$ single crystals in which the site symmetry of the complexes is known to be D_3.[39] In this compound the molecular $\vec{C}_3$ axes are parallel to the crystallographic $\vec{c}$-axis. Thus, the compound is well suited for the application of polarized spectroscopy, since the group-theoretical selection rules can *directly* be used.

4.1.1. Temperature dependence of the broad-band emission

Fig. 4 shows the polarized emission of single-crystal $[Ru(bpy)_3](PF_6)_2$ at various temperatures.[40,41] Temperature increase produces considerable changes in the differently polarized emission spectra with respect to the energy-positions of the resolved bands, the intensities of the transitions, their shapes and half-widths. Starting at $T = 1.6$ K these changes are due to the successive appearance of new transitions by thermal activation of higher-lying excited states through temperature increase. These transitions grow in according to a Boltzmann distribution and due to the fact that the pre-exponential factors (= ratios of radiative rates) exceed the exponential Boltzmann factors. The application of a Boltzmann distribution for the determination of the activation energies requires the occurrence of a fast thermal equilibration between the states.[44] Wherein 'fast' means, that the equilibration should occur in a much shorter time than the corresponding emission lifetimes. Since these are relatively long in the whole investigated temperture range $\{\tau(T = 1.6\text{ K}) \approx 2 \cdot 10^2 \mu s$ [27]; $\tau(T = 300\text{ K}) \approx 3\mu s$ [45]$\}$ this condition seems to be fulfilled (but should be checked independently, see below).

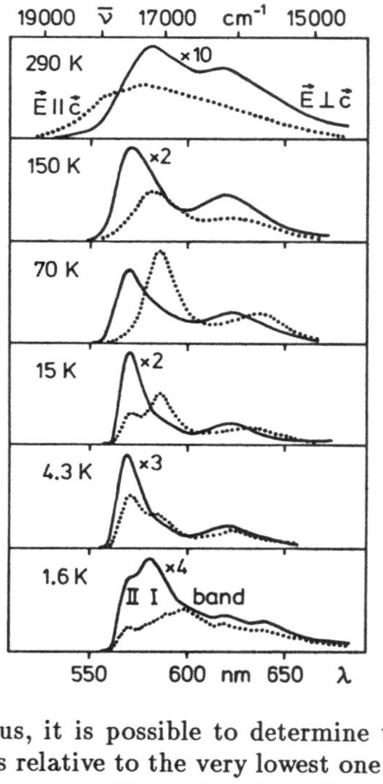

Figure 4. Polarized emission spectra of single-crystal $[Ru(bpy)_3](PF_6)_2$ at different temperatures.[41,40] The electric field vector $\vec{E}$ is either parallel or perpendicular to the $\vec{c}$-axis. The intensities of the differently polarized spectra are comparable after multiplication with the given factors. For different temperatures the intensities are not comparable. The luminescence measurements were carried out with a spectrophotometer, especially constructed for measurements of the polarized emission of microcrystals.[42,43] The laser power of the excitation ($\lambda_{exc} = 364$ nm) was held below 1 mW to avoid sample heating. The size of the laser spot and the registered luminescing area were adjusted to be smaller than the size of the crystal face under observation. (For more details see [22].)

Thus, it is possible to determine the activation energies of the excited (and emitting) states relative to the very lowest one just from the temperature development of the broad-band spectra. From these investigations an energy level scheme consisting of five different excited states can be deduced. The increase of the pre-exponential factors (see Tab. IV) from lower to higher excited states is explained on the basis of an increasing amount of singlet character mixed in by spin-orbit coupling, which is expected from the discussions presented in section 3.. Nevertheless, these five states should still be considered to be mainly of triplet character since the emission lifetime is relatively long up to 300 K.

The group-theoretical assignments of these states are also possible from the broad-band spectra due to the large ratios of the differently polarized intensities. Applying the selection rules for D_3 one can directly determine the total symmetries (double group representations) of the involved states. The result is the energy level diagram shown in Fig. 5.[41]

The emission behavior below $T = 10$ K is determined by the properties of the two lowest excited states. Fig. 4 shows that at $T = 1.6$ K the emission maximum ($\vec{E} \perp \vec{c}$) lies at ≈ 582 nm (band I). With temperature increase band II grows in at ≈ 569 nm and is dominant at 4.3 K. The growing in of band II is accompanied by an intensity increase by a factor of about four.[41] This behavior is explained by the relatively large difference in the radiative rates k_{II_r} [for the second excited state $| II > (2E')$] compared to k_{I_r} [for the lowest excited state $| I > (1E')$], with $k_{II_r}/k_{I_r} >> 1$. The blue-shift of about 400 cm^{-1} results from different vibronic patterns coupled to the corresponding transitions giving displacements of the centers of the unresolved vibronic distributions.[18,41] (See also 4.1.2..) An approximate decomposition of the emission spectra into two separate components resulting from the two states is possible, and an Arrhenius plot of the corresponding emission intensities allows to determine the activation energy from state $| I >$ to state $| II >$ to $\Delta E = (7.5 \pm 1)$ cm^{-1} and k_{II_r}/k_{I_r} to $\approx 10^2$.[46]

A comparison of these single-crystal data to emission data obtained from diluted complexes in a $[Zn(bpy)_3]SO_4$ matrix[47] shows that the described temperature dependence of the spectra is identical and that the quantitative results are comparable. Therefore, one can conclude that even in the neat material the emission properties can be described on the basis of excited complexes which are not subject to significant cooperative solid-state interactions (see also [22]). This conclusion is in contrast to assumptions presented in [48].

The emission of the third excited state ($1A_2'$ see Fig. 5) is appreciably red-shifted compared to the emission from the lower lying states and its maximum occurs with a slight delay of about one μs.[18] This delay presumably is connected to energy transfer processes involving the A_2' state, possibly making the group-theoretical assignment somewhat less confident. But the existence of this state is not in doubt due to the observation of the corresponding zero-phonon line in the excitation spectra (see Tab. IV).

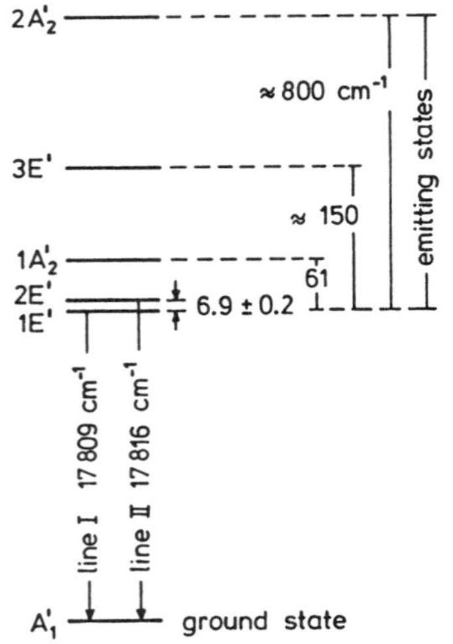

Figure 5. Experimentally determined energy level diagram for the emitting states of single-crystal $[Ru(bpy)_3](PF_6)_2$.[41] The group-theoretical classification is given for the D_3 point group. We use here a notation of primed representations to distinguish the numbering of the experimentally determined states from that given in Tab. I. The very accurate energy-positions of the three lowest excited states are determined from zero-phonon lines [18,20], see also 4.1.2..

It is interesting to compare the blue emission flank of $[Ru(bpy)_3](PF_6)_2$ at low temperature with the corresponding flanks of salts with other counter ions. In most cases one finds a much flatter slope than for the PF_6^--salt. This indicates that the $[Ru(bpy)_3]^{2+}$ chromophores feel different electron-phonon coupling strengths and/or different inhomogeneous broadening effects in the various salts.[49] Motivated by this finding we investigated a large number of $[Ru(bpy)_3]X_2$ salts hoping to be able to "tune" the broadening effects to become still smaller through variation of X^-. Indeed, the blue flank of the emission of the ClO_4^--salt exhibits a slightly steeper rise than that of the PF_6^--salt and, what is important, additionally shows a fine structure at the high-energy side (zero-phonon lines, see below).[19,49,50] Especially this result stimulated us to grow single crystals of very good quality of the PF_6^--salt and to re-investigate these crystals more carefully with high-resolution spectroscopy, and in fact a fine structure can also be observed.

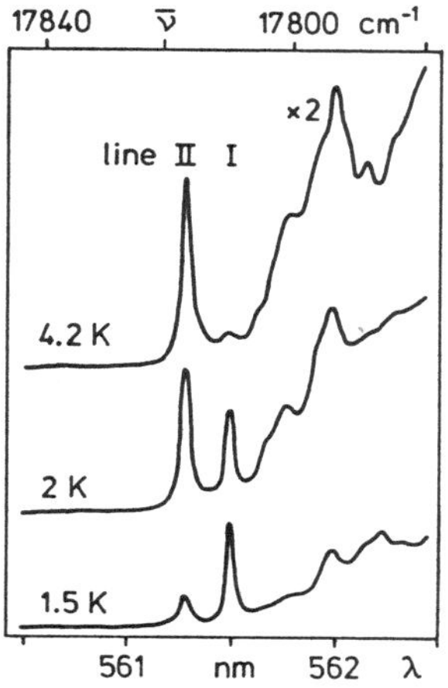

Figure 6. High-energy part of the highly resolved $\vec{E} \perp \vec{c}$-polarized emission of single-crystal $[Ru(bpy)_3](PF_6)_2$.[46,18,20,22] The intensities are comparable after multiplication with the given factor. The maximum of the broad-band emission is by a factor of about twenty higher than the peaks of the fine structure. For $T \gtrsim 10$ K the fine structure is smeared out.

4.1.2. Electronic zero-phonon lines and vibronic fine structure

Below $T = 10$ K the blue side of the emission spectrum of a high-quality $[Ru(bpy)_3](PF_6)_2$ single crystal exhibits a fine structure which is dominated by a pair of narrow lines, denoted as I (17809 cm^{-1}) and as II (17816 cm^{-1}) in Fig. 6. Their energy separation is (6.9 ± 0.2) cm^{-1} (halfwidth ≈ 2 cm^{-1}) and they are $\vec{E} \perp \vec{c}$-polarized. (The intensity in $\vec{E} \parallel \vec{c}$ is at least by a factor of 20 weaker.) Further, the intensities of the two lines change drastically with temperature increase, leading to a growing in of line II on the expense of line I.

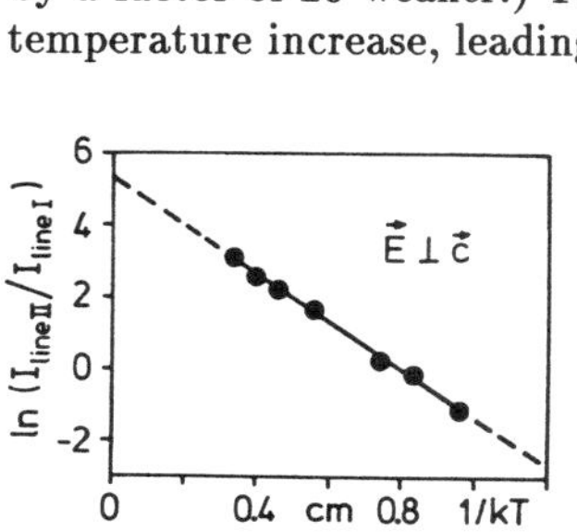

Figure 7. Arrhenius plot of the intensity ratio of the two zero-phonon lines for $[Ru(bpy)_3](PF_6)_2$.[46] The ratio of radiative rates k_{II_r}/k_{I_r} is calculated (from the intercept) to 210 ± 20 by assuming equal electron-phonon coupling strengths for the two transitions, which allows to replace the total emission intensities by the zero phonon lines. The slope gives $\Delta E = (6.8 \pm 0.4)$ cm^{-1}

It is near at hand to assign these features of highest energy (in the emission spectrum) to zero-phonon lines.[18,20,22,46] In this case and under assumption of a thermal equilibration an Arrhenius plot of the intensity ratio should deliver an activation energy equal to the observed spectral separation of the two lines. In fact, this result is found as Fig. 7 demonstrates! Moreover, zero-phonon lines should occur at exactly the same energies in absorption and excitation spectra as observed in emission. Fig. 8 shows that this condition is fulfilled for line II, which appears – as expected – only in $\vec{E} \perp \vec{c}$. The situation seems to be somewhat less clear for line I, for which no absorption could be registered. However, this may be explained by a very low molar extinction coefficient of the corresponding transition. From Fig. 8 we find that ϵ (line II) $\approx$ 5 l/mol cm and Fig. 7 gives $k_{II_r}/k_{I_r} = \epsilon_{II}/\epsilon_I \approx 210$. Thus, ϵ (line I) should be about 0.02 l/mol cm. This value is far too small to be detected. On the other hand, line I becomes more allowed under application of high magnetic fields and then can be observed in absorption, too (see 4.1.3. and [46,51-53]). Especially by this result the zero-phonon character of line I is further evidenced and no "marked gap between absorption and emission" occurs in neat $[Ru(bpy)_3](PF_6)_2$, as has been assumed to exist for $[Ru(bpy)_3]^{2+}$ in $[Zn(bpy)_3](PF_6)_2$ and attributed to an excited state relaxation[54].

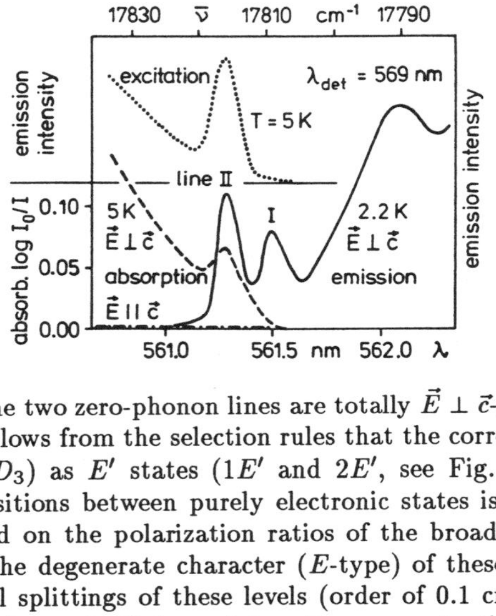

Figure 8. Polarized absorption, polarized emission and excitation of neat single-crystal $[Ru(bpy)_3](PF_6)_2$ for the low-energy zero-phonon transitions. No absorption for $\vec{E} \parallel \vec{c}$ is registered in this energy range, crystal thickness (80 ± 10) μm. (For further arguments concerning the interpretation of the zero-phonon transitions see [22].)

The two zero-phonon lines are totally $\vec{E} \perp \vec{c}$-polarized in absorption and emission. Thus, it follows from the selection rules that the corresponding excited states should be classified (in D_3) as E' states ($1E'$ and $2E'$, see Fig. 5).[18,22] This assignment deduced from transitions between purely electronic states is much more reliable than the classification based on the polarization ratios of the broad emission bands (4.1.1.). Further evidence for the degenerate character (E-type) of these states is deduced from the fact that very small splittings of these levels (order of 0.1 cm^{-1} recorded as ODMR-signals, see 4.1.4.) are observed. It is an interesting property that these two states of same representation do not couple strongly despite the small energy separation. This effect might result if the two states came from different, only very weakly interacting configurations, as for example $[d(e)]^3[d(a_1)]^2[\pi_1^*(a_2)]$ and $[d(e)]^4[d(a_1)]^1[\pi_1^*(e)]$ (see also 4.1.3.).

It is an important result that one finds a close parallelism between the properties of the broad-band spectra and the lines (of zero-phonon character) with respect to their temperature dependence (and magnetic-field behavior, see 4.1.3.)[46] This shows that the fine structure and the broad bands are connected to the same chromophores and the same excited states. Thus, one must reject an assignment to impurities [55] as well as a more recent alternative interpretation[33,56], which contests the single complex origin and relates the sharp structure in the luminescence to the band structure of the neat material.

From the presented data a further interesting conclusion is possible. The result of being able to determine exactly the energy separation between the two lowest excited states from an Arrhenius plot implicitly expresses the occurrence of a fast thermal equilibration [44] (with respect to the predominant part of the integrated intensity). A thermal equilibration between the lowest excited states that has been assumed by Crosby [26-30] and our group [40,41,46] was at first contested by Ferguson and Krausz [31,33,9] but later – after becoming aware of the importance of sample heating effects for $[Ru(bpy)_3]^{2+}$ through the exciting laser power [22] – also accepted by Krausz and Moran [57] on the basis of new measurements.

The simultaneous appearance of sharp lines and broad bands in a spectrum is a common phenomenon in low-temperature solid-state spectroscopy.[58-61] On the one hand, the sharp lines represent purely electronic transitions without involvement of phonons or intra-molecular vibrations (Fig. 6). On the other hand, these transitions can couple to a small number of specific phonons and/or vibrations. Thus sharp vibronic lines may also appear. In contrast the broad bands result from the corresponding multi-phonon transitions (= multi-phonon side bands). These can smear out the whole structure. The amount of this effect is characterized by the electron-phonon coupling strength, commonly described by the Huang-Rhys factor S. It expresses the strength of the interaction of a specific electronic state with the surroundings. S can be estimated from the ratio of the intensity of the zero-phonon line I_{0-0} to the intensities of I_{0-0} plus that of the multi-phonon sideband I_{band}, giving [58,19]:

$$\frac{I_{0-0}}{I_{0-0}+I_{band}} \approx exp\{-S[1+6.6(\frac{T}{\Theta_D})^2]\}$$

The Debye temperature Θ_D for this class of compounds is of the order of 10^2 K.[19,62] Therefore, one obtains for the low-temperature limit the simple expression:

$$\frac{I_{0-0}}{I_{0-0}+I_{band}} \approx exp(-S)$$

Table III: Vibrational satellites to the transition $| I > \rightarrow | 0 >$ (line I) at 17809 cm^{-1} of $[Ru(bpy)_3](PF_6)_2$ observed in emission (T = 1.5 K) compared to IR and Raman data [cm^{-1}]

Energy separation[a)] to line I	IR[b)] vibrations[64]	Resonance Raman vibrations[65]
345	337	337
373		372
442		
478	474	
766	765	765
816		
1028	1027	1025
1173		1173
1273	1271	1278
1317	1313	1317
1441	1446	
1491		1490

a) Experimental error less than ± 2 cm^{-1}; b) Measured with $[Ru(bpy)_3](PF_6)_2$ at T = 300 K

Using this expression we evaluated the Huang-Rhys factor for the lowest excited states of $[Ru(bpy)_3](PF_6)_2$ to about 6.[20] This value means that the electron-phonon coupling is quite strong for the PF_6^--salt. From this result it can be understood, why the emission (or absorption) spectrum of $[Ru(bpy)_3](PF_6)_2$ is so badly resolved in the vibronic range: A large number of close-lying vibrations couple to the electronic transitions (see [19,63] and 5.2.) and each (zero-phonon) vibronic transition is connected to a broad multi-phonon side band. A superposition of all these components gives the observed smeared out emission or absorption structure (apart from the zero-vibron range). Nevertheless a very careful registration of the emission at $T = 1.5$ K (applying a zero-point suppression technique) allows to determine several vibrational components even for the $[Ru(bpy)_3](PF_6)_2$-salt.[63] Tab. III summarizes these vibrational energies and presents a comparison with IR and Raman data. The observed vibrational energies also correspond well to those found from the (better resolved) emission spectrum of the ClO_4^--salt[19,50].

Table IV. Photophysical parameters for $[Ru(bpy)_3]^{2+}$ in different environments[a)]

	Lowest excited and emitting states[b)]					
$[Ru(bpy)_3]^{2+}$	I	II	III	IV	V	Ref.
$[\,](PF_6)_2$ [c)]						[40,41]
Classification[d)]	$1E'$	$2E'$	$1A_2'$	$3E'$ [e)]	$2A_2'$ [f)]	
Energy E_{0-0}[cm^{-1}]	17 809	17 816	17 870[g)]	≈ 17 960	≈ 18 600	
ΔE_{0-0} [cm^{-1}]	-	6.9 ± 0.2	61 ± 1	≈ 150	≈ 800	
k_{ir}/k_{Ir}	1	210 ± 20	≈ 10^3	≈ 10^4	≈ 10^4	[46,62]
ϵ_{0-0}[l/mol cm]	≈ 0.02	5				[22]
$[\,](ClO_4)_2$ [c)]						[19,50]
Energy E_{0-0} [cm^{-1}]	17 605	17 613	17 662			
ΔE_{0-0} [cm^{-1}]	-	8.2 ± 0.2	57 ± 1			
k_{ir}/k_{Ir}	1	60 ± 20	≈ 10^3			
$[\,]^{2+}$ in PMMA[h)]						[28]
ΔE [cm^{-1}]	-	10	61			
τ [μs]	180	20	0.7			
$[\,]^{2+}$ in $[Zn(bpy)_3](BF_4)_2$						[57]
ΔE [cm^{-1}]	-	8	68			
τ[μs]	250	20	1			

a) Zero-phonon lines have also been detected for $[Ru(bpy)_3](SbF_6)_2$ (at $E_{0-0} = 17\,817$ and 17 824 cm^{-1}) and for $[Ru(bpy)_3](AsF_6)_2$ (at $E_{0-0} = 17\,812$ and 17 819 cm^{-1}). b) Higher lying states do not contribute to the emission below $T = 350$ K (see 4.2.). c) Data for single crystals. d) The classification is given in the D_3 double group to take into account the spin-orbit coupling. No experimental evidence is found for a symmetry reduction in the very lowest excited states of this single-crystal compound, see also 4.1.4.. e) 95 % of the emission intensity results from this state at $T = 300$ K. f) Ground state and excited state have different equilibrium positions, see section 4.2. and [21,67]. g) Observed in an unpolarized excitation spectrum [62]. h) PMMA = polymethylmethacrylate

Tab. IV summarizes a series of photophysical parameters which are known for the different lowest excited states of crystalline $[Ru(bpy)_3](PF_6)_2$. Some data for other salts or dissolved complexes are also included. It is interesting to note that the energy separations (but not the group-theoretical classifications) between the three lowest excited states determined from single crystals fit quite nicely to those found by Crosby's group [26-30] about fifteen years ago. In a very recent investigation Krausz and Moran[57] confirmed these energy separations, thus showing that an older interpretation[31,33] is incorrect.

Further, it is emphasized that only for a small number of $[Ru(bpy)_3]X_2$-salts it is possible to register the purely electronic zero-phonon lines (Tab. IV), due to relatively small inhomogeneous broadening effects and electron-phonon coupling strengths. On the other hand most of the investigated salts and all dissolved complexes give only broad bands.[49,57] This points to the fact – as has already been mentioned (4.1.1.) – that the second coordination, surrounding the $[Ru(bpy)_3]^{2+}$ chromophores, influences their photophysical properties. Therefore a certain "tuning" of the properties is possible by changing the second coordination.[49] Moreover, the identified different excited states react differently with respect to these changes.[66,67] Consequently, one must be very careful in presenting conclusions on properties of *the excited state* without well-defining it and without clearifying how a specific, possibly very unsymmetric environment (e.g. a polar solvent) influences this specific state. For example – as has been pointed out already several years ago[41] – the interaction with solvent molecules surrounding the $[Ru(bpy)_3]^{2+}$ chromophore might be responsible for a distortion or symmetry reduction, that is observed for the excited complex. Up to now we have no indication, which requires – in the limit of the usual optical resolution – a classification of the very lowest excited states of single-crystal $[Ru(bpy)_3](PF_6)_2$ in a lower than the ground state symmetry D_3. Especially, the identification of (nearly) degenerate E-states in single crystals (with small energy splittings of the order of 0.1 cm^{-1}, see 4.1.4.) supports the presented model.

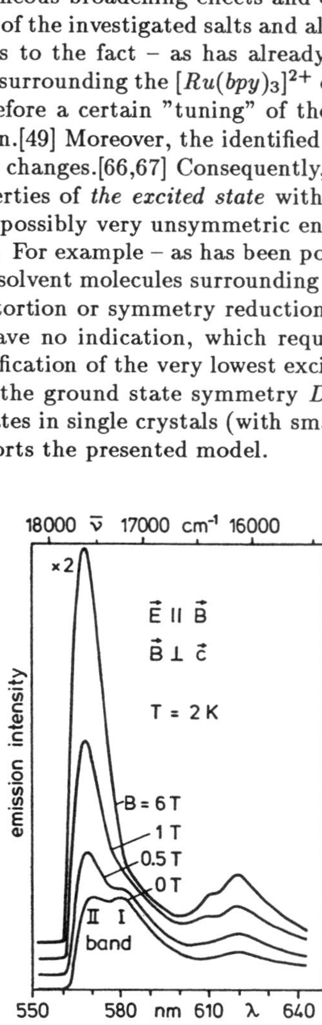

Figure 9. Broad-band emission spectra of single-crystal $[Ru(bpy)_3](PF_6)_2$ for different magnetic-field strengths.[52] The magnetic field $\vec{B}$ is oriented perpendicular to the needle axis of the crystal ($\vec{c}$-axis) and the emission is detected with $\vec{E} \parallel \vec{B} (\vec{E} \perp \vec{c})$. The emission intensities of the $\vec{E} \parallel \vec{c}\ (\vec{E} \perp \vec{B})$ polarized spectra are at least by a factor 10 weaker. (The spectrum for $B = 6$ T has to be multiplied by the given factor to be comparable.)

4.1.3. Magnetic-field effects

Magnetic fields influence the low-temperature luminescence of many transition metal complexes.[68,69] Also $[Ru(bpy)_3]^{2+}$ exhibits obvious changes of the emission under high magnetic field strengths B.[47,70] Since no investigations concerning different orientations of $\vec{B}$ relative to the molecular $\vec{C}_3$-axis nor studies of the effect of $\vec{B}$ on the zero-phonon lines had been carried out we studied the magnetic field effects on the spectra of $[Ru(bpy)_3](PF_6)_2$ single crystals.[50-53] In a more recent investigation Krausz also addressed to this subject.[56]

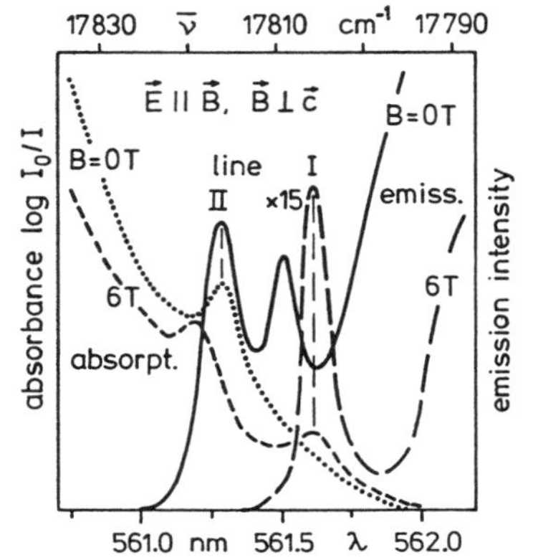

Figure 10. Absorption and emission ($T = 2$ K) in the range of the zero-phonon lines of single-crystal $[Ru(bpy)_3](PF_6)_2$ (orientation of $\vec{B}$ like in Fig. 9).[52] The emission intensities as well as the absorbance of the $\vec{E} \parallel \vec{c}$ ($\vec{B} \perp \vec{c}$) polarized spectra are by a factor of at least 20 weaker. The Zeeman shift is the same in absorption and emission and the energies of the corresponding lines coincide within $\pm$ 0.5 cm^{-1}. (The emission intensity at $B = 6$ T has to be multiplied by 15 to be comparable.)

Fig. 9 shows the $\vec{E} \perp \vec{c}$-polarized broad-band emission at $T = 2$ K for different magnetic fields with $\vec{B} \perp \vec{c}$ ($\vec{E} \parallel \vec{B}$, observation $\vec{0} \perp \vec{c}$, $\vec{0} \perp \vec{B}$). In Fig. 10 the magnetic-field dependence of the zero-phonon lines in emission and absorption is compared at $T = 2$ K. The lines are in resonance for all applied fields. Finally, the field dependence of the energy separation of the zero-phonon lines (being clearly non-linear with B) is summarized in Fig. 11. All these results are obtained for the orientations given above. A magnetic field with $\vec{B} \parallel \vec{c}$ produces only very small effects, which are – within limits of experimental error – identical to those for $\vec{B} \perp \vec{c}$ at low magnetic field. Therefore, these weak $\vec{B} \parallel \vec{c}$-effects may result from small misalignments of the crystals relative to the orientation of the field.[52] It is an important result that within an experimental uncertainty of 0.5 to 1 cm^{-1} no splitting of the zero-phonon lines is observed up to $B = 6$ T in any orientation.[50-53 and 56]

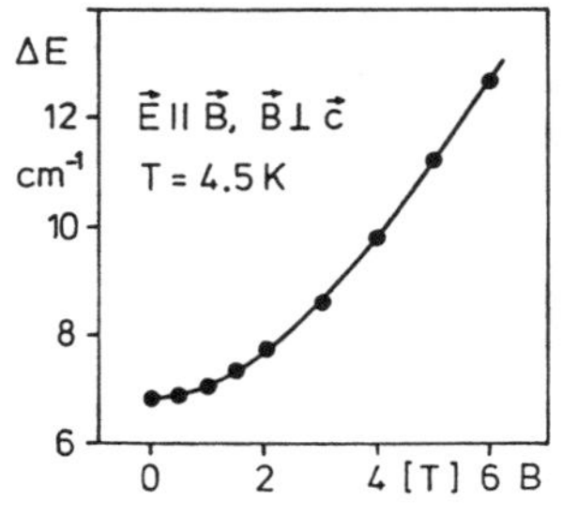

Figure 11. Energy separation of the two lowest zero-phonon emission lines versus magnetic field strength for single-crystal $[Ru(bpy)_3](PF_6)_2$.[46,52]

The results presented in the Figs. 9 to 11 may be explained phenomenologically by discussing only the properties of the transitions from the two lowest excited states $| I >$ and $| II >$ (The next state lies $\approx$ 54 cm^{-1} higher, Tab. IV): The emission behavior without magnetic fields is shortly summarized on the basis of the energy level diagram shown on the left hand side of Fig. 12. (see also 4.1.2.) At low temperature we observe mainly an emission from $| I >$ as line I and band I. With increasing temperature the emission from $| II >$ grows in as line II and band II, whereby the band II maximum appears about 400 cm^{-1} blue-shifted compared to the band I maximum while line II lies only 6.9 cm^{-1} higher. The intensity ratio of the emission components is governed by a Boltzmann factor and the ratio of the radiative rates ($k_{II_r}/k_{I_r} \approx 2 \cdot 10^2$). Under magnetic fields ($\vec{B} \perp \vec{c}$) the (unperturbed) wavefunctions $| I >$ and $| II >$ can mix, which may be discussed in a perturbation approach in this two-state model. The perturbed wavefunctions may be written as: [52,68-73]

$$| I' >= \alpha \, | I > + \beta \, | II > \quad \text{and} \quad | II' >= \alpha \, | II > - \beta \, | I >$$

with the normalization condition $\alpha^2 + \beta^2 = 1$ and the mixing coefficient

$$\beta = \mu_B B \frac{< II \mid k\mathbf{L} + \mathbf{g}\mathbf{S} \mid I >}{\Delta E}$$

μ_B and ΔE are the Bohr magneton and the energy difference between the two states, respectively. **L** and **S** are the operators for the total angular and the total spin momentum, respectively. **g** represents the phenomenological anisotropic g-factor (tensor) and k is a reduction factor for the angular momentum.

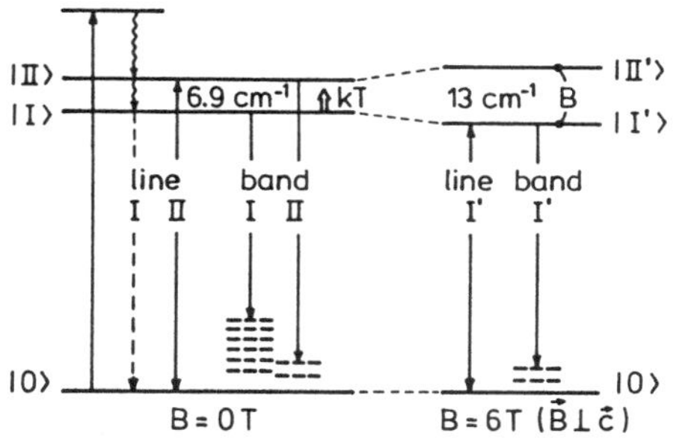

Figure 12. Energy level diagram for the two lowest excited states of $[Ru(bpy)_3](PF_6)_2$. The lines are observed as zero-phonon transitions. The bands represent superpositions of different vibronic transitions with different vibrational coupling properties for $| I > \rightarrow | 0 >$ and $| II > \rightarrow | 0 >$, respectively. Under magnetic fields one observes a Zeeman effect (for $\vec{B} \perp \vec{c}, \vec{E} \| \vec{B}$).

Thus, one can express the magnetic field dependence of the ratio of radiative rates as:[52]

$$\frac{k_{I'_r}}{k_{II'_r}}(B) - \frac{k_{I_r}}{k_{II_r}}(0) \approx \frac{\beta^2}{1-\beta^2}$$

For weak fields, where $\beta^2 << 1$, one obtains a B^2-dependence of the relative rates. Indeed, this is found experimentally as is seen in Fig. 13. Further, using this expression one can roughly estimate the extent to which the wavefunctions are mixed at high magnetic fields (although the perturbational approach is developed only for weak distortions). For $B = 6$ T this estimate shows that the wavefunction of the lowest excited state $| I' >_{6T}$ consists of about two-thirds of its unperturbed wavefunction $| I >$ and one third of $| II >$ being mixed in. Since the unperturbed states $| I >$ and $| II >$ behave very differently at zero magnetic field the properties of the transition $| I' > \leftrightarrow | 0 >$ should change drastically with increasing magnetic fields by adopting properties of the transition $| II > \leftrightarrow | 0 >$.[52,53]

Consequently, the zero-phonon absorption corresponding to line I, which is strongly forbidden without magnetic fields, gets transition probability mixed in. This leads to a growing in of the corresponding absorption. With this behavior a powerful support for the intrinsic zero-phonon character of line I is given. (See 4.1.2.) Further, due to the fact that the vibrational coupling properties of the two unperturbed transitions differ strongly the mixing of the wavefunctions should also produce drastic changes in the vibronic structure. Indeed, the emission from the lowest excited state develops with increasing magnetic field strength from the band I envelop to the band II envelop. (Fig. 9 and right part of Fig 12.)

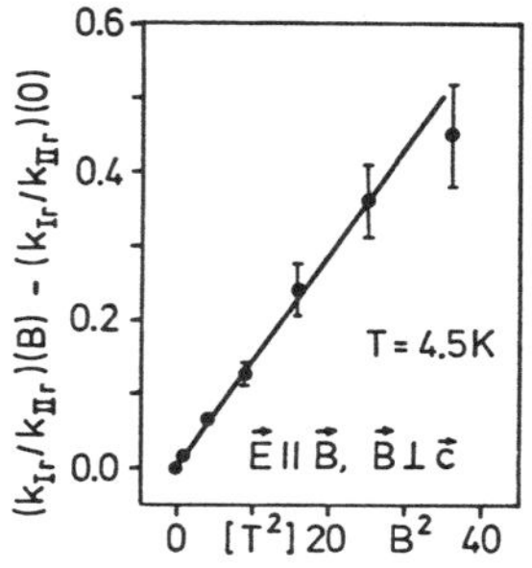

Figure 13. Magnetic-field dependence of the difference of the ratio of radiative rates k_{Ir}/k_{IIr} for the two lowest zero-phonon transitions of $[Ru(bpy)_3](PF_6)_2$, plotted versus B^2.[52]

Moreover, it is expected that the marked magnetic-field induced coupling of the wavefunctions $| I >$ and $| II >$ is connected with an increase of the energy separation ΔE (Zeeman effect) which should develop approximately quadratically with increasing field strengths.[73] In fact, this tendency is also observed.(Fig. 11)

The presented discussion of the influence of magnetic fields on $[Ru(bpy)_3](PF_6)_2$ allows a deeper insight into the properties of this compound, mainly with respect to the lowest excited states and even a fairly well quantitative description is possible although the model is certainly phenomenological. However, several questions arise immediately:

The two lowest excited states have been classified as degenerate representations. If the electron spin g value is assumed to be ≈ 2, one expects a splitting of the order of 5 cm^{-1} ($\vec{B} \parallel \vec{c}$, $B = 6$ T). But within limits of an experimental uncertainty of 0.5 to 1 cm^{-1} no splitting is observed. Mainly for that reason Krausz[56] proposed (in dropping the E' assignment [31]) that the corresponding states cannot be degenerate. On the other hand, one observes distinct ODMR signals on the zero-phonon lines (4.1.4.), which strongly supports the given E' classification. Therefore, one should ask, if it is possible to have E' representations with a g-value much smaller than 2.

For the $\vec{B} \perp \vec{c}$-orientation no first order splitting is expected due to symmmetry restrictions ($g_\perp = g_{xy} = 0$).[72,73,47] Hence, we have to focus on the case of the $\vec{B} \parallel \vec{c}$-orientation. It is the intention of the following discussion to outline that also for this orientation not every degenerate E state must split by serveral cm^{-1}, including states which result from triplet parents .

The lowest excited states represent mainly triplet components. This is concluded from the long emission lifetimes and the very low transition probabilities.(4.1.1. and 4.1.2.) Also theoretical considerations ([13], section 3.) show that the very lowest states might indeed be regarded as fairly pure triplets, even when spin-orbit coupling is included. These triplets can be described by products of orbital and spin wavefunctions. Now, let us focus on a state of total E representation which is built up from the non-degenerate A_2 spin component ($m_S = 0$) and the orbitally degenerate E component. (This combination can only result from ^3E parent states, see Tab. I.) The corresponding wavefunctions may be written in the following form:

$$| \psi_{1,2}(E) > = | A_2 >_{spin} \cdot | E_{1,2} >_{orb.}$$

wherein $| E_{1,2} >_{orb.}$ represents the two functions spanning the orbital E representation. Hence, the order of magnitude of the splitting through magnetic fields with $\vec{B} \parallel \vec{c}$ is characterized by a matrix element of the following type ($i, j = 1, 2$)

$$\begin{aligned} &< \psi_i(E) \ | k\mathbf{L}_z + g_z\mathbf{S}_z \,|\, \psi_j(E) > \\ &= < \psi_i(E) \,|\, k\mathbf{L}_z \,|\, \psi_j(E) > \ + g_z < \psi_i(E) \,|\, \mathbf{S}_z \,|\, \psi_j(E) > \\ &= < A_2 \,|\, A_2 >_{spin} \cdot < E_i \,|\, k\mathbf{L}_z \,|\, E_j >_{orb.} + \\ &\quad g_z < A_2 \,|\, \mathbf{S}_z \,|\, A_2 >_{spin} \cdot < E_i \,|\, E_j >_{orb.} \end{aligned}$$

Since $\mathbf{S}_z \,|\, A_2 >= m_S \,|\, A_2 >$ and $m_S = 0$ for the A_2 representation of the spin (section 3.) the second term of the matrix element is zero. Consequently the splitting under magnetic fields is only given by the matrix elements of the angular momentum. Thus, due to the known effects of quenching of the angular momentum [16,73,74] one expects only relatively small splittings.

An application of these model-considerations to the *two* lowest excited states ($1E'$, $2E'$) is also possible. Due to the fact that every ^3E parent term gives only one $| A_2 >_{spin} \cdot | E_{1,2} >_{orb.}$ combination, two E' representations being constructed in this way have to result from two different ^3E terms. As Tab. I shows, these are only found in different one-electron configurations. This is consistent with the considerations presented in 4.1.2., concerning the different configurations which might give the two E' states. The second lowest excited state ($2E'$) seems to contain a small amount of singlet character ($\epsilon \approx 5$ l/mol cm) but we believe that this would not invalidate the estimate presented above. Thus, the described model shows that there is *no need* to drop the (well-founded) E'-classification on the basis that a Zeeman splitting of the individual E' states is not observed in the optical spectra.

4.1.4. Zero-field splittings, ODMR-signals

The two lowest excited states of $[Ru(bpy)_3](PF_6)_2$ have been assigned within a D_3 point group symmetry to degenerate E' states mainly due to the fact that the corresponding optical zero-phonon transitions are totally $\vec{E} \perp \vec{c}$-polarized (4.1.2.). However, in a real system one expects at least small distortions of the D_3 symmetry which should lead to small zero-magnetic-field splittings of the degenerate states. But the optical spectra of highest obtained resolution (in the zero-phonon lines) show that any splitting must be smaller than 0.5 cm^{-1}. Therefore, it is near at hand to investigate this system with techniques of microwave resonances in the excited states. This allows a detection of splittings in the GHz-energy range ($1\ cm^{-1} \,\hat{=}\, 30 GHz$). An adequate method represents the zero-field ODMR spectroscopy (optically detected magnetic resonance). In this technique one monitors the luminescence (excited as usual) at a fixed wavelength and tunes the microwave frequency (in a repetitive way).[75-77] In the case of resonance between sublevels of an excited state one may observe intensity changes in the emission as positive or negative signals. The amount of these changes depends on the differences in the radiative rates of the sublevels and on the extent of microwave-induced changes in the relative population of these levels (which in general are not thermally equilibrated at low temperatures).[75,78]

Indeed, ODMR signals have been observed for $[Ru(bpy)_3](PF_6)_2$ single crystals at $T = 1.5$ K. The emission was detected on the zero-phonon line II at 17816 cm^{-1} ($2E' \rightarrow A_1'$) and at the corresponding broad band. In the investigated frequency range ($1.5 \leq \nu \leq 6GHz$) resonance occurred at several frequencies (1.89, 2.81, 3.26, 4.16, 4.35 GHz). For all resonances we found an increase of the emission intensity. But no signals could be observed in the investigated range when the emission was detected on the zero-phonon line I at 17809 cm^{-1} ($1E' \rightarrow A_1'$). On the other hand at $T = 1.5$ K the analogous $[Ru(bpy)_3](ClO_4)_2$ exhibits resonances at 3.41 and 4.75 GHz (investigated range: $3 \leq \nu \leq 5\ GHz$) on line I and the corresponding broad band, however, with a decrease in the emission intensity. A detection on line II (which is very weak at $T = 1.5$ K) shows only a very weak signal at $\approx$ 3.4 GHz with intensity increase.[62,79]

These results are consistent with the presented model of two nearly degenerate doublets which are split in the order of 0.1 cm^{-1}. The microwave resonances occur between these doublet sublevels. An appearance of several resonance frequencies for the same emission line is not unusual[80] and is ascribed to different sites. Especially with respect to the established occurrence of different $[Os(bpy)_3]^{2+}$ sites in the $[Ru(bpy)_3](PF_6)_2$ lattice this assignment is near at hand (see 5.2.). The fact that we could only find resonances (positive signals) for the second excited state of the PF_6^- compound and not for the very lowest one seems – at first sight – to be puzzling. But the close analogy of the properties of the PF_6^--salt compared to the ClO_4^- compound[19,50,63] and the finding of negative ODMR signals for the very lowest excited state as well as a positive although very weak one for the second excited state in $[Ru(bpy)_3](ClO_4)_2$ allows to conclude on the existence of a doublet also for the PF_6^--salt. Possibly, the radiative rates of the corresponding sublevels are nearly the same or the splitting is smaller than 1.5 GHz. This second assumption is not unreasonable since the expected distortion of the $[Ru(bpy)_3]^{2+}$ sites is less pronounced in the PF_6^- salt compared to the ClO_4^- compound.[79,64]

ODMR signals have also been found for the broad-band emission of neat and dissolved $[Ru(bpy)_3](BF_4)_2$.[81] In this reference it is reported that five microwave resonances occur in the investigated energy range. These lie at somewhat different energies than for the PF_6^- and ClO_4^- salts as is expected due to the changed environment. Three of the microwave transitions are arbitrarily assigned – in analogy to the situation found in purely organic compounds [75-78, 80] – to zero-field split components of a normal T_1 spin triplet, while the other transitions are also ascribed to different sites. Following this T_1 model DeArmond and his group[82] tried to explain also the optical properties of $[Ru(bpy)_3]^{2+}$ on the basis of a single orbitally non-degenerate spin triplet. This model, however, ignores the well-established $d\pi^*$-CT-character of the excited many-electron states of $[Ru(bpy)_3]^{2+}$. It further disregards the extensive investigations on zero-field splittings in compounds like the metalloporphyrins[83] and the metallophthalocyanines[83,73]. In these compounds even the $^3\pi\pi^*$ states with a small d-admixture exhibit relatively large zero-field splittings through the enhanced spin-orbit coupling. Thus, one finds T_1-splittings of several cm^{-1}, e.g. 2 cm^{-1} and 26 cm^{-1} for $Pd(II)$- and $Pt(II)$-phthalocyanines, respectively.[84,85] Consequently, we suggest that the interpretations given in [81] and [82] should not be used.

4.2. Higher lying excited states

In most cases it is very difficult to obtain accurate and complete data of higher lying states since the superposition of vibronic structures belonging to different electronic transitions smears out the spectra. Further, the usual optical absorption spectroscopy with neat single crystals fails for measurements of highly allowed transitions due to the large absorbances even of very thin crystalline foils. Consequently, for these transitions one has to investigate doped systems which – especially for $[Ru(bpy)_3]^{2+}$ – introduce (the discussed) problems of extensive inhomogeneous broadenings. Thus, we reproduce in Fig. 14 two kinds of spectra to cover the polarized absorption of $[Ru(bpy)_3]^{2+}$. These spectra are obtained with neat single-crystal $[Ru(bpy)_3](PF_6)_2$-foils for lower ϵ-values (up to about $3 \cdot 10^3 \ell mol^{-1} cm^{-1}$, giving the total $\vec{E} \parallel \vec{c}$-polarized MLCT range) and with complexes doped into single-crystal $[Zn(bpy)_3]SO_4 \cdot 7H_2O$ (taken from [86]) for larger molar extinctions, respectively.

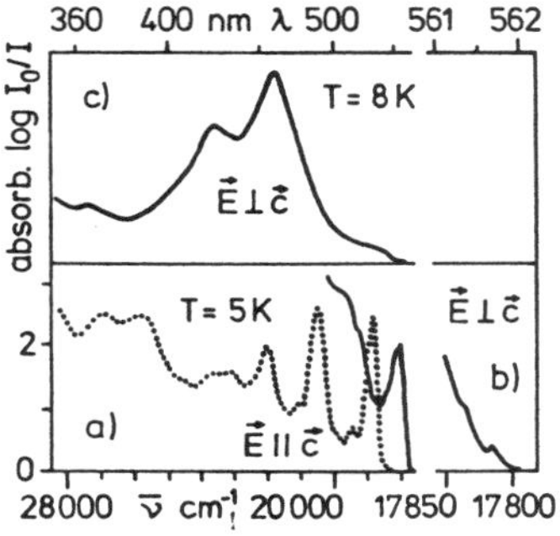

Figure 14. Polarized absorption of single-crystal $[Ru(bpy)_3](PF_6)_2$ at $T = 5$ K.[21] Crystal thickness: a) (5 ± 2) μm; b) (80 ± 10) μm. For comparison we reproduced the $\vec{E} \perp \vec{c}$-polarized spectrum for the energy range of the MLCT transitions of high absorbances (c) measured at $T = 8$ K with $[Ru(bpy)_3]^{2+}$ in $[Zn(bpy)_3]SO_4 \cdot 7H_2O$ (from [86]).

The $\vec{E} \perp \vec{c}$ spectrum exhibits a weak and sharp peak at 17816 cm^{-1} which represents the zero-phonon line II ($A_1' \rightarrow 2E'$, 4.1.2.). The observed structure near 18000 cm^{-1} may be related to the multi-phonon side band of this transition or may be connected to a further E' state (possibly $3E'$, Fig. 5). A definite assignment cannot be given from the actual absorption spectra. However, from the ϵ values and the transition energy this peak might be regarded to consist largely of triplet character. The resolvable structure in the energy range between $\approx$ 18000 and 20000 cm^{-1} is assigned to intra-molecular vibrations (160, 330, 730, 1030, and 1270 + 160 cm^{-1}) which couple to the transition near 18000 cm^{-1}. Within experimental error the vibrational energies are the same as observed in ground state Raman experiments.[21,65] The absorbance increases drastically above $\approx$ 20000 cm^{-1} to ϵ-values of several $10^4 \ell mol^{-1} cm^{-1}$. Although the obtained resolution is much lower for this energy range one clearly sees three transitions which are assigned to different E' states being mainly of singlet character. For more details concerning these transitions see [12, 13, 65, 86].

The $\vec{E} \parallel \vec{c}$ spectrum does not exhibit any absorption of $\epsilon \geq 0.3$ $\ell mol^{-1} cm^{-1}$ ($\hat{=}$ experimental uncertainty) up to about 17860 cm^{-1}. The first dominating peak at 18770 cm^{-1} ($\epsilon \approx 2.4 \cdot 10^3 \ell mol^{-1} cm^{-1}$) represents the zero-vibron multi-phonon component of the $A_1' \rightarrow 2A_2'$ transition.[21] The zero-phonon component could not be determined in absorption but from the temperature dependence of the polarized emission it is concluded that the electronic origin should lie near 18600 cm^{-1} (Fig. 5), which lies just in the low-energy rise of the first dominating peak as is expected for the position of the zero-phonon line.

It is an interesting further aspect that this $A_1' \rightarrow 2A_2'$ transition is connected to a distinct $\approx 1600\ cm^{-1}$ progression (that additionally couples to several other intra-molecular vibrations as 160, 660, and 1180 cm^{-1}). One may discover about four members of this progression (which previously had been assigned erroneously – on the basis of much less well resolved absorption spectra – to different electronic A_2' origins[13]). The intensity distribution of the progression can be used to calculate bond length changes of *bpy* for the equilibrium position of the excited $2A_2'$ state relative to the ground state. We obtained an inter-ring distance decrease ($\approx$ 4 pm) and an inner-ring expansion (average $C-C$ increase $\approx$ 2 pm).[67] Moreover, a band analysis using the knowledge about the members of the observed progression allows to determine two further zero-vibron origins at $\approx$ 23000 and 25600 cm^{-1}.[67,21] For completeness we want to point out that we could not detect a similar progression from the well resolved vibronic emission spectra resulting from the two lowest excited states. This behavior points to the different properties of the excited states, also with respect to distortions, bond length changes, and vibronic coupling.

5. Spectroscopy of $[Os(bpy)_3]^{2+}$ doped into single-crystal $[Ru(bpy)_3](PF_6)_2$

The spectroscopic properties of $[Os(bpy)_3]^{2+}$ have also been investigated by several research groups[3, 13, 15, 54, 87-93], but the progress in identifying and/or assigning the lowest excited states and the vibronic structures was very moderate due to the relatively broad and unresolved optical spectra obtained so far. The lowest excited states, in which one is mainly interested, are classified throughout the cited literature to triplets of $Os5d - bpy\pi^*$ CT character. More detailed assignments are partly contradictory to each other or sometimes even inconsistent in themselfs. Therefore, it is desirable to investigate the $[Os(bpy)_3]^{2+}$ complex on the basis of better resolved spectra. But up to now, even spectra recorded at $T < 2$ K with neat single crystals did not clearly reveal sharp zero-phonon or vibronic transitions. However, a small amount of $[Os(bpy)_3]^{2+}$ doped into selected $[Ru(bpy)_3]X_2$ matrices (e.g. with $X^- = ClO_4^-, PF_6^-, AsF_6^-, SbF_6^-$) led to an enormous enhancement in the resolution of the $[Os(bpy)_3]^{2+}$ spectra.[94-98,66] Further, these investigations uncover a very interesting additional feature of these doped systems. One observes a radiationless energy transfer from the matrices to the dopant. This allows to conclude on important properties of the host material. It is subject of this section to discuss at first some aspects of the energy transfer and then to consider the lowest excited states of $[Os(bpy)_3]^{2+}$ also with respect to vibrational coupling properties.

5.1. Radiationless energy transfer

The emission spectrum of single-crystal $[Ru_{1-x}Os_x(bpy)_3](PF_6)_2$ ($x \approx 0.01$) excited at $\lambda_{exc} = 363.8$ nm (corresponding to an absorption of the host) consists of two distinctly separated components. The high-energy one, lying between $\approx$ 560 and $\approx$ 680 nm results from the $[Ru(bpy)_3](PF_6)_2$ host while the low-energy part (with $\lambda > 680$ nm) comes from the $[Os(bpy)_3]^{2+}$ guests. The low-energy emission spectrum is identical irrespectively if the host is excited or directly the guests ($\lambda_{exc} = 632.8$ nm). This component is absent if no guest molecules have been doped in. Consequently, one can conclude on the occurrence of a radiationless energy transfer from the host (donor) to the guests (acceptor). This is further substantiated by a drastic reduction of the emission intensity and a shortening of the emission lifetime of the donor compared to the neat material. Moreover, the energy transfer is evidenced by the observed fast rise and the prolonged decay of the acceptor emission.[96, 97]

A more detailed study of the decay properties show that the donor emission exhibits (at low acceptor concentrations) a relatively fast decay component (order of 10^2 ns) and a long-living component (order of many μs). These properties can be rationalized if one takes into account two different processes of energy transfer. The fast process is ascribed to a relatively efficient resonance transfer between nearest donor-acceptor neighbors, presumably being dominated by an exchange mechanism. On the other hand, the long-living component is assigned to be connected with a donor-donor excitation transport and a subsequent transfer to the guests. This is concluded from the fact that the long-living decay of the donor is just the same as the (prolonged) acceptor decay and that both decays become (equally) shorter with increasing guest concentrations. Relatively high acceptor concentrations are needed (order of 10^{-2} mole/mole) to effectively quench the donor emission (at low temperatures). These concentrations should be compared to other single-crystal compounds in which amounts of impurities that are several orders of magnitude smaller drastically quench the donor emission.[99-103] Thus, it follows that the donor-donor transfer is rather ineffective and that the inter-molecular interaction between the $[Ru(bpy)_3]^{2+}$ complexes is relatively small (see also [22]). Presumably, this behavior is connected to an inhomogeneous distribution of the donors which leads to an energy mismatsch between adjacent $[Ru(bpy)_3]^{2+}$ centers and thus to a reduced donor-donor transfer efficiency.[96]

5.2. Lowest excited states and vibrational coupling

The emission spectrum of $[Os(bpy)_3]^{2+}$ doped into single-crystal $[Ru(bpy)_3](PF_6)_2$ exhibits a large number of very sharp transitions (halfwidths ≈ 2 cm^{-1}). But the assignment of the lines seems to be complicated due to the occurrence of a triple pattern which, on the one hand, has been ascribed to a phonon progression [104] and, on the other hand, to a

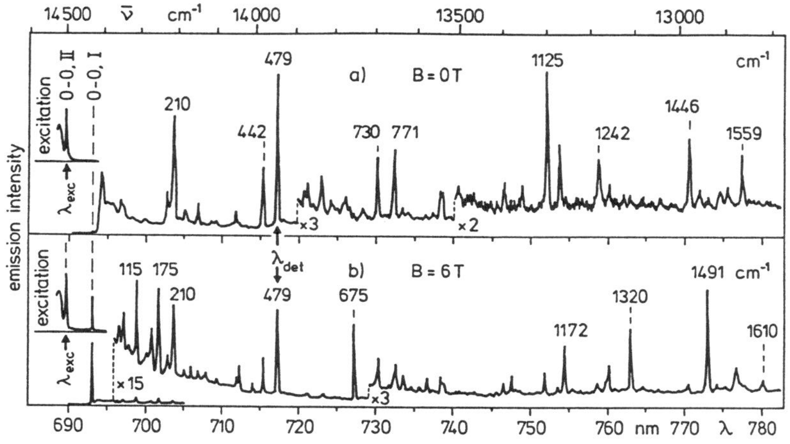

Figure 15. Site selective emission and excitation spectra for the site of lowest energy of $[Os(bpy)_3]^{2+}$ (≈ 1 mol%) in single-crystal $[Ru(bpy)_3](PF_6)_2$ at $T \leq 2$ K at different magnetic fields ($\vec{B} \perp \vec{c}$). The excitation spectra are detected at $\lambda_{det} = 717.2$ nm and the emission spectra are excited ate $\lambda_{exc.} = 689.9$ nm. The spectra are $\vec{E} \perp \vec{c}$-polarized. The $\vec{E} \parallel \vec{c}$-polarized emission intensity is at least by a factor of 10^2 weaker. The states $\mid I >$ and $\mid II >$ have an energy separation of 72 cm^{-1}.

superposition of emission spectra resulting from three different guest sites [94, 96, 97]. The assignment to sites could be proven by recording the selectively excited spectrum of the site A of lowest energy. This spectrum is much simpler than the non-selective ones since it does no longer exhibit the triple pattern.[66, 98] Fig. 15a reproduces this site-selective emission and the corresponding excitation spectrum.

For temperatures below $T \approx 2$ K the emission results from the lowest excited state $\mid I >$, but the corresponding zero-phonon transition (line I) is hardly to observe in emission and absent in absorption as well as in excitation. However, a vibronic analyses (using the known vibrational energies of $[Os(bpy)_3]^{2+}$ [105, 95, 66] and of $[Ru(bpy)_3]^{2+}$ [65, 106, 19, 50]) fits very well, when the zero-phonon line I corresponding to site A is assigned to lie at (14423 ± 1) cm^{-1}. This assignment is proven through the application of a magnetic field which leads to a growing in of the zero-phonon line I at exactly the same energy by about three orders of magnitude in emission, excitation, and absorption. (See Fig. 15b and [94, 98, 66].) This behavior is similar to the situation found for $[Ru(bpy)_3](PF_6)_2$. (4.1.3.) Magnetic fields induce a mixing-in of wavefunctions of higher lying states, providing the enormous increase of the transition probability. As expected, one finds a B^2 dependence for this coupling.[98] The most probable canditate for the perturbing state seems to be state $\mid II >$ which lies 72 cm^{-1} above $\mid I >$. Due to the relatively large energy separation between these states no energy shifts could be observed up to $B = 6$ T. (In an alternative interpretation[92] the occurrence of the "energy gap" between absorption and emission at zero magnetic field is taken as indication of a relaxation process in the excited state. However, this model is proven to be invalid on the basis of the presented data.)

Finally, we want to focus to a further very interesting magnetic-field effect. The vibronic structure in emission is drastically changed by magnetic fields. (Fig. 15)[97,98] At low temperatures ($T \approx 2$ K), the vibronic structure of $[Os(bpy)_3]^{2+}$ in $[Ru(bpy)_3](PF_6)_2$ is dominated for $B = 0$ T by IR-aktive [65,66,106] vibrations coupling to $\mid I > \rightarrow \mid 0 >$ (with prominent vibronic components lying e.g. at 479, 1125, 1154, 1242, 1446, and 1559 cm^{-1}). With increasing magnetic fields one finds that (resonance enhanced) Raman-active [65,105] vibrations (e.g. at 675, 1172, 1270, 1370, 1491, 1552, and 1610 cm^{-1}) gradually grow in until they are dominant at $B = 6$ T. This effect is explained by the mixing-in of the wavefunction of $\mid II >$ into $\mid I >$, whereby the two states are connected to the different properties in vibrational coupling. Thus, magnetic fields allow a *tuning* of the vibrational coupling. For completeness it is mentioned that $[Os(bpy)_3]^{2+}$ in $[Ru(bpy)_3](CO_4)_2$ exhibits a similar change in the vibronic coupling by application of hydrostatic pressure.[64] At atmospheric pressure IR-active as well as Raman-active vibrations couple to $\mid I > \rightarrow \mid 0 >$. Pressure increases to ≈ 10 kbar ($T = 2$ K) leads to a growing in of the (resonance enhanced) Raman modes while the intensities of the IR-active modes decrease.

6. Conclusion

In connection with extensive investigations of processes of solar energy conversion the interest in $[Ru(bpy)_3]^{2+}$ and related complexes was mainly focussed on chemical and photochemical aspects. In this paper it is shown that the title compounds also exhibit very interesting physical properties. From the obtained highly resolved spectra, it is possible to deduce an information about electron-phonon coupling strengths, inhomogeneous broadenings, radiationless energy transfer properties, as well as characteristics of the vibrational coupling. With the tunability of this coupling through magnetic fields a significant feature has been observed for these compounds.

Acknowledgments. The authors thank the "Verband der Chemischen Industrie", the "Deutsche Forschungsgemeinschaft" and the "Stiftung Volkswagenwerk" for financial support. Prof. Dr. G. Gliemann is acknowledged for support of this work.

a) Present address: Bayerisches Landesamt für Umweltschutz, 8000 München, FRG
b) Present address: BASF AG, 6700 Ludwigshafen, FRG

References

1 A. Juris, V. Balzani, F. Barigelletti, S. Campagna, P. Belser, A. von Zelewsky; Coord. Chem. Rev. **84**, 85 (1988)
2 R. A. Krause; Structure and Bonding **67**, 1 (1987)
3 T. J. Meyer; Pure and Appl. Chem. **58**, 1193 (1986)
4 N. Sutin, C. Creutz; Adv. Chem. Ser. **168**, 1 (1978)
5 J. N. Demas; J. Chem. Educ. **60**, 803 (1983)
6 R. J. Watts; J. Chem. Educ. **60**, 834 (1983)
7 K. Kalyanasundaram; Coord. Chem. Rev. **46**, 159 (1982)
8 H. Yersin, A. Vogler; "Photochemistry and Photophysics of Coordination Compounds"; Springer Verlag Berlin (1987)
9 E. Krausz; Comments Inorg. Chem. **7**, 139 (1988)
10 G. A. Crosby; Acc. of Chem. Res. **8**, 231 (1975)
11 L. E. Orgel; J. Chem. Soc. **1961**, 3683
12 A. Ceulemans, L. G. Vanquickenborne; J. Am. Chem. Soc. **103**, 2238 (1981)
13 E. M. Kober, T. J. Meyer; Inorg. Chem. **21**, 3967 (1982)
14 K. W. Hipps, G. A. Crosby; J. Am. Chem. Soc. **97**, 7042 (1975)
15 J. Ferguson, F. Herren; Chem. Phys. **76**, 45 (1983)
16 H. L. Schläfer, G. Gliemann; "Basic Principles of Ligand Field Theory", Wiley-Interscience, London (1969)
17 G. F. Koster, J. O. Dimmock, R. G. Wheeler, H. Statz; "Properties of the thirty-two point groups", M. I. T. Press, Cambridge, Mass. (1963)
18 H. Yersin, E. Gallhuber, G. Hensler; J. de Physique (France) **46**, Suppl., C7-453 (1985)
19 E. Gallhuber, G. Hensler, H. Yersin; Chem. Phys. Lett. **120**, 445 (1985)
20 G. Hensler, E. Gallhuber, H. Yersin; Inorg. Chim. Acta **113**, 91 (1986)
21 H. Yersin, E. Gallhuber, G. Hensler; Chem. Phys. Lett. **134**, 497 (1987)
22 H. Yersin, G. Hensler, E. Gallhuber; Inorg. Chim. Acta **132**, 187 (1987)
23 C. M. Carlin, M. K. DeArmond; Chem. Phys. Lett. **89**, 297 (1982)
24 P. S. Braterman, G. A. Heath, L. J. Yellowless; J. Chem. Soc. Dalton **1985**, 1081
25 G. Gliemann, H. Yersin; Structure and Bonding **62**, 87 (1985)
26 R. W. Harrigan, G. D. Hager, G. A. Crosby; Chem. Phys. Lett **21**, 487 (1973)
27 R. W. Harrigan, G. A. Crosby; J. Chem. Phys. **59**, 3468 (1973)
28 G. D. Hager, G. A. Crosby; J. Am. Chem. Soc. **97**, 7031 (1975)
29 G. D. Hager, R. J. Watts, G. A. Crosby; J. Am. Chem. Soc. **97**, 7037 (1975)
30 W. H. Elfring, G. A. Crosby; J. Am. Chem. Soc. **103**, 2683 (1981)
31 J. Ferguson, E. R. Krausz; Chem. Phys. Lett. **93**, 21 (1982)
32 J. Ferguson, E. Krausz; Inorg. Chem. **26**, 1383 (1987)
33 J. Ferguson, E. Krausz; Chem. Phys. **112**, 271 (1987)

34 P. G. Bradley, N. Kress, B. A. Hornberger, R. F. Dallinger, W. M. Woodruff; J. Am. Chem. Soc. **103**, 7441 (1981)
35 M. Forster, R. E. Hester; Chem. Phys. Lett. **81**, 42 (1981)
36 S. M. Angle, M. K. DeArmond, R. J. Donohoe, K. W. Hanck, D. W. Wertz; J. Am. Chem. Soc. **106**, 3688 (1984)
37 C. V. Kumar, J. K. Barton, N. J. Turro, I. R. Gould; Inorg. Chem. **26**, 1455 (1987)
38 P. J. Carroll, L. E. Brus; J. Am. Chem. Soc. **109**, 7613 (1987)
39 D. P. Rillema, D. S. Jones, H. A. Levy; J. Chem. Soc. Chem. Commun., **1979**, 849
40 H. Yersin, E. Gallhuber, A. Vogler, H. Kunkely; J. Am. Chem. Soc. **105**, 4155 (1983)
41 H. Yersin, E. Gallhuber; J. Am. Chem. Soc. **106**, 6582 (1984)
42 H. Yersin, G. Gliemann; Messtechnik (Braunschweig) **80**, 99 (1972)
43 M. Stock, H. Yersin; Chem. Phys. Lett. **40**, 423 (1976)
44 H. Yersin, H. Otto, G. Gliemann; Theor. Chim. Acta **33**, 63 (1974)
45 H. Yersin, G. Hensler, E. Gallhuber; IIIrd European Conference on Solid State Chem., Regensburg, Book of Abstracts, Vol. 2, 329 (1986)
46 E. Gallhuber, G. Hensler, H. Yersin; Reference [8] p. 93
47 D. C. Baker, G. A. Crosby; Chem. Phys. **4**, 428 (1974)
48 J. Ferguson, E. Krausz; Chem. Phys. Lett. **127**, 551 (1986)
49 H. Yersin, G. Hensler, E. Gallhuber, W. Rettig, L. O. Schwan; Inorg. Chim. Acta **105**, 201 (1985)
50 H. Yersin, E. Gallhuber, G. Hensler; Reference [8] p. 101
51 H. Yersin, E. Gallhuber, G. Hensler; Bunsentagung, Heidelberg, May 1986, Vol. of Abstracts p. D77; E. Gallhuber, G. Hensler, H. Yersin; XI IUPAC Symposium on Photochemistry, Lisbon, July 1986, Vol. of Abstracts p. 384
52 E. Gallhuber, G. Hensler, H. Yersin; J. Am. Chem. Soc. **109**, 4818 (1987)
53 G. Hensler, E. Gallhuber, H. Yersin; Inorg. Chem. **26**, 1641 (1987)
54 J. Ferguson, F. Herren; Chem. Phys. Lett. **89**, 371 (1982)
55 E. Krausz, T. Nightingale; Inorg. Chim. Acta **120**, 37 (1986)
56 E. Krausz; Chem. Phys. Lett. **135**, 249 (1987)
57 E. Krausz, G. Moran; J. Luminescence **42**, 21 (1988)
58 D. B. Fitchen, R. H. Silsbee, T. A. Fulton, E. L. Wolf; Phys. Rev. Lett. **11**, 275 (1963)
59 D. Haarer; J. Chem. Phys. **67**, 4076 (1977)
60 G. F. Imbusch; in: "Luminescence Spectroscopy", ed. M. D. Lumb, Acad. Press, London 1978, p. 1
61 R. G. Denning; see the paper in this volume
62 E. Gallhuber; Thesis, Universität Regensburg, in preparation
63 G. Hensler, E. Gallhuber, H. Yersin; XI IUPAC Symposium on Photochemistry, Lisbon (1986), Vol. of Abstracts p. 382; G. Hensler, Thesis, Universität Regensburg (1988)
64 H. Yersin, D. Braun, et al. unpublished results
65 O. Poizat, C. Sourisseau; J. Phys. Chem. **88**, 3007 (1984)
66 D. Braun, E. Gallhuber, G. Hensler, H. Yersin; Mol. Phys. (1989), in press
67 T. Schönherr, J. Degen, E. Gallhuber, G. Hensler, H. Yersin; submitted
68 G. Gliemann; Comments Inorg. Chem. **5**, 263 (1986)
69 G. Gliemann, Pure Appl. Chem., in press
70 R. J. Watts, R. W. Harrigan, G. A. Crosby; Chem. Phys. Lett. **8**, 49 (1971)
71 M. H. L. Pryce; Proc. Phys. Soc. A **63**, 25 (1950)
72 J. S. Griffith; Phys. Rev. **132**, 316 (1963)

73 G. W. Canters, J. H. van der Waals; in „The Porphyrins“, Vol. III, ed. D. Dolphin, Acad. Press, New York , p. 531 (1978)
74 R. M. Macfarlane; Phys. Rev. **B1**, 989 (1970)
75 M. A. El-Sayed; J. Chem. Phys. **54**, 680 (1971)
76 C. B. Harris, R. J. Hoover; J. Chem. Phys. **56**, 2199 (1972)
77 J. Zuclich, D. Schweitzer, A. H. Maki; Photochem. Photobiol. **18**, 161 (1973)
78 C. B. Harris, M. J. Buckley; in: "Adv. in Nuclear Quadrupole Resonance", Vol. 2, ed. J. A. S. Smith, Heyden, London 1975, p. 15
79 H. Yersin, E. Gallhuber, G. Hensler, D. Schweitzer, in preparation
80 D. Schweitzer, K. H. Hausser, H. Vogler, F. Diederich, H. A. Staab; Mol. Phys. **46**, 1141 (1982)
81 S. Yamauchi, Y. Komada, N. Hirota; Chem. Phys. Lett. **129**, 197 (1986)
82 M. L. Myrick, R. L. Blakley, M. K. DeArmond, M. L. Arthur; J. Am. Chem. Soc. **110**, 1325 (1988)
83 J. A. Kooter, G. W. Canters, J. H. van der Waals; Mol. Phys. **33**, 1545 (1975)
84 W.-H. Chen, K. E. Rieckhoff, E.-M. Voigt; Chem. Phys. **102**, 193 (1986)
85 W.-H. Chen, K. E. Rieckhoff, E.-M. Voigt; Mol. Phys. **59**, 355 (1986)
86 F. Felix, J. Ferguson, H. U. Güdel, A. Ludi; J. Am. Chem. Soc. **102**, 4096 (1980)
87 J. N. Demas, G. A. Crosby; J. Am. Chem. Soc. **93**, 2841 (1971)
88 F. Felix, J. Ferguson, H. U. Güdel, A. Ludi; Chem. Phys. Lett. **62**, 153 (1979)
89 B. J. Pankuch, D. E. Lacky, G. A. Crosby; J. Phys. Chem. **84**, 2061 **and** 2068 (1980)
90 S. Decurtins, F. Felix, J. Ferguson, H. U. Güdel, A. Ludi; J. Am. Chem. Soc. **102**, 4102 (1980)
91 J. Ferguson, F. Herren, G. M. McLaughlin; Chem. Phys. Lett. **89**, 376 (1982)
92 J. Ferguson, F. Herren, E. R. Krausz, M. Maeder, J. Vrbancich; Coord. Chem. Rev. **64**, 21 (1985)
93 E. M. Kober, J. V. Caspar, R. S. Lupmkin, T. J. Meyer; J. Phys. Chem. **90**, 3722 (1986)
94 G. Hensler, E. Gallhuber, H. Yersin; Reference [8], p. 107
95 H. Yersin, E. Gallhuber, G. Hensler; Chem. Phys. Lett. **140**, 157 (1987)
96 H. Yersin, D. Braun, E. Gallhuber, G. Hensler; Ber. Bunsenges. Phys. Chem. **91**, 1228 (1987)
97 H. Yersin, G. Hensler, E. Gallhuber; J. Luminesc. **40/41**, 676 (1988)
98 H. Yersin, D. Braun, E. Gallhuber, G. Hensler; XII IUPAC Symposium on Photochemistry, Bologna, July 1988, Vol. of Abstracts p. 804
99 H. Yersin, W. v. Ammon, M. Stock, G. Gliemann; J. Luminesc. **18/19**, 774 (1979)
100 W. Holzapfel, H. Yersin, G. Gliemann; J. Chem. Phys. **74**, 2124 (1981)
101 H. Yersin, M. Stock; J. Chem. Phys. **76**, 2136 (1982)
102 M. Köhler, D. Schmid, H. C. Wolf; J. Luminesc. **14**, 41 (1976)
103 M. Bettinelli; see the paper in this volume
104 E. Krausz, G. Moran; Adv Magn. Optics, J. Magn. Soc. Jpn. **11**, Suppl. S1, 23 (1987)
105 J. V. Caspar, T. D. Westmoreland, G. H. Allen, P. G. Bradley, T. J. Meyer, W. H. Woodruff; J. Am. Chem. Soc. **106**, 3492 (1984)
106 P. K. Mallick, G. D. Danzer, D. P. Strommen, J. R. Kincaid; J. Phys. Chem. **92**, 5628 (1988)

ORBITAL MODELS AND THE PHOTOCHEMISTRY OF TRANSITION–METAL COMPLEXES

A. CEULEMANS
Department of Chemistry
University of Leuven
Celestijnenlaan 200F
B–3030 Leuven
Belgium

ABSTRACT. This article examines the ligand field based orbital models, that are being used to explain photochemical properties of transition–metal complexes. Special attention is devoted to the formalism of dynamic ligand field theory, which constitutes a promising new tool in this respect. A case study is presented to illustrate the important role of the Jahn–Teller effect in inorganic photochemistry.

1. Introduction

Coupling of electronic and nuclear motion is an evident characteristic of all chemical processes, including photochemical reactions that are initiated by photon absorption. Nevertheless in the chemical literature the widespread importance of vibronic coupling is not generally recognized. The chemist's understanding of thermal and photochemical reaction mechanisms is often based on qualitative orbital diagrams, rather than on elaborate vibronic hamiltonians. Orbitals are pictorial tools, which tell where the electrons are, and eventually provide some indication as to how the nuclei may be put into motion. In the present article we deliberately use these orbital concepts as a starting point. Subsequently the connection with vibronic coupling matrix elements is developed.

2. Orbital Shapes and Symmetries

Orbital models for transition–metal complexes usually are based on the limited set of the five d–orbitals. Still these may be represented in an infinity of ways and therefore some criterion is needed to define the d–orbital shapes. A convenient definition may be based on a sequence of group and subgroup representation labels, which gradually fixes the orbital basis [1]. As an example the label sequence $d > e_g > a_{1g}$, based on the group chain $O(3) > O_h > D_{4h}^{z}$, provides an unequivocal definition for the d_{z^2} function, leaving only its phase undetermined. In contrast, when a sequence terminates in a degenerate representation, a further symmetry lowering is

C. D. Flint (ed.), Vibronic Processes in Inorganic Chemistry, 221–254.
© 1989 by Kluwer Academic Publishers.

needed to fix the degenerate components. Hence the d_{xz} and d_{yz} orbitals, which are both characterized by the sequence $d > t_{2g} > e_g$, may be distinguished by a twofold symmetry axis along the x direction C_2^x. This operation resolves the twofold degenerate e_g representation of D_{4h}^z into a symmetric and an antisymmetric part. These parts will be denoted as e_x and e_y respectively. The splitting of the e_g representation under C_2^x may then be represented in the following way [2] :

$$C_2^x(e_x,e_y) = (e_x,e_y)\begin{pmatrix}1 & 0\\ 0 & -1\end{pmatrix} \tag{1}$$

The d_{yz} function is symmetric under C_2^x and therefore matches the $d > t_{2g} > e_{gx}$ sequence, while d_{xz} corresponds to the $d > t_{2g} > e_{gy}$ antisymmetric component. As a further refinement the fourfold axis of D_{4h}^z may be used to connect the partners, thereby fixing their relative phase. This connection of the components by C_4^z will be defined as follows :

$$C_4^z(e_x,e_y) = (e_x,e_y)\begin{pmatrix}0 & -1\\ 1 & 0\end{pmatrix} \tag{2}$$

Symmetry adaptation of d_{xz} and d_{yz} with respect to equation (2) can be achieved by changing the sign of d_{xz}. The resulting basis $(d_{yz},-d_{xz})$ will obey precisely the symmetry requirements in equations (1) and (2). The full tetragonal form of the d–orbital basis is presented in equation (3).

$$d \begin{cases} \nearrow e_g \begin{cases} \nearrow a_{1g} : d_{z^2} \\ \searrow b_{1g} : d_{x^2-y^2} \end{cases} \\ \searrow t_{2g} \begin{cases} \nearrow b_{2g} : d_{xy} \\ \searrow e_g \begin{cases} \nearrow x : d_{yz} \\ \searrow y : -d_{xz} \end{cases} \end{cases} \end{cases} \tag{3}$$

Similarly a trigonal quantization scheme may be developed along the chain $O(3) > O_h > D_3^{xyz}$. The defining symmetry relations for the doubly degenerate e representation of D_3 are as follows :

$$C_2^{x\bar{y}}(e_\theta,e_\epsilon) = (e_\theta,e_\epsilon)\begin{pmatrix}1 & 0\\ 0 & -1\end{pmatrix}$$

$$C_3^{xyz}(e_\theta,e_\epsilon) = (e_\theta,e_\epsilon)\begin{pmatrix} -1/2 & -\sqrt{3}/2 \\ \sqrt{3}/2 & -1/2 \end{pmatrix} \qquad (4)$$

Here the splitting operation, $C_2^{x\bar{y}}$, is a twofold rotation axis in the xy plane at the x = –y direction; e_θ and e_ϵ denote respectively the symmetric and antisymmetric component. The connecting operation C_3^{xyz}, is the principal threefold rotation axis along the x = y = z direction. The resulting trigonal basis set is specified in equation (5). Computer drawings of the trigonal t_{2g} functions may be found in the recent literature [3,4].

$$d \rightarrow e_g \rightarrow e \begin{cases} \theta : d_{z^2} \\ \epsilon : d_{x^2-y^2} \end{cases}$$

$$d \rightarrow t_{2g} \begin{cases} a_1 : (d_{xz} + d_{yz} + d_{xy})/\sqrt{3} \\ e \begin{cases} \theta : (2d_{xy} - d_{xz} - d_{yz})/\sqrt{6} \\ \epsilon : (d_{yz} - d_{xz})/\sqrt{2} \end{cases} \end{cases} \qquad (5)$$

It is by no means imperative to follow the conventions outlined in equations (3) and (5). Depending on the problem at hand, other subduction schemes might be more appropriate. No matter which choice will be preferred, it always remains of utmost importance to specify explicitly the defining symmetry relations and to apply these in a consistent and uniform way. The benefit of using a well defined symmetry basis becomes apparent when constructing energy matrices. In general matrix elements between components of different symmetries or subsymmetries will be zero. As an example in a trigonal ligand field matrix elements between e_θ and e_ϵ functions yield zero.

$$<(e_g)e_\theta|\mathcal{V}^{D3}|(t_{2g})e_\epsilon> = 0$$
$$<(e_g)e_\epsilon|\mathcal{V}^{D3}|(t_{2g})e_\theta> = 0 \qquad (6)$$

Moreover, as a result of phase adaptation to a connecting symmetry operation, allowed matrix elements will be independent of the subsymmetries. As an example :

$$<(e_g)e_\theta|\mathcal{V}^{D3}|(t_{2g})e_\theta> = <(e_g)e_\epsilon|\mathcal{V}^{D3}|(t_{2g})e_\epsilon> \qquad (7)$$

Having defined a suitable orbital basis set, the next step requires the construction of many–electron wavefunctions describing excited states of potential photochemical interest.

3. Orbital Analysis of Many–Electron States

3.1 ORBITAL OCCUPATION NUMBERS

The d^n configuration of a transition–metal ion gives rise to a considerable number of ligand field eigenstates. The corresponding wavefunctions may easily be obtained by using symmetry arguments, in combination with the tried methods of ligand field theory [1,5]. In general such eigenfunctions will take the form of a linear combination of determinantal functions, as represented in equation (8).

$$\psi = \sum_k c_k D_k \tag{8}$$

Here D_k is a normalized Slater determinant, representing a single d^n configuration, and c_k is the corresponding expansion coefficient. At first sight the multideterminantal form of ψ tends to blur the simple orbital picture. However this picture can be restored through the concept of orbital occupation numbers. The occupation of the ith d–orbital, $n(d_i)$, is defined as in equation (9).

$$n(d_i) = \sum_k c_k \bar{c}_k n_k(d_i) \tag{9}$$

In this equation $n_k(d_i)$ represents the occupation number (0, 1, or 2) of d_i in D_k. In this way a given many–electron function, ψ, can be associated with a set of five fractional numbers between 0 and 2, $\{n(d_i)\}_{i=1,5}$, describing its d–orbital composition.

When considering an orbitally degenerate state, the orbital occupation numbers for the individual components do not offer a reliable description, since the components may be represented in an infinity of ways. However the state averages of these numbers may be verified to remain invariant under unitary transformations of the degenerate basisfunctions. Hence for a degenerate state meaningful orbital occupation numbers can be obtained by averaging the results for the individual components.

As an example, table 1 lists the three degenerate components of the excited $^4T_{2g}$ state in octahedral d^3 complexes. The table also includes the d–orbital occupation numbers for the individual components. The state averages of these numbers yield the orbital composition of the $^4T_{2g}$ state as a whole.

In a photochemical context, d–orbital occupation numbers, in combination with the orbital shapes, may provide a first indication as to the potential photochemical activity of a given excited state. As an obvious example, the $^4T_{2g}$ state of octahedral d^3 complexes is likely to be substitutionally labile, due to the partial occupation of the d_{z^2} and $d_{x^2-y^2}$ orbitals, which are antibonding along the metal–ligand interatomic axes. For more detailed predictions of this type, we refer to the energy expressions in section 4.

TABLE 1. Orbital analysis of the $^4T_{2g}$ state in octahedral d^3 complexes. Tetragonal component labels follow the defining relations in equations (1) and (2)

$^4T_{2g}(t_{2g}^2 e_g^1)$	n: d_{z^2}	d_{yz}	d_{xz}	d_{xy}	$d_{x^2-y^2}$
$^4B_{2g}$: $\vert(yz)(xz)(x^2\text{-}y^2)\vert$	0	1	1	0	1
$^4E_{gx}$: $-\frac{\sqrt{3}}{2}\vert(xz)(xy)(z^2)\vert - \frac{1}{2}\vert(xz)(xy)(x^2-y^2)\vert$	3/4	0	1	1	1/4
$^4E_{gy}$: $\frac{\sqrt{3}}{2}\vert(yz)(xy)(z^2)\vert - \frac{1}{2}\vert(yz)(xy)(x^2-y^2)\vert$	3/4	1	0	1	1/4
State average :	1/2	2/3	2/3	2/3	1/2

3.2 HALF–FILLED SHELL CONFIGURATIONS

A treatment of orbital occupation numbers would be incomplete without paying attention to the exceptional configurational structure of half–filled shell multiplets [6,7]. Examples are the $(t_{2g})^3$ multiplets in octahedral Cr^{3+} and Mn^{4+} complexes, the $(t_{2g})^3(e_g)^2$ states in high–spin Mn^{2+} or Fe^{3+} complexes and the $(t_{2g})^6(e_g)^2$ states in Ni^{2+}. Other examples may be found in the lanthanide series and in organic diradicals [8].

As an example table 2 lists the $(t_{2g})^3$ states of octahedral d^3 systems. A trigonal quantization scheme was chosen [3], following the standard symmetry relations of equations (4) and (5). If one calculates the t_{2g} orbital occupation numbers for the individual components in table 2, one finds that all nine *components* have exactly the same orbital composition : $n(a_1) = n(\theta) = n(\epsilon) = 1$. Hence, in contrast to the $^4T_{2g}$ state in table 1, the orbital composition for degenerate half–filled shell states is found to be the same for all degenerate components. This proves to be a general property of half–filled shell multiplets in real representational form. As has been pointed out by Griffith [1], it is related to a special form of hole–particle complementarity and may be shown to have far–reaching consequences. In the following we will substantiate this complementarity concept, by defining an exchange symmetry operation which commutes with spatial and permutational symmetry operators.

The starting point is the totally symmetric closed–shell determinant. For a $(t_{2g})^6$ shell this determinant will be written in standard order as $|a_1\bar{a}_1\theta\bar{\theta}\epsilon\bar{\epsilon}|$. An open–shell determinant may be looked upon as a *minor* of this function. Clearly exchange of holes and particles will turn this minor into its complement. A somewhat more precise definition of the exchange operator may be based on the Laplace expansion theorem for determinants

TABLE 2. The $(t_{2g})^3$ multiplet components in a trigonal quantization scheme. The labels a_1, θ, ϵ refer to the t_{2g} orbitals, defined in equation (5) of the text.

$${}^4A_{2g}(M_S = 3/2)$$

$A_2 : |a_1\theta\epsilon|$

$${}^2E_g\,(M_S = 1/2)$$

$$E\theta : \frac{1}{\sqrt{6}}(\sqrt{2}|a_1\bar{a}_1\theta| - \sqrt{2}|\theta\epsilon\bar{\epsilon}| + |a_1\theta\bar{\theta}| - |a_1\epsilon\bar{\epsilon}|)$$

$$E\epsilon : \frac{1}{\sqrt{6}}(\sqrt{2}|a_1\bar{a}_1\epsilon| - \sqrt{2}|\theta\bar{\theta}\epsilon| + |a_1\bar{\theta}\epsilon| - |a_1\theta\bar{\epsilon}|)$$

$${}^2T_{1g}\,(M_S = 1/2)$$

$$A_2 : \frac{1}{\sqrt{6}}(2|\bar{a}_1\theta\epsilon| - |a_1\bar{\theta}\epsilon| - |a_1\theta\bar{\epsilon}|)$$

$$E\theta : \frac{1}{\sqrt{6}}(-\sqrt{2}|a_1\theta\bar{\theta}| + \sqrt{2}|a_1\epsilon\bar{\epsilon}| - |\theta\epsilon\bar{\epsilon}| + |a_1\bar{a}_1\theta|)$$

$$E\epsilon : \frac{1}{\sqrt{6}}(\sqrt{2}|a_1\theta\bar{\epsilon}| - \sqrt{2}|a_1\bar{\theta}\epsilon| + |a_1\bar{a}_1\epsilon| - |\theta\bar{\theta}\epsilon|)$$

$${}^2T_{2g}\,(M_S = 1/2)$$

$$A_1 : \frac{1}{\sqrt{2}}(|a_1\epsilon\bar{\epsilon}| + |a_1\theta\bar{\theta}|)$$

$$E\theta : \frac{1}{\sqrt{2}}(|a_1\bar{a}_1\theta| + |\theta\epsilon\bar{\epsilon}|)$$

$$E\epsilon : \frac{1}{\sqrt{2}}(|a_1\bar{a}_1\epsilon| + |\theta\bar{\theta}\epsilon|)$$

[9]. In this scheme the hole–electron exchange operator, $\mathcal{E}$, will be defined as the operator which turns a given minor into its *algebraic* complement. This means that the action of $\mathcal{E}$ on a minor creates a *signed* complement, the sign being determined by the sum of the row and column numbers of the argument minor in the closed–shell determinant. As an example, for a $(t_{2g})^3$ string, such as $|a_1\epsilon\bar{\epsilon}|$, one has :

$$\mathcal{E}|a_1\epsilon\bar{\epsilon}| = (-1)^{1+2+3+1+5+6}|\bar{a}_1\theta\bar{\theta}| = |\bar{a}_1\theta\bar{\theta}| \tag{10}$$

From the phase factor in this equation, one recognizes that the argument minor $|a_1\epsilon\bar{\epsilon}|$ is formed from rows 1,2,3 and columns 1,5,6 in the parent $|a_1\bar{a}_1\theta\bar{\theta}\epsilon\,\bar{\epsilon}|$ determinant [10].
Clearly, if a shell is half–filled, it will be mapped onto itself under the

exchange of holes and electrons. In this instance $\mathscr{E}$ can be identified as a new invariance operation, besides the established spatial and permutational symmetries. As a function operator in the space of the half-filled shell determinants, $\mathscr{E}$ exhibits the following properties [7] :

a) $\mathscr{E}$ is an antilinear operator
b) $\mathscr{E}$ commutes with spatial symmetry operators
c) $\mathscr{E}^2 = \pm 1$, applying $\mathscr{E}$ twice yields the identity operator when the half-filled shell contains an even number of fermions, and minus the identity operator when the number of fermions is odd.

$\mathscr{E}$ shares all these properties with the time reversal operator, $\mathscr{T}$. Thus we are led to the conclusion that *exchange of holes and electrons implies reversal of time* ! In consequence, by multiplying $\mathscr{E}$ with $\mathscr{T}$, one describes a loop in time leaving as a net result a pure configurational transformation. Hence the true generator of the hole–electron exchange symmetry group for half-filled shell states is given by the product $\mathscr{T}\mathscr{E}$ [11]. This operator product is linear, commutes with spatial and spin symmetry operators, and may be shown to be its own inverse, as equation (11) illustrates [7].

$$\begin{aligned} \mathscr{T}\mathscr{E}\mathscr{T}\mathscr{E}|\, a_1\epsilon\bar{\epsilon}\,| &= \mathscr{T}\mathscr{E}\mathscr{T}|\,\bar{a}_1\theta\bar{\theta}\,| = -\,\mathscr{T}\mathscr{E}|\,a_1\theta\bar{\theta}\,| \\ &= -\mathscr{T}|\,\bar{a}_1\epsilon\bar{\epsilon}\,| = |\,a_1\epsilon\bar{\epsilon}\,| \end{aligned} \tag{11}$$

Therefore $\mathscr{T}\mathscr{E}$ generates a group of order two with only two representations (or *parities*) : a symmetric one, labeled (+), and an antisymmetric one, labeled (–). Since $\mathscr{T}\mathscr{E}$ is *linear* and commutes with spatial and spin symmetry operators, half-filled shell multiplets must have a definite parity under the $\mathscr{T}\mathscr{E}$ operation. One may indeed verify that the $^4A_{2g}$, ^2E_g and $^2T_{1g}$ states in table 2 are (+) states, while the $^2T_{2g}$ state is a (–) state. «The fact that we have two different kinds of terms in the half-filled shell is not ... merely a matter of definition and is of interest and importance.» [12]. It affects the interaction matrix elements. In the case of a one-electron hermitian operator, $\mathscr{H}$, the following *selection rules* may be derived [7] :

a) Interaction elements between half-filled shell states of opposite parity will be zero if $\mathscr{H}$ is antisymmetric under time reversal.
b) Off-diagonal interaction elements between half-filled shell states of identical parity will be zero if $\mathscr{H}$ is symmetric under time reversal.
c) Diagonal interaction elements between half-filled shell states will be zero if $\mathscr{H}$ is symmetric under time reversal and non-totally symmetric under the spatial symmetry operations.

A case in point is the linear Jahn–Teller (JT) interaction. The corresponding operator is a one-electron operator, which is symmetric under time reversal, and non-totally symmetric under spatial symmetry operations. Linear JT interaction elements for degenerate half-filled shell states will thus be zero. *The implication is that degenerate half-filled shell states cannot exhibit a first-order Jahn-Teller effect* [7] ! The observed equality of the t_{2g}-orbital occupation numbers for the $(t_{2g})^3$ terms in table 2 concurs with this result. In spectroscopy, the absence of first-order JT effects in half-filled shell terms is a well known feature, which facilitates the observation of weak higher-order effects [6,13]. Further experimental manifestations of self-complementarity concern the photochemical and photophysical properties of these half-filled shell multiplets. As an example the low-lying $^2E_g(t_{2g}^3)$ state in octahedral Cr^{3+} complexes is probably photochemically inert, has a long lifetime and gives rise to narrow emission spectra, that are not very sensitive to solvent changes. This relative robustness of the ^2E_g state is a typical consequence of the fact that each *component* of this state has exactly the same orbital composition as the $^4A_{2g}$ ground state [3,7].

While the hole–electron exchange is an important symmetry characteristic of half–filled shell states, it must be kept in mind that it is not an invariance group of the total hamiltonian. Accordingly, it may be broken by configuration interactions, which infringe upon the configurational uniformity of these states. Usually – as far as the lower multiplets of a half–filled shell manifold are concerned – these interactions are not very pronounced. However, they may be intensified by introducing heteroligands, which lower the symmetry of the first coordination sphere. The resulting relaxation of the parity constraints may give rise to the appearance in the spectra of Franck–Condon progressions and appreciable Stokes and solvent shifts [3,14,15,16,17]. Eventually strong substituent effects may even induce direct photochemical reactivity.

4. Orbital Energies

In the previous section excited states were characterized by a set of fractional numbers, which specified the occupation of the five d–orbitals. These numbers may be combined with the orbital shapes to produce a pictorial description of electronic structure. However when it comes to lower symmetry complexes, it is more appropriate to combine the orbital occupation numbers directly with orbital energies. In this way a quantitative description of the photochemical activity of an excited state may be obtained. Two types of energy expressions will now be discussed.

4.1 LIGAND FIELD STABILIZATION ENERGY

The Ligand Field Stabilization Energy (LFSE) is defined as the energy difference between the actual configuration of a state and an average reference configuration [1]. For an octahedral complex the energy of a $t_{2g}^m e_g^n$ configuration may be expressed in the Angular Overlap Model (AOM) parameters $e_\sigma, e_\pi, e_\delta$ as follows :

$$E(t_{2g}^m e_g^n) = m(4e_\pi + 2e_\delta) + n(3e_\sigma + 3e_\delta) \quad (12)$$

In the reference configuration the m + n electrons are equally distributed over the five d–orbitals. The reference orbital occupation number of each d–orbital therefore corresponds to (m + n)/5. Hence one has :

$$E(d^{m+n}) = \frac{m+n}{5}(6e_\sigma + 12e_\pi + 12_\delta) \quad (13)$$

The LFSE is then given by :

$$\mathrm{LFSE} = E(t_{2g}^{m}e_{g}^{n}) - E(d^{m+n})$$

$$= \frac{3n-2m}{5}\,[3(e_\sigma - e_\delta) - 4(e_\pi - e_\delta)] \qquad (14)$$

As may be seen from equation (14) the LFSE is proportional to the spectrochemical strength parameter $10\,Dq = 3(e_\sigma - e_\delta) - 4(e_\pi - e_\delta)$, which describes the splitting of the d–shell in an octahedral ligand field. By preferentially occupying the lower lying t_{2g} orbitals the ground state achieves an extra stability, since these orbitals keep the metal electrons away from the repulsive interactions of the ligand charges. The LFSE concept is perfectly suited to rationalize the variations across the transition series of several thermodynamic and structural properties. However as far as ligand substitution reactions are concerned, its utility is far more restrained [18]. This is because the LFSE is a typical *global* concept. It measures the stabilization energy of the complex as a whole, but does not consider the bonding of the individual ligands. The study of ligand substitution processes requires a *local* bond descriptor. The concept, that is fitted for this purpose, is the ligand field bond index, which will be discussed below.

4.2 BOND INDICES

A bond index is a semi–quantitative measure for the strength of an individual metal–ligand bond [19,20]. Its calculation is based on the particularly simple view of the coordination bond, that is present in the work of McClure [21], Kettle [22], Burdett [23] and others. In this view it is assumed that a coordination bond is formed because occupied ligand orbitals interact with unoccupied d–orbitals; the stabilization of the bonding combination is assumed to outweigh exactly the destabilization of the antibonding combination. In this mirror–image model the total bond energy, I, is simply the sum of the energies of the unoccupied d–orbitals.

$$I = \sum_i [2 - n(d_i)] < d_i | \mathscr{V} | d_i > \qquad (15)$$

In this expression $2 - n(d_i)$ represents the degree of vacancy of the ith d–orbital, while the matrix element $<d_i|\mathscr{V}|d_i>$ refers to the energy of d_i under the ligand field potential $\mathscr{V}$. The summation runs over the five d–orbitals, *that diagonalize the ligand field perturbation matrix.* Remark that I is defined as a positive quantity.

In an additive ligand field formalism, the potential $\mathscr{V}$ is written as a sum over individual ligand potentials, $\mathscr{V}_L$. This allows a natural partitioning of I into local bond indices $I(M-L)$:

$$I = \sum_L I(M-L) \qquad (16a)$$

with

$$I(M-L) = \sum_i [2 - n(d_i)] \langle d_i | \mathcal{V}_L | d_i \rangle \tag{16b}$$

These formulae may be applied to any ground or excited state, by simply inserting the relevant orbital occupation numbers and energy expressions. The AOM parametrization scheme seems particularly well designed for this type of application since it uses the local M–L interactions as its elementary building blocks [24,25].

As a straightforward illustration of equation (16), we calculate the bond indices for the $^4A_{2g}$ and $^4T_{2g}$ states of a d^3 complex. In the $^4A_{2g}(t_{2g}^3)$ ground state there are four holes in the e_g shell with energy $3e_\sigma + 3e_\delta$, and three holes in the t_{2g} shell with energy $4e_\pi + 2e_\delta$. Neglecting the e_δ parameter, one obtains for the $^4A_{2g}$ state :

$$\begin{aligned} I &= 12e_\sigma + 12e_\pi \\ I(M-L) &= 2e_\sigma + 2e_\pi \end{aligned} \tag{17}$$

In the $^4T_{2g}(t_{2g}^2e_g)$ excited state, one electron is transferred from t_{2g} to e_g. The excited state bond indices, symbolized as I*, now become :

$$\begin{aligned} I^* &= 9e_\sigma + 16e_\pi \\ I^*(M-L) &= \frac{3}{2}e_\sigma + \frac{8}{3}e_\pi \end{aligned} \tag{18}$$

Clearly, since $e_\sigma \gg e_\pi$, $I^*(M-L)$ will be less than $I(M-L)$ thus facilitating ligand loss in the excited state. The net ligand labilization energy upon excitation can easily be determined by making the difference between $I(M-L)$ and $I^*(M-L)$:

$$I(M-L) - I^*(M-L) = \frac{1}{2}e_\sigma - \frac{2}{3}e_\pi = {}^1/_6(10Dq) \tag{19}$$

Hence the bond weakening depends on the energy gap between the t_{2g} and e_g shells, as was to be expected. More detailed information may be obtained from the study of lower symmetry cases, such as a tetragonal *trans*-CrL_4X_2 complex with four equatorial L ligands and two axial X ligands. In this D_{4h} complex the $^4T_{2g}$ state splits into two tetragonal components $^4B_{2g}$ and ^4E_g. The zeroth-order wavefunctions and orbital occupation numbers for these components were given in table 1. The relevant AOM orbital energy expressions and the tetragonal splitting of the $^4T_{2g}$ state are given in equations (20a) and (20b) respectively :

$$E(xz) = E(yz) = 2e_\pi^L + 2e_\pi^X$$

$$E(xy) = 4e_\pi^L$$

$$E(z^2) = 2e_\sigma^X + e_\sigma^L$$

$$E(x^2 - y^2) = 3e_\sigma^L \qquad (20a)$$

$$E(^4B_{2g}) - E(^4E_g) = \frac{1}{2}(10Dq_L - 10Dq_X) \qquad (20b)$$

Combination of orbital energies and occupation numbers yields the following I* values :

$$^4B_{2g} \begin{cases} I_L^* = \frac{5}{4} e_\sigma^L + 3e_\pi^L \\ I_X^* = 2e_\sigma^X + 2e_\pi^X \end{cases}$$

$$^4E_g \begin{cases} I_L^* = \frac{13}{8} e_\sigma^L + \frac{5}{2} e_\pi^L \\ I_X^* = \frac{5}{4} e_\sigma^X + 3e_\pi^X \end{cases} \qquad (21)$$

As equation (21) shows, the I* model makes a clear distinction between the labilization of the axial and equatorial sites in the $^4B_{2g}$ and ^4E_g states. In the $^4B_{2g}$ state the bond index of the X ligand remains unchanged as compared to the ground state value, whereas the four equatorial L ligands are strongly labilized. In the ^4E_g state the labilization has an opposite polarization and is mainly concentrated on the axial ligands.

As a rule the component at lower energy will be the dominant photoactive level (Photochemical Kasha rule [26]). Hence, if the tetragonal splitting is positive, i.e. if $10Dq_X < 10Dq_L$, the ^4E_g component is lower and a dominant axial labilization will be expected. If the tetragonal splitting is reversed, i.e. if $10Dq_L < 10Dq_X$, the equatorial $^4B_{2g}$ component will be the lower level and equatorial ligand labilization should occur. In either case the site of weakest spectrochemical strength is preferentially labilized. This prediction precisely coincides with Adamson's first empirical rule regarding the site selectivity of the photosubstitution reactions in mixed Cr^{3+} complexes [27]. In the past decade the empirical test of the I* model has been extended to a wide range of photolysis data for various transition–metal ions [20,28–32]. In most cases the ligand with the lowest I*(M – L) value for the photochemically active excited state is indeed found to be the ligand that is most easily expelled. This remarkable success of the bond index model is quite surprising when compared with the crude assumptions on which this model is based.

5. Coordinate derivatives of orbital energies

In the previous section parametric expressions for metal–ligand bond strengths were derived within the framework of the Angular Overlap Model. These quantities allowed a rationalization of the observed site selectivity in the photochemistry of low-symmetry complexes. However for a more detailed understanding of photochemical reactivity, it is necessary to focus on the vibrationally equilibrated excited states or "thexi" states, which are the immediate precursors of the reactive decay [29]. In recent years there has been some progress towards a direct structural characterization of these thexi-states. This progress has mainly been based on high-resolution spectroscopy of the vibronic structure of ligand field transitions [33,34]. In principle Resonance Raman studies could also contribute to the identification of the force field and structure of the excited state.

From a theoretical point of view the relaxation from a Franck-Condon state to a thexi-state is a typical example of vibronic coupling. The principal coupling operator, $\frac{\partial \mathcal{H}}{\partial Q}\Big|_0 Q$, involves the first derivative of the electronic hamiltonian $\mathcal{H}$ with respect to the nuclear displacement coordinates Q, taken at the ground state equilibrium position. Matrix elements of this operator represent the forces that act upon the system after excitation. In combination with a restoring force field for the excited state, these quantities allow a characterization of the excited state potential energy surface, including deactivation channels and metastable minima. Ligand field theory can play an important role in the evaluation of the vibronic coupling elements. Indeed, as far as the d-electrons are concerned, the vibronic coupling operator effectively reduces to the coordinate derivative of the ligand field potential, $\frac{\partial \mathcal{V}}{\partial Q}$, multiplied by Q. The part of ligand field theory that is concerned with evaluating matrix elements of the type $\langle d_i|\frac{\partial \mathcal{V}}{\partial Q_\alpha}|d_j\rangle$, and $\langle d_i|\frac{\partial^2 \mathcal{V}}{\partial Q_\alpha \partial Q_\beta}|d_j\rangle$ is called *Dynamic Ligand Field Theory*. In the following this theory will be presented within the AOM formalism.

5.1 THE DYNAMIC AOM

The dynamic extension of the AOM is due to Bacci [35-37]. Here we review the essential features of the formalism, keeping in line with Bacci's notation. In AOM the general expression for a matrix element of the ligand field potential is given by [24] :

$$\langle d_i|\mathcal{V}|d_j\rangle = \sum_L \sum_{\lambda\omega} F_{\lambda\omega}(d_i;\theta_L\varphi_L)F_{\lambda\omega}(d_j;\theta_L\varphi_L)\, e_\lambda^L \qquad (22)$$

with : $\lambda\omega = \sigma, \pi y, \pi x, \delta xy, \delta x^2-y^2$

In this equation $F_{\lambda\omega}(d_i;\theta_L\varphi_L)$ is an element of the AOM rotation matrix, θ_L and φ_L being the angular coordinates of ligand L; e_λ^L is the λ-type AOM parameter for ligand L. For convenience the σ and π columns of $\mathbb{F}$ are listed in table 3.

TABLE 3. The $\lambda\omega = \sigma, \pi y, \pi x$ part of the AOM rotation matrix

$F_{\lambda\omega}(d_i;\theta\varphi)$	σ	πy	πx
d_{z^2}	$\frac{1}{4}(1 + 3\cos 2\theta)$	0	$-\frac{1}{2}\sqrt{3}\sin 2\theta$
d_{yz}	$\frac{1}{2}\sqrt{3}\sin 2\theta \sin\varphi$	$\cos\theta\cos\varphi$	$\cos 2\theta \sin\varphi$
d_{xz}	$\frac{1}{2}\sqrt{3}\sin 2\theta\cos\varphi$	$-\cos\theta\sin\varphi$	$\cos 2\theta\cos\varphi$
d_{xy}	$\frac{1}{4}\sqrt{3}(1-\cos 2\theta)\sin 2\varphi$	$\sin\theta\cos 2\varphi$	$\frac{1}{2}\sin 2\theta\sin 2\varphi$
$d_{x^2-y^2}$	$\frac{1}{4}\sqrt{3}(1-\cos 2\theta)\cos 2\varphi$	$-\sin\theta\sin 2\varphi$	$\frac{1}{2}\sin 2\theta\cos 2\varphi$

One must realize that AOM – in contrast to the pure crystal field model – does not specify the operator $\mathcal{V}$. As such direct calculations of matrix elements of the derivative operator $\partial\mathcal{V}/\partial Q$ cannot be performed and must be replaced by differentiation of the parametric AOM expressions in equation (22) with respect to nuclear displacements. Since AOM uses spherical polar coordinates to specify the ligand positions, it is indicated to carry out the differentiation in the same coordinate system. The spherical polar form of the gradient operator for the displacement of ligand L is given by :

$$\vec{\nabla}_L = (\nabla_L^R, \nabla_L^\theta, \nabla_L^\varphi) \tag{23}$$

with
$$\nabla_L^R = \frac{\partial}{\partial R_L}$$
$$\nabla_L^\theta = \frac{1}{R_L}\frac{\partial}{\partial\theta_L}$$
$$\nabla_L^\varphi = \frac{1}{R_L\sin\theta_L}\frac{\partial}{\partial\varphi_L}$$

The differential operator for an internal coordinate Q_α can be expanded over the individual ligand displacements in the following way :

$$\frac{\partial}{\partial Q_\alpha} = \sum_L\sum_\ell U_{L\ell,\alpha}\nabla_L^\ell \tag{24}$$

with : $\ell = R, \theta, \varphi$

Combination of equations (22) and (24) finally yields the parametric expression for the first–order dynamic ligand field matrix.

$$<d_i|\frac{\partial \mathscr{V}}{\partial Q_\alpha}|d_j> = \sum_L \sum_\ell \sum_{\lambda\omega} U_{L\ell,\alpha} \nabla_L^\ell [F_{\lambda\omega}(d_i;\theta_L\varphi_L) F_{\lambda\omega}(d_j;\theta_L\varphi_L)e_\lambda^L] \quad (25)$$

For a pure bending mode, the dynamic coupling elements will be expressed in exactly the same e_λ^L parameters as the ordinary static AOM elements. For a stretching mode new $\partial e_\lambda^L/\partial R_L$ parameters appear, which incorporate the radial dependence of the AOM parameters.

More recently, the dynamic ligand field theory has been extended to evaluate second-order terms of the general type $<d_i|\frac{\partial^2 \mathscr{V}}{\partial Q_\alpha \partial Q_\beta}|d_j>$ [38]. The second-order differential operators are given by :

$$\frac{\partial^2}{\partial Q_\alpha \partial Q_\beta} = \sum_L \sum_\ell \sum_{L'} \sum_{\ell'} U_{L\ell,\alpha} U_{L'\ell',\beta} (\nabla\nabla)_{LL'}^{\ell\ell'} \quad (26)$$

Since the energy expressions are additive in the ligands, there is no need for inter-ligand terms in this operator. Consequently equation (26) reduces to :

$$\frac{\partial^2}{\partial Q_\alpha \partial Q_\beta} = \sum_L \sum_\ell \sum_{\ell'} U_{L\ell,\alpha} U_{L\ell',\beta} (\nabla\nabla)_L^{\ell\ell'} \quad (27)$$

Here the intra-ligand term $(\nabla\nabla)_L^{\ell\ell'}$ is not an ordinary product of linear differential operators, but it is a symmetrical spherical tensor, as specified in equation (28).

$$(\nabla\nabla)_L = \begin{pmatrix} \frac{\partial^2}{\partial R_L^2} & \frac{\partial}{\partial R_L}\left(\frac{1}{R_L}\frac{\partial}{\partial \theta_L}\right) & \frac{\partial}{\partial R_L}\left(\frac{1}{R_L \sin\theta_L}\frac{\partial}{\partial \varphi_L}\right) \\ \frac{\partial}{\partial R_L}\left(\frac{1}{R_L}\frac{\partial}{\partial \theta_L}\right) & \frac{1}{R_L}\frac{\partial}{\partial R_L} + \frac{1}{R_L^2}\frac{\partial^2}{\partial \theta_L^2} & \frac{1}{R_L^2}\frac{\partial}{\partial \theta_L}\left(\frac{1}{\sin\theta_L}\frac{\partial}{\partial \varphi_L}\right) \\ \frac{\partial}{\partial R_L}\left(\frac{1}{R_L \sin\theta_L}\frac{\partial}{\partial \varphi_L}\right) & \frac{1}{R_L^2}\frac{\partial}{\partial \theta_L}\left(\frac{1}{\sin\theta_L}\frac{\partial}{\partial \varphi_L}\right) & \frac{1}{R_L}\frac{\partial}{\partial R_L} + \frac{1}{R_L^2 \sin^2\theta_L}\frac{\partial^2}{\partial \varphi_L^2} + \frac{\cot\theta_L}{R_L^2}\frac{\partial}{\partial \theta_L} \end{pmatrix} \quad (28)$$

For a detailed derivation of this tensor we refer to Stone [39]. The surface tensor, which applies to a purely curvilinear bending motion, can easily be obtained from equation (28) by imposing a constant value of R_L. Obviously second–order dynamic ligand field elements may contain three types of parameters : e_λ^L, $\frac{\partial e_\lambda^L}{\partial R_L}$ and $\frac{\partial^2 e_\lambda^L}{\partial R_L^2}$.

5.2 VIBRONIC COUPLING ELEMENTS IN AN OCTAHEDRON

As a case in point we consider the linear vibronic coupling elements for an octahedral complex. In O_h the d–orbitals span the reducible representation $e_g + t_{2g}$. The symmetrized square of this representation determines the symmetries of the allowed normal modes.

$$[e_g + t_{2g}]^2 = 2a_{1g} + 2e_g + t_{1g} + 2t_{2g} \tag{29}$$

Since there is no normal mode of t_{1g} symmetry, linear vibronic coupling interactions in an octahedron are limited to displacements of a_{1g}, e_g and t_{2g} symmetry. In equation (29) the two a_{1g} terms refer to the coupling of the e_g and t_{2g} orbitals to the totally symmetric breathing mode. The two e_g terms and one of the t_{2g} terms represent the E x e type JT activity of the e_g orbitals and the T x (e + t_2) type JT activity of the t_{2g} orbitals. Finally the remaining t_{2g} interaction is due to the pseudo–linear coupling between the e_g and t_{2g} shell. Expressions for all these coupling elements may be found in the literature [36,37]. By way of exercise, the derivation of the pseudo–linear coupling elements will be repeated here in great detail. The relevant t_{2g} bending mode comprises three components, which will be denoted in the standard Griffith conventions [1] as Q_ζ, Q_η, Q_ξ. The positive Q_ζ coordinate and the ligand numbering convention are as follows :

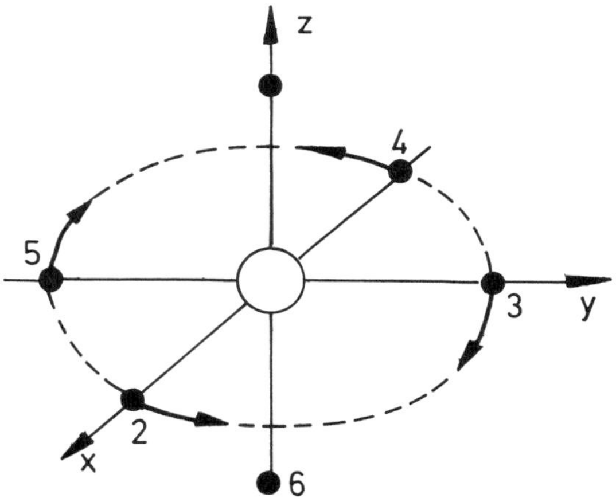

The corresponding normalized curvilinear $\partial/\partial Q_\zeta$ displacement operator is given by :

$$\frac{\partial}{\partial Q_\zeta} = \frac{1}{2R}\left(\frac{\partial}{\partial\varphi_2} - \frac{\partial}{\partial\varphi_3} + \frac{\partial}{\partial\varphi_4} - \frac{\partial}{\partial\varphi_5}\right) \tag{30}$$

R is the M – L distance. In principle the Q_η and Q_ξ expressions may be obtained from equation (30) by applying a threefold symmetry axis. However since these displacements involve ligands 1 and 6 at the poles, the polar singularity of the spherical polar coordinates must be taken into account. This means that one must refrain from substituting the θ coordinates of the polar ligands by their octahedral values until the actual derivatives have been calculated. Hence the η and ξ differential operators are given by :

$$\frac{\partial}{\partial Q_\eta} = \frac{1}{2R}\left(\cos\theta_1\cos\varphi_1\frac{\partial}{\partial\theta_1} - \frac{\sin\varphi_1}{\sin\theta_1}\frac{\partial}{\partial\varphi_1} - \cos\theta_6\cos\varphi_6\frac{\partial}{\partial\theta_6} + \frac{\sin\varphi_6}{\sin\theta_6}\frac{\partial}{\partial\varphi_6} - \frac{\partial}{\partial\theta_2} + \frac{\partial}{\partial\theta_4}\right)$$

$$\frac{\partial}{\partial Q_\xi} = \frac{1}{2R}\left(\cos\theta_1\sin\varphi_1\frac{\partial}{\partial\theta_1} + \frac{\cos\varphi_1}{\sin\theta_1}\frac{\partial}{\partial\varphi_1} - \cos\theta_6\sin\varphi_6\frac{\partial}{\partial\theta_6} - \frac{\cos\varphi_6}{\sin\theta_6}\frac{\partial}{\partial\varphi_6} - \frac{\partial}{\partial\theta_3} + \frac{\partial}{\partial\theta_5}\right) \tag{31}$$

As an example, using the static AOM expressions of equation (22) and the $\mathbb{F}$ matrix in table 3, the matrix element $<d_{z^2}|\mathcal{V}|d_{xz}>$ is found to be :

$$<d_{z^2}|\mathcal{V}|d_{xz}> = \sum_L\left[\frac{\sqrt{3}}{8}(1 + 3\cos2\theta_L)\sin2\theta_L\cos\varphi_L\right]e_\sigma^L - \left[\frac{\sqrt{3}}{2}\sin2\theta_L\cos2\theta_L\cos\varphi_L\right]e_\pi^L \tag{32}$$

The Q_η derivative of this matrix element in the octahedral origin is then given by :

$$<d_{z^2}|\frac{\partial\mathcal{V}}{\partial Q_\eta}|d_{xz}>_0 = \frac{\sqrt{3}\ e_\sigma}{2\ R}(\sin^2\varphi_1 + \cos^2\varphi_1 + \sin^2\varphi_6 + \cos^2\varphi_6 - 1) - \frac{\sqrt{3}\ e_\pi}{2\ R}(\sin^2\varphi_1 + \cos^2\varphi_1 + \sin^2\varphi_6 + \cos^2\varphi_6 - 2) \tag{33}$$

At this point the undeterminate φ-coordinates of the polar ligands are eliminated easily to yield :

$$<d_{z^2}|\frac{\partial \mathcal{V}}{\partial Q_\eta}|d_{xz}>_0 = \frac{\sqrt{3}\, e_\sigma}{2\,R} \qquad (34)$$

The corresponding vibronic coupling element is obtained by multiplying this result with the coordinate Q_η. Proceeding in this way the total pseudo-linear coupling matrix of the octahedral d-system may easily be derived. The results are displayed in table 4.

TABLE 4. The pseudo–linear JT interaction elements in an octahedral complex. In units of e_σ/R.

	d_{yz}	d_{xz}	d_{xy}
d_{z^2}	$\frac{\sqrt{3}}{2}Q_\xi$	$\frac{\sqrt{3}}{2}Q_\eta$	$-\sqrt{3}\,Q_\zeta$
$d_{x^2-y^2}$	$-\frac{3}{2}Q_\xi$	$\frac{3}{2}Q_\eta$	0

Obviously the matrix elements in table 4 may be written as a product of a reduced matrix element and the standard coupling coefficients for the $E\,T_2T_2$ triple, as given by Griffith [40]. Higher–order coupling coefficients may be found in appendix IV of Englman's book [41].

5.3 APPLICATIONS

For applications of dynamic ligand field theory to excited state distortions of octahedral complexes, the reader is referred to a recent publication by Schmidtke and Degen [42]. These authors have carried out a detailed analysis of photochemically active states, representing d–electron coupling with a_{1g}, e_g and t_{2g} vibrations, for which well resolved vibronic spectra have been reported. Both the crystal field and AOM versions of the theory were used. In spite of their approximate nature, these simple methods performed quite well in rationalizing typical vibronic features such as Huang–Rhys factors and excited state distortions. Other examples in the literature mainly concern the evaluation of JT coupling constants for complexes with degenerate ground states [43–47]. Table 5 lists a few examples of T_d and O_h symmetry.

In conclusion dynamic ligand field theory appears to be an interesting additional tool in the study of vibronic coupling in transition–metal complexes. More detailed treatments of the thexi–states in low–symmetry complexes should be within reach. Further interesting aspects of this theory will be presented in the next section.

TABLE 5. Applications of dynamic ligand field theory in octahedral and tetrahedral complexes

Metal	Ion	Symmetry	State	Ref.
d^1	Ti^{3+}	O_h	$^2T_{2g}$,^2E_g	[35]
	V^{4+}	T_d	^2E	[43]
d^3	Cr^{3+}	O_h	$^4T_{2g}$	[42]
	Re^{4+}	O_h	$^4A_{2g}$,$^2T_{2g}$	[37]
d^5	Mn^{2+}	T_d	$^4T_1$,$^4T_2$	[44,45]
d^6	Co^{3+}	O_h	$^3T_{1g}$,$^1T_{1g}$	[42]
	Rh^{3+}	O_h	$^3T_{1g}$	[42]
	Pt^{4+}	O_h	$^3T_{1g}$	[42]
d^8	Fe^0	T_d	$^3T_1$	[38]
	Ni^{2+}	T_d	$^3T_1$	[46]
d^9	Co^0	T_d	$^2T_2$	[38]
	Cu^{2+}	T_d	$^2T_2$	[46,47]
	Cu^{2+}	O_h	^2E_g	[36]

6. Photostereochemistry and the Jahn–Teller effect

Photochemical reactions often exhibit a remarkable stereochemistry, which is different from the ground–state stereochemical behavior [29]. Obviously any attempt to explain such an observation must start off from the reaction mechanism. When interpreting the mechanistic relevance of thermodynamic or kinetic studies, it must be realized that such studies report macroscopic properties. These properties are of course related to the intimate reaction mechanism, but the link is usually far from simple or direct. A considerable body of concurring evidence has to be accumulated before a mechanism can be considered to be well established.

As far as substitution reactions are concerned, mechanistic proposals range from a limiting associative to a limiting dissociative mechanism. An investigation of such alternatives with simple orbital models will usually be centered on the highly reactive species of higher or lower coordination number, that are generated upon ligand association or ligand loss [20]. In the present section this model strategy will be illustrated for the case of the coordinatively unsaturated Cr(III) fragments, that result from photolabilization of hexacoordinate Cr(III) complexes. The treatment will entirely be based on the dynamic ligand field theory, which was exposed in the previous section.

The *trans*–CrL_4XY complexes (L is a strong σ donor such as NH_3, X and Y are weak–field acido ligands such as Cl^-) undergo photosubstitution reactions with almost complete stereomobility. Hence if Y is the leaving ligand, the photoproduct in aqueous solution is found to adopt a *cis*–configuration.

$$trans\text{–}CrL_4XY + H_2O \xrightarrow{h\nu} cis\text{–}CrL_4X(H_2O) + Y \qquad (35)$$

Kirk has summarized this observation as follows : "The entering ligand will stereospecifically occupy a position corresponding to entry into the coordination sphere *trans* to the leaving ligand" [48]. Strictly speaking this *trans*–attack rule implies an associative reaction mechanism, involving the formation of a 7–coordinate species. Although such a mechanism seems to be supported by pressure studies, it begs the important question why the heptacoordinate structure should be formed by *trans*–attack only. Indeed from an orbital point of view both *cis* and *trans* entries should be equally probable and a mixture of stereoretentive and stereomobile products would be anticipated. Moreover, after the formulation of the *trans*–attack rule new information has been gathered on several complexes containing fluoro– and cyano–substituents [49,50,51]. Some of these complexes exhibit a more diversified photostereochemistry, which does not comply with the stereomobility pattern of *trans*–attack. In this respect a dissociative mechanism offers more fruitful prospects. For this reason we will concentrate on the pentacoordinate Cr(III) fragment.

6.1 THE PENTACOORDINATE CrL_5 FRAGMENT

The stereochemistry of an idealized CrL_5 fragment with five identical ligands is based on the interplay of only two alternative geometries : the square pyramid (SP) and the trigonal bipyramid (TBP), respectively with C_{4v} and D_{3h} symmetry. One way of exploring the stereochemistry of this fragment is by performing point to point static ligand field calculations along the various pathways that interconvert the two structures of maximal symmetry. This method, which we have been using previously [20,52], yields Woodward–Hoffmann type correlation diagrams for one–dimensional sections of the reaction surface. At present we will follow an alternative procedure, which is based on the dynamic ligand field theory. This formalism provides a direct functional description of the multidimensional potential energy surface around the SP and TBP structures. As such it allows to overlook the reaction surface as a whole. In the end such a *panoramic view* should lead to more comprehensive models of photostereochemistry. Tables 6 and 7 collect the basic ingredients of a dynamic orbital model for TBP and SP fragments : orbital functions, symmetries and energies, and their coordinate derivatives [53]. Only bending modes have been considered, because these are the only modes that give rise to stereochemical changes. In addition bending modes are known to be more flexible than stretchings.The information, that is contained in the tables, will now be analyzed in detail for each of the two geometries.

6.1.1 *The Trigonal Bipyramid.* The high–spin ground state of the TBP is a $^4E'$ state. Its two $M_S = 3/2$ components will be denoted as follows :

$$|{}^4E' \; x^2\text{-}y^2> \; = \; |(yz)(xz)(xy)|$$

$$|{}^4E' \; xy> \; = \; - \; |(yz)(xz)(x^2-y^2)| \qquad (36)$$

In D_{3h} the $x^2 - y^2$ and xy labels point to different symmetries with respect to a C_2^x axis. The connecting operation is the principal C_3^z axis.

$$C_2^x(x^2 - y^2, xy) = (x^2 - y^2, xy)\begin{pmatrix} 1 & 0 \\ 0 & -1 \end{pmatrix} \qquad (37a)$$

$$C_3^z(x^2 - y^2, xy) = (x^2 - y^2, xy)\begin{pmatrix} -1/2 & \sqrt{3}/2 \\ -\sqrt{3}/2 & -1/2 \end{pmatrix} \qquad (37b)$$

The $^4E'$ state is orbitally degenerate and therefore should exhibit a $E \times e$ type JT activity, involving the e' in–plane bending mode. This mode comprises the $Q_{x^2-y^2}$ and Q_{xy} components, defined in table 6. To first–order the corresponding JT surface may be described by the standard formula in equation (38).

$$\begin{array}{cc} |E' \; x^2 - y^2> & |E'xy> \end{array}$$

$$E = 1/2 \; K(Q_{x^2\text{-}y^2}^2 + Q_{xy}^2) + \left\| \begin{array}{cc} -F Q_{x^2\text{-}y^2} & F Q_{xy} \\ F Q_{xy} & F Q_{x^2 - y^2} \end{array} \right\| \qquad (38)$$

A transformation to cylindrical coordinates will convert this equation into the familiar expression for the *Mexican hat* potential [54].

$$E_{\pm} = 1/2 \; K\rho^2 \pm F\rho$$

$$\text{with} : \begin{cases} Q_{x^2\text{-}y^2} = \rho\cos\varphi \\ Q_{xy} = \rho\sin\varphi \end{cases} \qquad (39)$$

Here K is the bending mode force constant and F is the linear coupling constant. By comparing equation (38) to the dynamic ligand field matrix in the table, F can readily be identified as a simple function of the 10Dq parameter :

$$F = -\frac{\sqrt{3}}{2\sqrt{2}} \frac{10Dq}{R} \qquad (40)$$

TABLE 6. Dynamic orbital model for a D_{3h} trigonal bipyramidal fragment. Vibronic coupling is limited to the in–plane e' bending mode

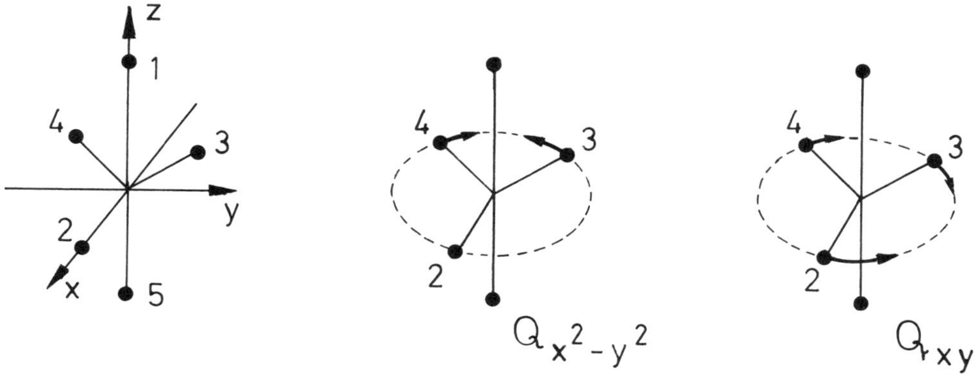

orbital energies ($e_\delta = 0$)

$$a_1'(d_{z^2}) : \frac{11}{4} e_\sigma$$

$$e'(d_{xy}, d_{x^2-y^2}) : \frac{9}{8} e_\sigma + \frac{3}{2} e_\pi$$

$$e''(d_{yz}, d_{xz}) : \frac{7}{2} e_\pi$$

e' displacement operators

$$\frac{\partial}{\partial Q_{x^2-y^2}} = \frac{1}{R\sqrt{2}} \left(\frac{\partial}{\partial \varphi_3} - \frac{\partial}{\partial \varphi_4} \right)$$

$$\frac{\partial}{\partial Q_{xy}} = \frac{1}{R\sqrt{6}} \left(2 \frac{\partial}{\partial \varphi_2} - \frac{\partial}{\partial \varphi_3} - \frac{\partial}{\partial \varphi_4} \right)$$

non–zero vibronic coupling elements [53]

$$<x^2 - y^2 | \frac{\partial \mathscr{V}}{\partial Q_{x^2-y^2}} | x^2 - y^2>_o = - \frac{\sqrt{3}}{2\sqrt{2}} \frac{10Dq}{R}$$

$$<xy | \frac{\partial \mathscr{V}}{\partial Q_{x^2-y^2}} | xy>_o = \frac{\sqrt{3}}{2\sqrt{2}} \frac{10Dq}{R}$$

$$<xy | \frac{\partial \mathscr{V}}{\partial Q_{xy}} | x^2 - y^2>_o = \frac{\sqrt{3}}{2\sqrt{2}} \frac{10Dq}{R}$$

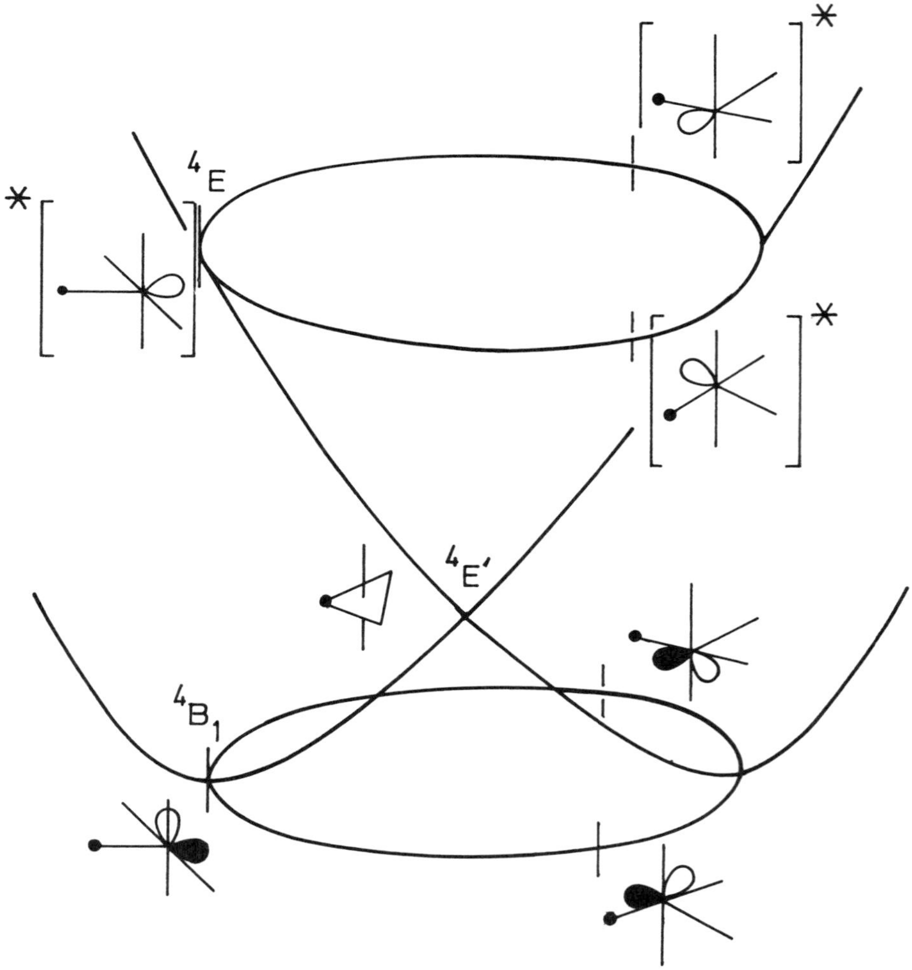

Figure 1. Mexican hat potential surface around a $^4E'$ state of a TBP. The three peripheral SP's are formed by a concerted rearrangement of the triangular TBP equator to a T-shaped form. The orbital lobes on the SP fragments refer to the singly occupied frontier orbitals. The black circle represents a substituent. The figure illustrates how a transverse SP* → TBP → SP decay may interchange apical and basal sites of the square pyramids.

Since F is proportional to 10Dq, a strong JT instability will result. Hence a d^3 TBP will be surrounded by an equipotential energy trough at lower energy. When second–order terms are included the Mexican hat will become warped, giving rise to the formation of three local minima on the bottom of the trough. As required by symmetry, these minima will lie along the three possible C_{2v} distortions of the D_{3h} structure. Interestingly these paths precisely correspond to the nuclear displacements that carry the central D_{3h} structure over into a square pyramid. In this way three SP isomers are located around a conical intersection on a TBP origin. This is schematically represented in figure 1. We will return to this figure when discussing the photostereochemistry of Cr(III) complexes.

6.1.2 *The Square Pyramid.* The most relevant orbital characteristics of the square pyramidal geometry are listed in table 7. A symmetry analysis confirms that the C_{2v} paths on the lower surface of the Mexican hat in figure 1 connect the $^4E'$ state of the central TBP to the $^4B_1$ ground state of the peripheral SP structures. Using the conventions of the table, this $^4B_1$ state may be denoted as follows :

$$^4B_1 = |(yz)(xz)(xy)| \tag{41}$$

The lowest unoccupied d–orbital of this configuration is the d_{z^2} orbital, which points towards the vacant coordination site below the basal plane. This site will thus be particularly susceptible to nucleophilic attack, restoring the octahedral skeleton. As a result the square pyramidal structures in the bottom of the Mexican hat will very easily be trapped by solvent molecules to yield the stable end products.

On the other hand the upper surface connects the $^4E'$ TBP state to the first excited SP quartet state. This state is orbitally degenerate and has ^4E signature. Its $M_S = 3/2$ components are given by :

$$|^4Ex> = |(yz)(xy)(z^2)|$$

$$|^4Ey> = |(xz)(xy)(z^2)| \tag{42}$$

In the convention of equation (42), the symmetry operation which discriminates Ex and Ey is the vertical xz reflection plane. The connecting operation is the C_4^z axis, which was defined earlier in equation (2).

Clearly the octahedral parent state of these ^4E components is the photoactive $^4T_{2g}$ state, which has been described in table 1. The twofold degeneracy of ^4E stems from the degeneracy of the xz and yz orbitals. Accordingly the non–rigidity of the excited square pyramid will in essence be determined by the JT couplings of the degenerate e orbitals. As table 7 shows, two bending modes can in principle be active. The corresponding coupling case is commonly denoted as the E x $(b_1 + b_2)$ case [54]. For both modes the JT activity is a sensitive function of the bond angle θ between the apical and basal bonds [36]. When the metal lies in the basal plane ($\theta = 90°$), the vibronic coupling reduces to zero for the b_1 mode and to a

TABLE 7. Dynamic orbital model for a C_{4v} square pyramidal fragment. Vibronic coupling is limited to the b_1 and b_2 bending modes.

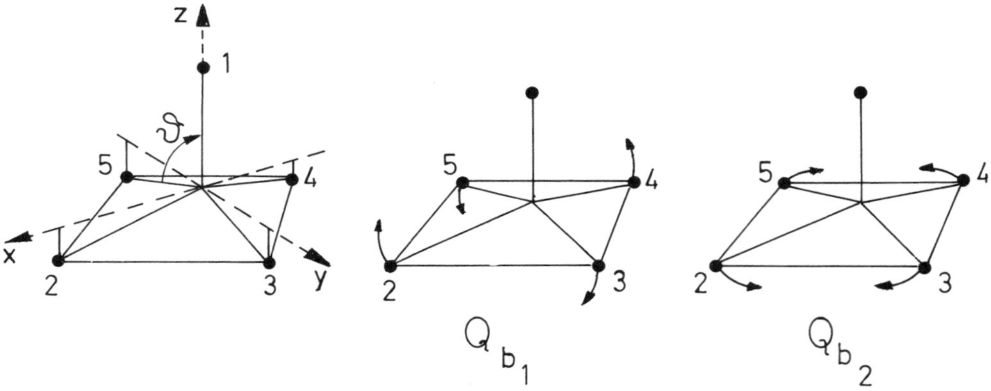

Orbital energies ($\theta = 0$; $e_\delta = 0$)

$b_1(d_{x^2-y^2}) : 3e_\sigma$

$a_1(d_{z^2}) : 2e_\sigma$

$e(d_{yz},d_{xz}) : 3e_\pi$

$b_2(d_{xy}) : 4e_\pi$

b_1 and b_2 displacement operators

$$\frac{\partial}{\partial Q_{b_1}} = \frac{1}{2R}\left(-\frac{\partial}{\partial\theta_2} + \frac{\partial}{\partial\theta_3} - \frac{\partial}{\partial\theta_4} + \frac{\partial}{\partial\theta_5}\right)$$

$$\frac{\partial}{\partial Q_{b_2}} = \frac{1}{2R}\left(\frac{\partial}{\partial\varphi_2} - \frac{\partial}{\partial\varphi_3} + \frac{\partial}{\partial\varphi_4} - \frac{\partial}{\partial\varphi_5}\right)$$

non–zero vibronic coupling elements ($e_\delta = 0$) [53]

$$<yz|\frac{\partial \mathcal{V}}{\partial Q_{b_1}}|yz>_o = \frac{1}{R}\left(\frac{3}{2}\sin 4\theta\, e_\sigma + (\sin 2\theta - 2\sin 4\theta)e_\pi\right)$$

$$<xz|\frac{\partial \mathcal{V}}{\partial Q_{b_1}}|xz>_o = -\frac{1}{R}\left(\frac{3}{2}\sin 4\theta\, e_\sigma + (\sin 2\theta - 2\sin 4\theta)e_\pi\right)$$

$$<xz|\frac{\partial \mathcal{V}}{\partial Q_{b_2}}|yz>_o = \frac{1}{2R}\left(3\sin^2 2\theta\, e_\sigma + 4(\cos^2 2\theta - \cos^2\theta)e_\pi\right)$$

small value of 2 e_π/R for the b_2 mode. However, when the metal is raised above the basal plane ($\theta > 90°$), the coupling abruptly increases. For $\theta = {}^5/_8\,\pi$ the e_σ contribution in the matrix elements of table 7 amounts to 1.5 e_σ/R for b_1, and 0.75 e_σ/R for b_2. Hence at a bond angle of 112.5° the b_1 mode appears to be more active by a factor of two.

From a stereochemical point of view the b_2 mode is unimportant, since it does not lead to the interchange of apical and basal ligands. In contrast, the b_1 mode has the proper C_{2v} symmetry to induce stereomobility. The actual bending mode, which will convert the SP into a TBP, indeed corresponds to a combination of the b_1 mode with the a_1 *umbrella*–motion. Two equivalent rearrangements from SP to TBP are conceivable, matching $Q_{a_1} + Q_{b_1}$ and $Q_{a_1} - Q_{b_1}$ distortions. The a_1 mode will be activated by interligand repulsion, while – for $\theta > 90°$ – the b_1 mode will be promoted by vibronic coupling. *Hence we predict that photoinduced ligand loss of an octahedral complex will favor access to the upper level of a TBP centered Mexican hat potential.*

Obviously the two–dimensional representation of the potential energy surface in figure 1 is too simplified to allow a precise description of the excited state entrance point. Photodissociation of a ligand from the $^4T_{2g}$ state appears to be a complicated multimode process, involving concomitant ligand dissociation and bending of the remaining fragment [55]. Perhaps a more detailed study of thexi–state geometries in low–symmetry complexes might clarify this problem. In any case the decay of the initial five–coordinated fragment through a conical intersection point at a TBP geometry is symmetry allowed and energetically favorable.

6.2 THE EFFECT OF SUBSTITUENTS

Before the stereochemical aspects of the model can be discussed, attention must be given to the symmetry lowering effect of substituents. Indeed, in addition to the study of typical reaction parameters such as steric hindrance and solvent changes, photochemists have mainly been concerned with the effect of substituents on ligand labilization and product distribution [49]. In this section, three aspects of the model will be investigated more closely in the light of the symmetry lowering effect of substituents, viz. : the excited state entrance point, the conical intersection, and the minimal energy trough.

6.2.1. *The Excited State Entrance Point.* It will be remembered that the excited state square pyramid can collapse to a TBP in two different ways, either by a $Q_{a_1} + Q_{b_1}$ distortion or by its $Q_{a_1} - Q_{b_1}$ counterpart. In the high–symmetry case these two processes are equivalent. However in low–symmetry complexes this equivalence may be broken. As we have demonstrated in a previous publication [20], the choice between the two pathways can be expressed by the following electronic selection rule :

Consider the plane of excitation. In the lowest excited quartet, this is the plane formed by the two axes of weakest average field. Upon removal of the leaving ligand from this plane, the resulting T–shaped structure will rearrange to a triangle. The perpendicular axis is conserved. If there are two equivalent weak–field planes, the rule yields the same result when applied to either one of them.

This rule was derived from a comparison of orbital correlation diagrams for the two possible rearrangements. It can equally well be obtained from dynamic ligand field theory. We will demonstrate this for the case of the *cis*–$Cr(NH_3)_4F_2^+$ complex. In this complex the plane formed by the two axes of weakest average spectrochemical strength is the equatorial plane containing the two F^- ligands. According to Adamson's rule (*vide supra*, section 4.2) this plane will be labilized upon irradiation. Removal of a NH_3 ligand from this plane leads to a square pyramid with one basal and one apical fluoro–ligand, as shown in figure 2. Using the coordinate system of table 7 the substituents will be located on ligand positions 1 and 2, i.e. in the xz plane. The lowest $^4T_{2g}$ component, responsible for the labilization of this plane, comprises a $d_{xz} \rightarrow d_{x^2-z^2}$ excitation, which is totally symmetric with respect to the σ_{xz} reflection operation. Therefore, upon ligand removal, this state will correlate with the equisymmetric Ex component of the square pyramidal ^4E as specified in equation (42). In ^4Ex the d_{yz} orbital is singly occupied. The detailed form of the vibronic coupling element between the ^4Ex component and the b_1 mode can easily be obtained from dynamic ligand field theory. One has :

$$<^4Ex|\frac{\partial \mathcal{V}}{\partial Q_{b_1}}|^4Ex> =$$

$$\frac{3}{2R}\sin 4\theta\, e_\sigma^{NH_3} + \frac{1}{2R}\sin 2\theta (e_\pi^F + e_\pi^{NH_3}) - \frac{2}{R}\sin 4\theta\, e_\pi^{NH_3} \qquad (43)$$

Clearly, for $\theta > 90°$, this coupling element will be positive. The corresponding energy *lowering* force thus will point in the direction of the $-Q_{b_1}$ distortion coordinate. As may be seen from table 7, this $-Q_{b_1}$ coordinate in combination with Q_{a_1} forms a triangle in the xz–plane. Hence decay towards a TBP with two F^- ligands in the equatorial plane will be allowed, while decay to the alternative TBP with one equatorial and one axial fluoride will be forbidden. The equatorial plane of the allowed TBP thus coincides with the original plane of excitation, exactly as predicted by the electronic selection rule.

In the case of cis–$Cr(NH_3)_4F_2^+$, this rule has observable stereochemical consequences. Indeed as may be seen from figure 2, it precludes the formation of the *facial*–$Cr(NH_3)_3(H_2O)F_2^+$ isomer, since this isomer can only be generated via the forbidden pathway. Kirk and Frederick have observed that in a photolyzed solution of *cis*–$Cr(NH_3)_4F_2^+$ the amount of *fac*–product is marginal and moreover wavelength dependent [28]. This seems to indicate that the small quantity of the *fac*–isomer originates from the higher excited

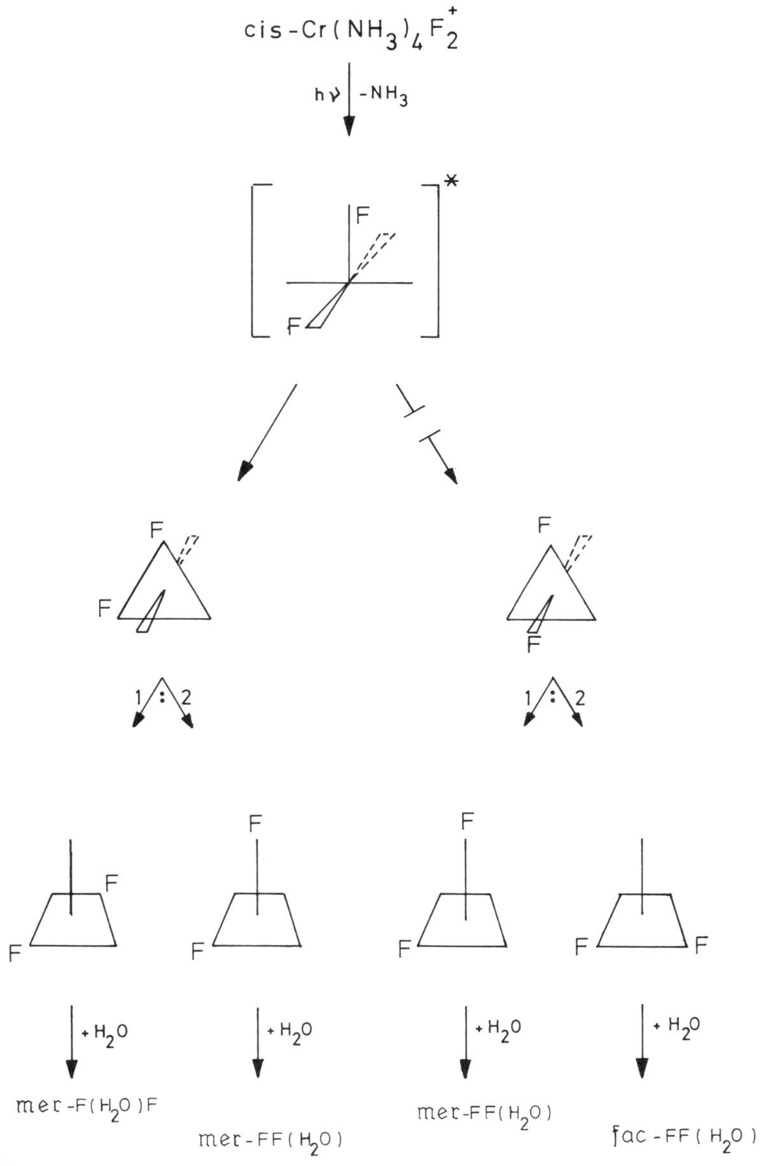

Figure 2. Possible reaction pathways for cis–$Cr(NH_3)_4F_2^+$, following equatorial NH_3 loss. The central square pyramid has one apical and one basal substituent. Of the two alternative deactivation routes of this fragment, only one is allowed by the electronic selection rule, as indicated on the figure. This decay indeed gives rise to the observed meridional isomers.

state, while the lowest photoactive component only gives rise to *meridional*-isomers, as required by the selection rule.

6.2.2. *The Conical Intersection.* The effect of symmetry lowering substituents on the potential energy surface near the TBP degeneracy point can be described in general by adding a symmetrical tensor S to the first-order formula in equation (38). The potential thus becomes :

$$E = {}^1/_2\, K(Q^2_{x^2-y^2} + Q^2_{xy}) + \begin{Vmatrix} -FQ_{x^2-y^2} + S_{11} & FQ_{xy} + S_{12} \\ FQ_{xy} + S_{12} & FQ_{x^2-y^2} + S_{22} \end{Vmatrix} \quad (44)$$

This general formula has been used to model the potential of substituted hexacoordinate $Cu^{II}L_{6-k}X_k$ complexes [56,57]. Exactly the same formalism may be used to introduce the effect of strain in solid state problems [58]. By taking out the trace of the coupling matrix, one obtains :

$$E = {}^1/_2\, K(Q^2_{x^2-y^2} + Q^2_{xy}) + \frac{S_{11} + S_{22}}{2} + \begin{Vmatrix} -FQ_{x^2-y^2} - \frac{S_{22} - S_{11}}{2} & FQ_{xy} + S_{12} \\ FQ_{xy} + S_{12} & FQ_{x^2-y^2} + \frac{S_{22} - S_{11}}{2} \end{Vmatrix} \quad (45)$$

Again this expression may be simplified by a coordinate transformation. This transformation will depend on the S tensor. One defines :

$$Q_{x^2-y^2} + \frac{S_{22} - S_{11}}{2F} = \rho\cos\varphi$$

$$Q_{xy} + \frac{S_{12}}{F} = \rho\sin\varphi \quad (46)$$

Combination of equations (45) and (46) then yields :

$$E_{\pm} = {}^1/_2\, K(Q^2_{x^2-y^2} + Q^2_{xy}) + \frac{S_{11} + S_{22}}{2} \pm F\rho \quad (47)$$

The main conclusion from this formula is that for $S_{22} - S_{11}$ and S_{12} values that do not exceed the energetic effect of the vibronic coupling elements, the conical intersection is *not destroyed but only displaced*, thereby acquiring an oblique shape [59]. The distance between the new degeneracy point and the TBP coordinate origin is specified in equation (48).

$$Q_{x^2-y^2} = (S_{11} - S_{22})/2F$$

$$Q_{xy} = -S_{12}/F \quad (48)$$

These results may be illustrated for the TBP structure that is formed upon expulsion of the Y ligand in *trans*–CrL_4XY. The relevant bipyramid is shown in the center of figure 1. It has a twofold symmetry axis, containing the heteroligand X. If we let this axis coincide with the x–direction in table 6, the S_{12} tensor element will be zero, since the two $^4E'$ components have different symmetries with respect to this C_2^X axis (see eq. (37a)). In contrast the diagonal tensor elements may have non–zero values. Their difference can readily be calculated by means of the static AOM. One obtains :

$$S_{11} - S_{12} = {}^1/_4(10Dq_L - 10Dq_X) \tag{49}$$

Hence for the CrL_4X TBP the coordinates of the crossing point will depend on the difference in spectrochemical strength between X and L, as specified in equation (50).

$$Q_{x^2-y^2} = (10Dq_L - 10Dq_X)/8F$$
$$Q_{xy} = 0 \tag{50}$$

In all *trans*–CrL_4XY complexes, which obey Kirk's stereomobility rule (see eq. (35)), X turns out to be a weaker ligand than L. Combining equations (40) and (50) one finds that a weaker X ligand will yield a negative value for $Q_{x^2-y^2}$. Hence in all these cases the crossing point will be displaced on the C_{2v} path in the direction of the SP with X on the apex. This crossing point indeed appears on a Woodward–Hoffmann type correlation diagram of this particular C_{2v} path [20,52]. However the present method is more general than the Woodward–Hoffmann approach in that it is able to detect crossing points, that do not lie on the rearrangements paths between SP and TBP. A case in point is the TBP fragment with two F^- heteroligands in the equatorial plane, which plays a role in the photochemistry of cis–$Cr(NH_3)_4F_2^+$. (see figure 2). In this fragment $Q_{x^2-y^2}$ equals $(10Dq_F - 10Dq_{NH_3})/8F$, which is a positive quantity, while Q_{xy} remains zero. Such a crossing point will lie in the continuation of the SP ⟷ TBP path, past the TBP origin. Hence it will not show up in a SP to TBP correlation diagram, which ends in the TBP origin. An incomplete picture of the potential energy surface thus may result.
A discussion of the stereochemical relevance of the effect of substituents on the crossing point will be postponed till section 6.3.

6.2.3. *The minimal energy trough.* The warping of the minimal energy trough surrounding TBP in figure 1 may have important stereochemical implications. If the energy barriers for pseudorotation along this trough are very low, thermal interconversion between the square pyramidal isomers may successfully compete with solvent trapping. In this case the product distribution should reflect the relative stability of the three energy wells on the ground state surface. On the other hand if the pseudorotation is hindered by considerable energy barriers, the product distribution should reflect the stereoselectivity of the decay process from the upper to the lower sheet of the warped Mexican hat potential.

Detailed ligand field calculations on substituted fragments strongly suggest that the second alternative is the more likely. Energy barriers for the thermal interconversion of the SP ground states turn out to be too high to allow successful competition with solvent trapping, irrespective of the nature of the substituent. Furthermore substitution does not lead to appreciable energy differences between the SP isomers. Hence the product distribution cannot be viewed as the result of a thermal equilibration process along the bottom of the Mexican hat potential. It must rather be attributed to the decay process through the conical intersection.

6.3 DISCUSSION

In the preceding sections we have presented a detailed description of the potential energy surface of a substituted 5–coordinate Cr^{3+} fragment. On the basis of this description, the following mechanistic model may be proposed : ligand loss from the lowest excited quartet state of the 6–coordinate reactant gives access to the upper sheet of a Mexican hat potential around a TBP structure. Decay on this surface will populate square pyramidal potential energy wells on the lower surface. Finally solvent attack on these structures gives rise to the photoproducts.

All these processes are highly stereospecific. The bending motion which determines the excited state entrance point must involve the original plane of excitation, as required by the electronic selection rule of section 6.2.1. Likewise the "exit channels" are determined by specific solvent attack on the vacant coordination site of the SP ground states. Last but not least the decay process itself appears to be stereospecific. Indeed in order to explain the observed photostereochemistry by a dissociative mechanism, one must assume that the tunneling through the conical intersection will take a trajectory, that is closest to a straight line path, and thus leads to preferential population of the SP ground states at the opposite sides of the excited state entrance point. This would mean that the tunneling tends to conserve nuclear momentum.

As an example for the *trans*–CrL_4XY complexes, that obey the stereomobility rule in equation (35), the excited state entrance point is a CrL_4X square pyramid with the heteroligand in the apex. As shown in figure 1, a transverse decay will convert this SP* into square pyramidal ground states with X in the basal plane. These SP's indeed yield the stereomobile *cis*–product. As another example equatorial NH_3 loss from *cis*–$Cr(NH_3)_4F_2^+$ leads to a square pyramid with one apical and one basal F^- substituent. Transverse tunneling via the allowed TBP will produce the SP precursors of the two meridional $Cr(NH_3)_3(H_2O)F_2^+$ isomers, in equal amounts. This is in near agreement with the experimental findings of Kirk and Frederick [28] :

$$cis\text{–}Cr(NH_3)_4F_2^+ \xrightarrow{h\nu} \begin{cases} 45\ \%\ mer\text{–}F(H_2O)F \\ 55\ \%\ mer\text{–}FF(H_2O) \end{cases} \qquad (51)$$

In other cases of equatorial labilization the observed product ratios sometimes appear to be less selective. A case in point is the *trans*–$Cr(NH_3)_4F_2^+$ complex ion which decays via the same TBP as the *cis*–complex. For the *trans*–complex selective transverse decay would yield 100 % *mer*–$FF(H_2O)$ product. However the observed product distribution is closer to the statistical 1:2 ratio [28] :

$$trans\text{–}Cr(NH_3)_4F_2^+ \xrightarrow{h\nu} \begin{cases} 29\ \%\ mer\text{–}F(H_2O)F \\ 71\ \%\ mer\text{–}FF(H_2O) \end{cases} \tag{52}$$

In this respect the following interesting correlation can be made : in all cases where the crossing point of upper and lower surface is displaced *towards* the entrance point, the preference for selective transverse tunneling is most pronounced. In cases where the crossing point is displaced *away* from the entrance point, the directional information in the trajectory is partly lost and a more random decay will be observed.

This correlation is related to our earlier stereochemical rules for the TBP fragment, that were based on a Woodward–Hoffmann approach [20]. However, as we have already pointed out in section 6.2.2 the present analysis is more general in that it draws attention to the substituent–independent structure of the surface as a whole.

Clearly a theoretical treatment of the decay process on a Mexican hat potential is beyond the scope of dynamic ligand field theory and requires a physical model, including nuclear kinetic energy. Apparently such a detailed model is not yet available. Nonetheless it is interesting to note that in semi–classical treatments trajectories may indeed be selected on the basis of conservation of nuclear momentum [60].

7. Conclusion

In the present article we have demonstrated the use of dynamic ligand field theory to investigate the structural non–rigidity of a $Cr^{III}L_5$ fragment. Elsewhere we have applied the same method to the coordinatively unsaturated $Fe(CO)_4$ fragment, that is generated upon irradiation of $Fe(CO)_5$ in an inert matrix at low temperature [38,61]. Another interesting application involves the geometry of the Cu^{2+} ion in 5–coordination [62]. In all these cases the Jahn–Teller theorem provides a valuable clue to understand the global structure of the potential energy surface. This illustrates the important role of vibronic coupling in inorganic chemistry. The vibronic coupling phenomenon, that is specifically related to inorganic photochemistry, is the decay process from an upper to a lower sheet of a Jahn–Teller potential energy surface.

Acknowledgement

The author is indebted to the Belgian Government (Programmatie van het Wetenschapsbeleid) and to the Belgian National Science Foundation

(NFWO) for financial support.

REFERENCES

[1] Griffith, J.S. "The Theory of Transition–Metal Ions"; Cambridge University Press, Cambridge, 1961.

[2] As a general rule symmetry operations will be defined in an active sense, i.e. as leaving the Cartesian axes immobile while rotating or reflecting the orbital functions.

[3] Ceulemans, A.; Bongaerts, N.; Vanquickenborne, L.G. *Inorg. Chem.* **1987**, *26*, 1566. In table II of this reference the sign of the $|a_1\theta\bar{\epsilon}|$ determinant in the $|^2A_2>$ component of $^2T_{1g}$ should be negative.

[4] Dionne, G.F.; Palm, B.J. *J. Magn. Resonance* **1986**, *68*, 355.

[5] Ballhausen, C.J. "Introduction to Ligand Field Theory"; McGraw-Hill, New York, 1962.

[6] Ceulemans, A.; Beyens, D.; Vanquickenborne, L.G. *J. Am. Chem. Soc.* **1982**, *104*, 2988 (For a rigorous treatment, see [7]).

[7] Ceulemans, A. *Meded. K. Acad. Wet., Lett. Schone Kunsten Belg., Kl. Wet.* **1985**, *46*, 82. For a discussion of the physical background, see : Stedman, G.E. *J. Phys. A* **1987**, *20*, 2629.

[8] Ceulemans, A. *Chem. Phys.* **1982**, *66*, 169.

[9] Turnbull, H.W. "The Theory of Determinants, Matrices and Invariants", Dover Publications, New York, 1960 (3rd edition).

[10] Clearly a fixed standard order for the closed–shell parent determinant is required. If one changes the phase of this determinant, the phase factor in equation (10) will change as well, and so will the hole–electron exchange parities.

[11] The time reversal operator $\mathcal{T}$ turns spatial functions into their complex conjugates. Its effect on the electron spin is to turn $|\alpha>$ into $|\beta>$, and $|\beta>$ into $-|\alpha>$. Strictly speaking $\mathcal{T}$ is defined up to an arbitrary phase factor. See also : Wigner, E. "Group Theory" (translated from the german by J.J. Griffin); Academic Press, New York, 1964 (4th printing).

[12] Cited from [1], p. 247.

[13] Flint, C.D. *J. Mol. Spectrosc.* **1971**, *37*, 414.

[14] Forster, L.S.; Rund, J.V.; Fucaloro, A.F. *J. Phys. Chem.* **1984**, *88*, 5012.

[15] Forster, L.S.; Rund, J.V.; Fucaloro, A.F.; Lin, S.H. *J. Phys. Chem.* **1984**, *88*, 5017.

[16] Fucaloro, A.F.; Forster, L.S.; Glover, S.G.; Kirk, A.D. *Inorg. Chem.* **1985**, *24*, 4242.

[17] Ryu, C.K.; Endicott, J.F. *Inorg. Chem.* **1988**, *27*, 2203.

[18] Basolo, F.; Pearson, R.G. "Mechanisms of Inorganic Reactions"; Wiley, New York, 1967 (2nd edition).

[19] Vanquickenborne, L.G.; Ceulemans, A. *J. Am. Chem. Soc.* **1977**, *99*, 2208.

[20] Vanquickenborne, L.G.; Ceulemans, A. *Coord. Chem. Revs* **1983**, *48*, 157.

[21] McClure, D.S. "VIth International Conference on Coordination Chemistry" (Ed. S. Kirschner); MacMillan, New York, 1961, p. 498.

[22] Kettle, S.F.A. *J. Chem. Soc. A* **1966**, 420.

[23] Burdett, J.K. *J. Chem. Soc. Faraday Trans. 2* **1974**, *70*, 1599.

[24] Schäffer, C.E.; Jørgensen, C.K. *Mol. Phys.* **1965**, *9*, 401.

[25] Schäffer, C.E. "XIIth International Conference on Coordination Chemistry, Sydney 1969", IUPAC, Butterworths, London, 1970, p. 316.

[26] Zink, J.I. *J. Am. Chem. Soc.* **1972**, *94*, 8039.

[27] Adamson, A.W. *J. Phys. Chem.* **1967**, *71*, 798.

[28] Kirk, A.D.; Frederick, L.A. *Inorg. Chem.* **1981**, *20*, 60.

[29] Endicott, J.F.; Ramasami, T.; Tamilarasan, R.; Lessard, R.B.; Ryu, C.K. *Coord. Chem. Revs.* **1987**, *77*, 1.

[30] Riccieri, P.; Zinato, E.; Damiani, A. *Inorg. Chem.* **1987**, *26*, 2667.

[31] Mønsted, L.; Mønsted, O. *Acta Chem. Scand.* **1984**, *A38*, 679.

[32] Herbert, B.; Reinhard, D.; Saliby, M.J.; Sheridan, P.S. *Inorg. Chem.* **1987**, *26*, 4024.

[33] Wilson, R.B.; Solomon, E.I. *Inorg. Chem.* **1978**, *17*, 1729.

[34] Güdel, H.U.; Snellgrove, T.R. *Inorg. Chem.* **1978**, *17*, 1617.

[35] Bacci, M. *Chem. Phys. Lett.* **1978**, *58*, 537.

[36] Bacci, M. *Chem. Phys.* **1979**, 40, 237. In table 1 of this reference the matrix element for the b_2 mode in a square pyramidal MX_5 compound is erroneous. The correct matrix element is given in our Table 7.

[37] Warren, K.D. *Structure and Bonding* **1984**, *57*, 120. In table 2 of this reference the matrix elements $<z^2|\partial\mathscr{V}/\partial Q_{yz}|yz>$ and $<x^2\text{-}y^2|\partial\mathscr{V}/\partial Q_{yz}|yz>$ should read $\frac{1}{2}\sqrt{3}\,\sigma$ and $-\frac{3}{2}\,\sigma$ respectively.

[38] Ceulemans, A.; Beyens, D.; Vanquickenborne, L.G. *J. Am. Chem. Soc.* **1984**, *106*, 5824. Equation (10) of this reference contains some misprints : second row, the term $^1/_4\, Q_\zeta^2$ should read $^1/_4\, Q_\xi^2$; fourth (fifth) row, a minus sign in front of $^1/_2\, Q_\vartheta Q_\xi$ ($^1/_2\, Q_\vartheta Q_\eta$) should be added; sixth row, the term $X_t Q_\zeta Q_\eta$ should read $X_t Q_\xi Q_\eta$.

[39] Stone, A.J. *Mol. Phys.* **1980**, *41*, 1339. In equation (3.4) of this reference the matrix element in the third column should read : $-\cos\theta \mathbf{e}_\vartheta - \sin\theta \mathbf{e}_r$.

[40] Griffith, J.S. "The Irreducible Tensor Method for Molecular Symmetry Groups"; Prentice–Hall, Englewood Cliffs, NJ, 1962.

[41] Englman, R. "The Jahn–Teller Effect in Molecules and Crystals"; Wiley, New York, 1972.

[42] Schmidtke, H.–H.; Degen, J. *Structure and Bonding*, in press.

[43] Agresti, A.; Ammeter, J.H.; Bacci, M. *J. Chem. Phys.* **1984**, *81*, 1861. Erratum : *ibid.* **1985**, *82*, 5299.

[44] Parrot, R.; Naud, C.; Gendron, F.; Porte, C.; Boulanger, D. *J. Chem. Phys.* **1987**, *87*, 1463.

[45] Stavrev, K.K.; Kynev, K.D.; Nikolov G. St. *J. Chem. Phys.*, **1988**, *88*, 7027.

[46] Reinen, D.; Atanasov, M.; Nikolov, G.S.; Steffens, F. *Inorg. Chem.* **1988**, *27*, 1678.

[47] Bacci, M. *J. Phys. Chem. Solids* **1980**, *41*, 1267.
[48] Kirk, A.D. *Mol. Photochem.* **1973**, *5*, 127.
[49] Kirk, A.D. *Coord. Chem. Revs.* **1981**, *39*, 225.
[50] Zinato, E.; Riccieri, P.; Prelati, M. *Inorg. Chem.* **1981**, *20*, 1432.
[51] Riccieri, P.; Zinato, E. *Inorg. Chem.* **1983**, *22*, 2305.
[52] Vanquickenborne, L.G.; Ceulemans, A. *J. Am. Chem. Soc.* **1978**, *100*, 475.
[53] Bacci, M.; *Chem. Phys.* **1986**, *104*, 191. Equation (22) of this reference uses the erroneous value for the b_2 coupling element of ref. [36]. For the correct value, see our table 7.
[54] Bersuker, I.B. "The Jahn–Teller Effect and Vibronic Interactions in Modern Chemistry", Plenum, New York, 1984.
[55] Endicott, J.F. *Comments Inorg. Chem.* **1985**, *3*, 349.
[56] Dyachkov, P.N.; Levin, A.A. "Vibronic Theory of the relative isomeric Stability of Inorganic Molecules and Complexes", in : *Itogi Nauki Tekh.: Str. Mol. Khim. Svyaz*, Moscow, 1987, vol. 11.
[57] Riley, M.J.; Hitchman, M.A.; Reinen, D.; Steffen, G. *Inorg. Chem.* **1988**, *27*, 1924.
[58] Ham, F.S. *Phys. Rev.* **1968**, *166*, 307.
[59] Even if no symmetry is present, the crossing point may persist and is said to be topologically sustained. See : Longuet–Higgins, H.C. *Proc. Roy. Soc. Lond.* **1975**, *A344*, 147.
[60] Carpenter, B.K. *J. Am. Chem. Soc.* **1985**, *107*, 5730.
[61] Poliakoff, M.; Ceulemans, A. *J. Am. Chem. Soc.* **1984**, *106*, 50.
[62] Reinen, D.; Friebel, C. *Inorg. Chem.* **1984**, *23*, 791.

TWO-PHOTON SPECTROSCOPY OF THE $^4A_{2g} \rightarrow {}^4T_{2g}$ TRANSITION IN Mn^{4+} IMPURITY IONS IN CRYSTALS

C. Campochiaro and D. S. McClure
Princeton University
Frick Chemical Laboratory
Washington Road
Princeton, New Jersey 08544
USA

and

P. Rabinowitz and S. Dougal
Exxon Research and Engineering Company
Clinton Township
Route 22 E.
Annandale, New Jersey 08801
USA

1. Introduction

Our motivation for studying the $^4A_{2g} \rightarrow {}^4T_{2g}$ two-photon(TP) spectra of d^3 ions is to observe the ground to excited state transitions unblemished by odd parity enabling modes. Then we would see the Jahn-Teller e_g and t_{2g} modes which would help to understand the nature of the complicated excited state potentials. In this paper we report our findings on the Mn^{4+}

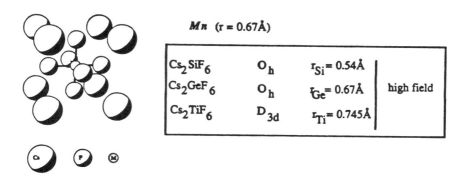

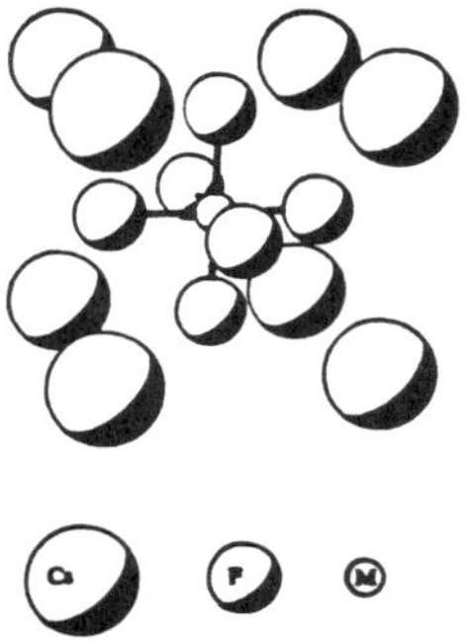

Mn (r = 0.67Å)

Cs_2SiF_6	O_h	r_{Si} = 0.54Å	high field
Cs_2GeF_6	O_h	r_{Ge} = 0.67Å	
Cs_2TiF_6	D_{3d}	r_{Ti} = 0.745Å	

Figure 1. Diagram of the Cs_2MF_6 lattice. Note that the MF_6 species is completely contained inside the cube of Cs ions.

C. D. Flint (ed.), *Vibronic Processes in Inorganic Chemistry*, 255–265.
© 1989 by Kluwer Academic Publishers.

ion which was incorporated in the tetravalent site in the cubic lattices Cs_2MF_6, where M is a Si or Ge ion, and a related trigonal host where M is a Ti ion as shown in figure 1.[1]

This family of host crystals has isolated MF_6 octahedra so the $[MnF_6]^{2-}$ impurity exhibits little sensitivity to the socket size of the host. These systems are high field so the ^2E_g and $^2T_{1g}$ states fall below the $^4T_{2g}$ state. The traditional emission and absorption or one-photon(OP) spectra have been reported and analyzed in detail by several authors.[2]

The major selection rule for TP spectroscopy is $g \rightarrow g$. All of the above systems have a center of inversion so this rule applies. Therefore we do not expect to see transitions to odd parity modes in the excited state, as these would be $g \rightarrow u$, nor did we see any such. There are intricate TP polarization selection rules for the $^4A_{2g} \rightarrow ^4T_{2g}$ transition which will be discussed in Section IV.[3]

In the previous work on Cs_2GeF_6: Mn^{4+} it was shown that the ^2E and $^2T_1$ states obeyed the rule of mutual exclusion which is a test for the presence of a center of symmetry and is familiar when comparing IR and Raman vibrational spectra.[3] Figure 2 shows a coincidence only at the ^2E origin where the OP transition is magnetic.

One- and Two-Photon Spectra of Cs_2GeF_6: Mn^{4+}

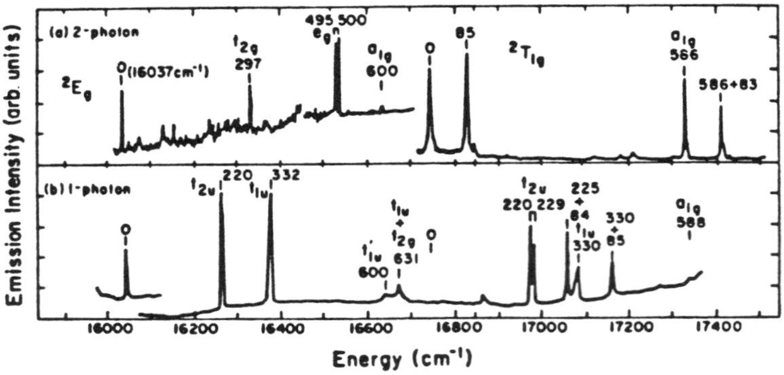

Figure 2. One- and two-photon spectra of the ^2E and $^2T_1$ states of Cs_2GeF_6: Mn^{4+} demonstrating that the even and odd parity vibrations are either one- or two-photon allowed--*not both* The ^2E origin is present in both spectra because it is magnetic dipole allowed.(From Chien, Berg, and McClure, reference 5.)

2. Experiment

Figure 3 shows the experimental method. It consists of passing radiation from a dye laser through a cell filled with H_2 to Raman shift it into the infrared by multiples of 4155 cm^{-1}.[4] The tunable IR laser beam is then focussed into the sample with power densities on the order of 1 GW/cm^2. The sample is packed on three sides with indium foil to assure good thermal contact to a He cooled cold finger. The fluorescence excited by the TP process is collected by a condenser lens pair with an effective f number of 1.5, and passed through filters to a photomultiplier tube. The laser beam passes through the crystal to a pyroelectric detector which serves as a reference. Each wavelength was sampled from 40 to 120 times and the average of the ratio[signal/(reference)^2] was recorded.

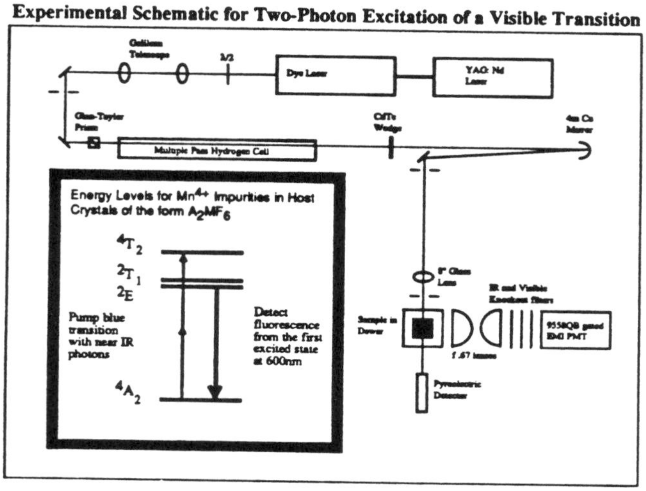

Figure 3. Experimental apparatus for TP excitation of states that emit visible light.

Almost all of the random noise was caused by shot to shot variations in the laser pulses. Both the Nd:YAG pump laser and the dye laser operate with multiple longitudinal modes. These modes interfere with each other over the 7ns duration of the pulse, resulting in a laser pulse that is not a smooth peaked function of time, and in fact has many sharp spikes of 10ps duration. The two-photon signal is proportional to the time integral of the square of the optical intensity over the pulse duration, so the highest spikes generate most of the signal. As the relative phases of the modes change from shot to shot, the peak intensities in the spikes can change dramatically, causing the observed shot to shot fluctuation in the signal. Furthermore, near the ends of a dye range these effects can be accentuated, resulting in an artificial rise at the juncture between two different dyes. One solution to this problem is to excite with single longitudinal mode lasers. Another is to reference the integrated square of the intensity rather than the energy of each pulse by using the second harmonic output of a doubling medium with sufficient conversion efficiency. This serving as a reference would extend the effective tuning range of the dye laser and improve the signal/noise ratio.

3. Results

Our current work with Cs_2GeF_6: Mn^{4+} shows better resolution than the previous work due to better heat sinking of the crystal and the resulting lower temperatures.[5] Figure 4 shows the first 550 cm^{-1} of the ${}^4A_{2g} \rightarrow {}^4T_{2g}$ transition. The origin quartet is better resolved and an interference by the 55 cm^{-1} lattice mode is revealed, so there has to be a change in the assignment of the third component. The t_{2g} vibrational multiplet is surprisingly strong, and is now shown to have many more components than at first thought. The interaction of the ${}^4T_{2g}$ multiplet with a t_{2g} mode should give 12 vibronic components, of which at least 8 seem to be present in figure 4. The lattice mode structure adds to these components, making for broad features in this region. At 487 and 525 cm^{-1} we observe the e_g and a_{1g} vibrations. Both have much lower intensities than the t_{2g} mode.

The complete TP spectrum of the ${}^4A_{2g} \rightarrow {}^4T_{2g}$ transition is shown in figure 5 for the three Mn^{4+} compounds we studied. The Si and Ge compounds exhibit progressions in both the e_g and a_{1g} modes. For the Ti compound both the perpendicular($\perp$) and parallel($\parallel$) spectra are shown. It is remarkable that for M = Si and Ge the t_{2g} mode seems to originate a progression of e_g and a_{1g} modes. It thus behaves like a promoting mode in a one-photon spectrum. The TP spectrum is simpler because there is only one octahedron promoting mode, while in the OP spectrum there are three($2t_{1u}$ and t_{2u}). In the trigonally distorted Ti compound the t_{2g} mode is very weak in the $\perp$ spectrum and the progression seen there appears to be in the e_g and a_{1g} modes beginning with the origin. It is not a regular progression however and several modes of similar frequency appear to be present.

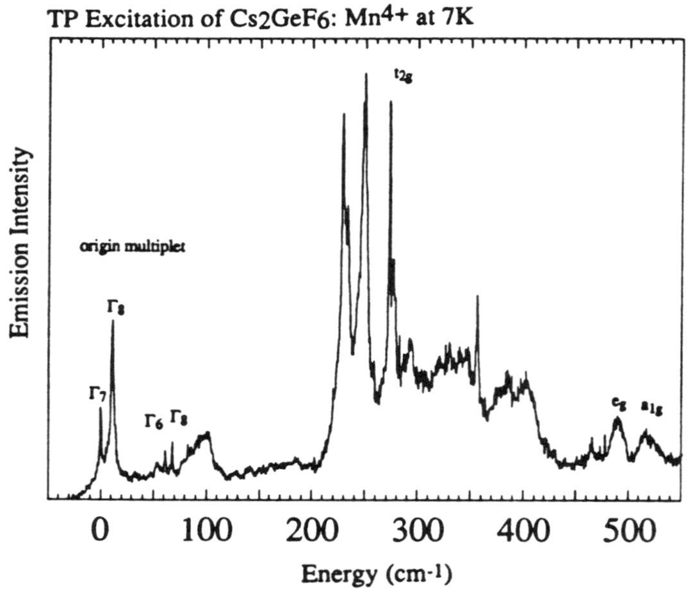

Figure 4. First 550 cm^{-1} of the TP excitation spectrum of Cs_2GeF_6: Mn^{4+} showing the origin multiplet, the t_{2g} vibrational multiplet, the e_g vibration, and the a_{1g} vibration.

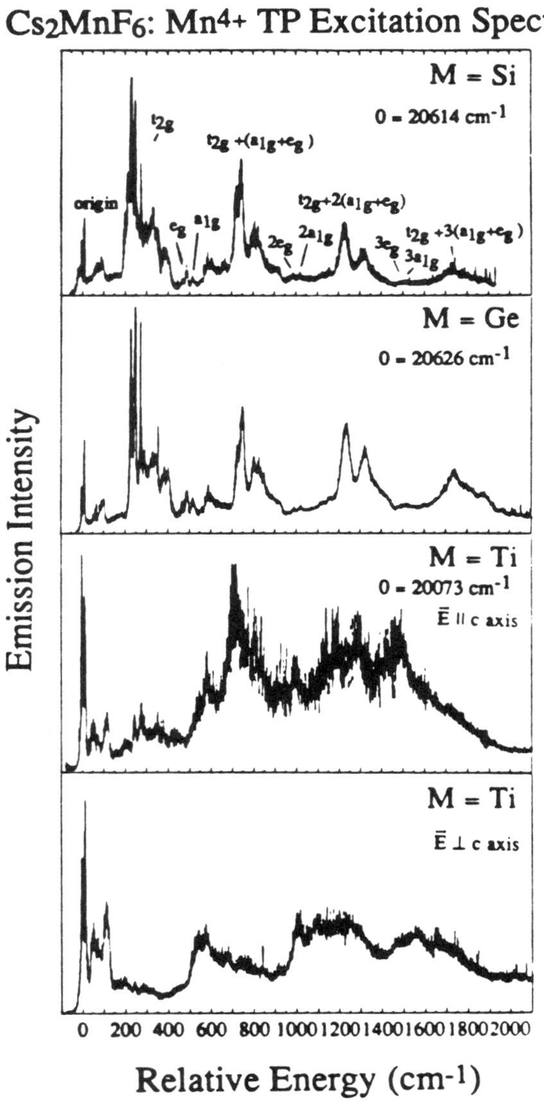

Figure 5. Low temperature TP excitation spectra of the entire $^4T_2$ states of Cs_2SiF_6: Mn^{4+} (15K), Cs_2GeF_6: Mn^{4+} (7K), and Cs_2TiF_6: Mn^{4+} (7K).

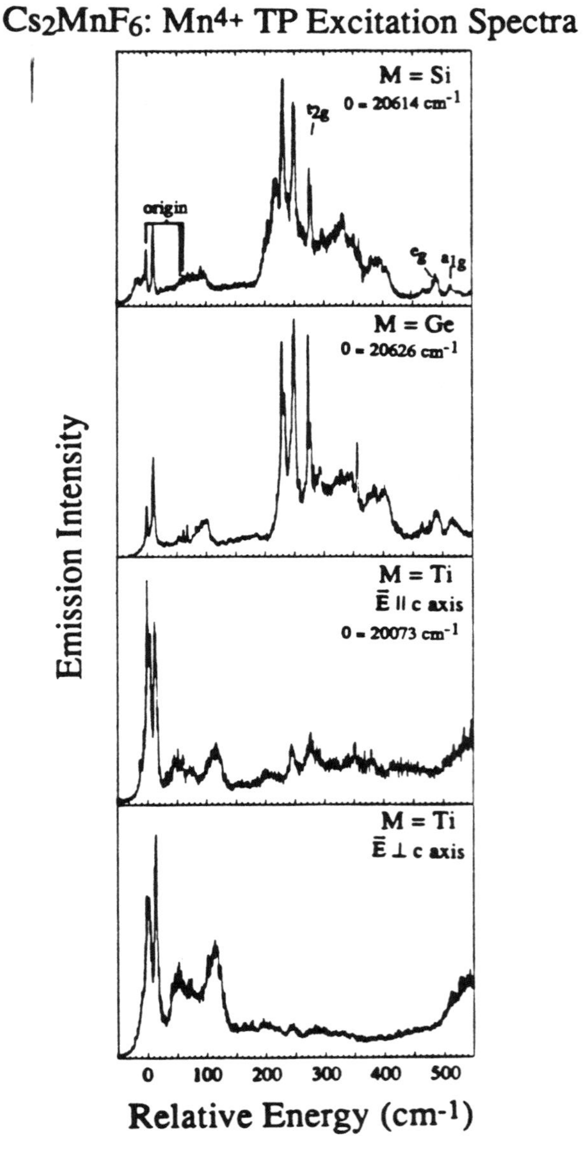

Figure 6. The same spectra shown in figure 5 are expanded in the region between 0 and 550 cm^{-1} so details of the origin and the t_{2g} vibration are revealed.

Figure 6 compares the first 550 cm^{-1} of the $^4A_{2g} \to ^4T_{2g}$ transition in the three crystals. Features appear in the $Cs_2SiF_6:Mn^{4+}$ crystal which seem to be due to Mn^{4+} pairs or clusters as shown by the broad weak band just before the first origin multiplet components and the first t_{2g} component. Otherwise the Si and Ge spectra are very similar. This figure also emphasizes the weakness of the t_{2g} mode in the Ti compound. The strong and broad features between 700 and 1700 cm^{-1} in the || spectrum do not look like "normal" Frank-Condon progressions often seen in molecular spectra, this may be due to the referencing problems discussed in Section 2.

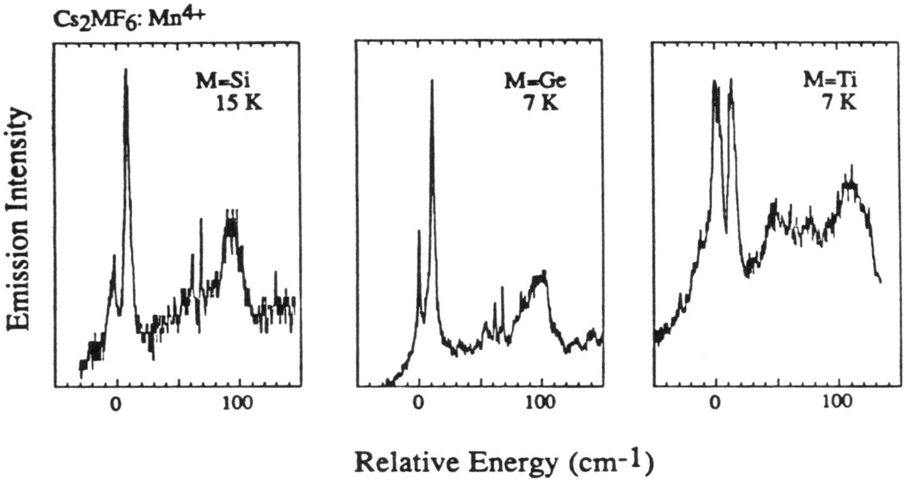

Figure 7. High resolution scans of the band origins. The four multiplet components are clearly observed at 0, 10, 64, and 72 cm^{-1} for Cs_2SiF_6: Mn^{4+}: Mn^{4+} and at 0, 11, 64, and 74 cm^{-1} for Cs_2GeF_6: Mn^{4+}; the combination of shot to shot noise and a phonon sideband obscures the higher energy features in the Cs_2TiF_6: Mn^{4+} case, but peaks are observed at 0, 5 and 13 cm^{-1}.

Figure 7 shows the zero phonon lines of the $^4T_{2g}$ state in the three crystals. These spectra were taken by averaging 140 laser pulses per point to reduce the noise due to the shot to shot fluctuations. In the Si and Ge salts the four components are clearly seen, but we cannot be sure of these for the Ti compound. The multiplet widths are 72 cm^{-1} for Si and 74 cm^{-1} for Ge.

4. Selection Rules and Analysis of OP and TP Spectra

The major features of the TP spectra can be explained with the help of the selection rules. First, it is illuminating to compare transition mechanisms for the $^4A_{2g} \to ^4T_{2g}$ transition in one- and two-photon spectroscopy. In OP transitions we find t_{1u} and t_{2u} vibrations acting as spectral origins, so the perturbation chain has the matrix elements:

$$< A_{2g} | \vec{r} | i >< i | \left(\frac{\partial H}{\partial Q_{t_{1u}}}\right) | T_{2g} > Q_{t_{1u}}$$

where i can be a T_{2u} intermediate state, and we are neglecting the spin part of the wavefunction for the moment.

There are nine components of this product which can be written as xV_y etc. where x is a component of the electric dipole transition operator, $e\cdot\vec{r}$, and V_y is a component of the vibronic operator, $\left(\frac{\partial H}{\partial Q_y}\right)Q_y$. The intermediate states |i><i| are omitted. In order for a $^4A_{2g}\rightarrow{}^4T_{2g}$ transition to occur the spin-independent transition operator must have the symmetry $A_{2g}\times T_{2g} = T_{1g}$. The vibronic operator with this symmetry has components

$$T^z_{1g}=xV_y\text{-}yV_x \text{ ,etc.,}$$

as can be seen by applying a T_1 projection operator to the set of nine components to get the antisymmetric direct product functions. These tensor components may be nonzero. Now compare this case to that for TP transitions.

In TP transitions we have two electric fields giving rise to :

$$< A_{2g} | \vec{r}_1 | i >< i | \vec{r}_2 | T_{2g} >$$

In a single beam experiment, $\vec{r}_1$ and $\vec{r}_2$ are parallel, but their direction with respect to the crystal axes can be controlled. But as $\vec{r}_1$ and $\vec{r}_2$ are parallel and indistinguishible, the antisymmetric direct product gives zero, and no T_{1g} operator is possible. Only A_{1g}, E_g, and T_{2g} two-photon operators can be formed, and none of these can cause an $A_{2g}\rightarrow T_{2g}$ transition. If we had two independent beams of different frequency such that two-photon transitions could only be caused by one photon from each beam, then the T_{1g} operator analogous to the T^z_{1g} vibronic operator of the last paragraph is $x_1y_2 - x_2y_1$ and would be non-zero. Thus the $A_{2g}\rightarrow T_{2g}$ transition would be permitted when the two beams are polarized along different cube axes.

Nevertheless we do see a direct TP transition as all four multiplet components are observed in several cases, and we now explain the reasons. Using the double group selection rules $^4A_2(\Gamma_8)\rightarrow{}^4T_2(\Gamma_6,\Gamma_7,\Gamma_8,\Gamma_8)$ the transitions are formally allowed by A_{1g}, E_g, and T_{2g} two-photon operators. The transition probability, however, comes entirely from the spin-orbit mixing of $^4T_{2g}$ and a^4T_{1g} states, where the a^4T_{1g} state is the one belonging to the same electron configuration, t^2e, as the $^4T_{2g}$ state. A complete diagonalization of the d^3 matrices shows that the admixture of these states is only a few percent for Mn^{4+}. But it is this small mixing which is responsible for the appearance of the origin quartet in the $A_{2g}\rightarrow T_{2g}$ two-photon spectrum.

The multiplet intensity patterns in TP are different from those in OP, and the difference depends on the spin-orbit perturbation routes into the quartet components, while the OP pattern is determined by the direct magnetic transition.

The vibrational structure of the TP band shows how the forbidden transition becomes allowed: the t_{2g} vibrational mode is very strong compared to the origin in Cs_2GeF_6: Mn^{4+}, and it is clearly the major spectral origin. The t_{2g} mode provides the symmetries needed as can be seen when tracing through the perturbation routes:

$$< A_{2g} | \vec{r}_1 | i >< i | \left(\frac{\partial H}{\partial Q_{t_{2g}}}\right) | j >< j | \vec{r}_2 | T_{2g} >$$

In the first matrix element $i = T_{2u}$. This means that j can be $T_{2u} \times T_{2g} = A_{2u} + E_u + T_{1u} + T_{2u}$. The last matrix element can be non-zero if j has any one of these symmetries.

It is illuminating to construct the triple tensor operator belonging to T_1 from the TP operators $\vec{r}_1$ and $\vec{r}_2$ and the vibronic coupling operator,

$$V_i = \left(\frac{\partial H}{\partial Q_i}\right) Q_i.$$

Here V_i belongs to the same partner of T_2 as Q_i, where Q_i is the ith partner of the t_{2g} vibrational mode, and $\left(\frac{\partial H}{\partial Q_i}\right)$ is the vibrational electronic coupling term for this mode. Both the E_g and T_{2g} combinations of the two photons can couple with V_i into a T_{1g} compound operator. For light propagating along the z axis of a cubic crystal we would have the components

$$T_1^x = y_1 y_2 V_{1x}, \quad \text{and}$$

$$T_1^y = x_1 x_2 V_{1y}$$

for the E_g intermediate couplings of the photons, and

$$T_1^x = (x_1 y_2 + x_2 y_1) V_y, \quad \text{and}$$

$$T_1^y = (x_1 y_2 + x_2 y_1) V_x$$

for T_{2g} intermediate coupling. It therefore appears that no marked polarizations should be observed in the vibronic sidebands.

This analysis shows that compound transition operators exist and could explain the presence of the t_{2g} vibronic sideband, but it does not explain its intensity relative to the origin quartet. Experimentally, the vibronic contribution to the intensity is at least ten times as large as the spin-orbit contribution. The vibronic perturbation borrows from the same $^4T_1$ states as does the spin-orbit coupling. This example provides a unique opportunity to compare and understand these two important perturbations.

5. Effect of the Trigonal Field

It is very interesting to see that the origin quartet in the Cs_2TiF_6: Mn^{4+} spectrum is considerably stronger than the t_{2g} vibronic mode. Figure 6 shows that the t_{2g} mode is almost absent from the perpendicular spectrum and definitely weaker than the origin region in the parallel spectrum. The trigonal field belongs to the t_{2g} representation in the cubic group so we can regard it as a frozen t_{2g} vibrational displacement. However, since it is present in both ground and excited states it intensifies the origin and the progressions built on it.

The compound tensor analysis of the last section could be carried out in the same way to understand the effect of the trigonal distortion, except that only one component of the trigonal distortion is present and active.

Both the E-type TP operator and the T_2-type can couple with V_{trig} to give a T_1 operator. If we carry out this coupling(cancelling out all but the z-component of V_{trig}) with coupling coefficients in the trigonal basis[6] we arrive at exactly the selection rules for the D_{3d} symmetry present in the trigonally distorted octahedron and given in reference 3. The transitions for a single beam propagating along the C_3-axis are isotropic and fully allowed for the $A_{2g} \rightarrow T_{2g}$ transition, but only for the E component of T_2: the A_1 component can only be seen in a two color experiment. The transition for light propagating perpendicular to C_3 and polarized parallel to it is strictly forbidden for both components in a single beam experiment. This forbidden transition is seen anyway, and as for the origin in the cubic crystals, it is the spin-orbit admixture of $^4T_{1g}$ which promotes it, since it is allowed under the double group selection rules.

6. The Jahn-Teller Effect

The $^4T_2$ state could have a T×e or a T×t Jahn-Teller effect. A previous study by Gudel and Snellgrove on the isoelectronic system Cs_2NaYCl_6: Cr^{3+} showed only a T×e distortion.[7] Since TP spectroscopy reveals the even parity vibrations, we directly observed the e_g progression in the excited state for the Si and Ge salts. Furthermore the t_{2g} vibration induces a progression in e_g. The pattern of the progression yields a Huang-Rhys factor of 1. The origin multiplet is also completely revealed and the Ham quenching calculation of Chien, Berg, and McClure(reference 5) also gives a Huang-Rhys factor of 1, confirming the value obtained from the vibrational progressions. We do not see a direct progression in t_{2g}, demonstrating that only the T×e Jahn-Teller distortion is active.

7. Conclusion

Since the $A_{2g} \rightarrow T_{2g}$ transition is forbidden in a single beam TP process, spin-orbit coupling and vibronic coupling become the dominant intensity generating mechanisms. As it turned out, the vibronic perturbation is stronger than the spin-orbit perturbation, so the t_{2g} progression is more intense than the origin bands

The single beam TP spectra are more complicated than we would like, but are nevertheless simpler than the OP spectra and provide new information about the excited T_2 state. A great deal of theoretical work needs to be done on the present spectra, and two-color TP experiments are also needed.

8. References

1. J.E. Huheey, *Inorganic Chemistry*, 3rd ed.(Harper & Row, New York, 1983) pp. 73-76.
2. a. L. Helmholz and M.E. Russo, *J. Chem. Phys.*, **59**(1973)5455. b. S.L. Chodos, A. M. Black, and C.D. Flint, *J. Chem. Phys*, **65**(1976)4816. c. W.C. Yeakel, R.W. Schwartz, H.G. Brittain, J.L. Slater, and P.N. Schatz, *Mol.Phys.*, **32**(1976)1751. d. A. G. Paulusz, *J. Lumin.*, **17**(1978)375. e. N.B. Manson, Z. Hasan, and C.D. Flint, *J. Phys. C*, **12**(1979)5483. f. Z. Hasan and N.B. Manson, *J. Phys. C*, **13**(1980)2325.
3. T. R. Bader and A. Gold, *Phys. Rev.*, **171**(1968)997.
4. P. Rabinowitz, B. N. Perry, and N. Levinos, *IEEE J. of Quant. Elec.*, **22**(1986)797.
5. R.-L. Chien, J.M. Berg, and D.S. McClure, *J. Chem. Phys.*, **84**(1986)4168.
6. S. Sugano, Y. Tanabe, and H. Kamimura, *Multiplets of Transition-Metal Ions in Crystals*, Academic Press, New York, 1970.
7. H.U. Gudel and T.R. Snellgrove, *Inorg. Chem.*, **17**(1978)1617.

THE JAHN-TELLER EFFECT OF Cu^{2+} IN SIX-, FIVE- AND FOUR-COORDINATION - EXAMPLES FOR THE IMPORTANCE OF VIBRONIC INTERACTIONS IN TRANSITION METAL CHEMISTRY

D. REINEN
Department of Chemistry
Hans-Meerwein-Strasse
D - 3550 Marburg/L.
Federal Republic of Germany

ABSTRACT. The stereochemistry of transition metal ions in high symmetry ligand fields is frequently determined by strong vibronic interactions between the electronic groundstate and certain nuclear motions, often leading to pronounced distortions of the environment as well as large term splittings and energetic groundstate lowering effects. These distortions may be of "dynamic" or "static" nature, depending on the ratio between the thermal energy and the vibronic interaction energies, which determine the minima in the groundstate potential surface. A prominent example is Cu^{2+}, which undergoes very pronounced and moderately strong vibronic couplings of the Jahn-Teller type in octahedral ($E \otimes \varepsilon$) and tetrahedral coordination ($T_2 \otimes \varepsilon$), respectively, but also exhibits rather strong vibronic interactions between groundstate and the first excited in trigonal-bipyramidal coordination ($A' \otimes \varepsilon' \otimes E'$: "pseudo-Jahn-Teller effect").

A. INTRODUCTION

Most d^n cations in high symmetry ligand fields possess orbitally degenerate groundstates (Table 1), which may - according to the Jahn-Teller theorem [1] - undergo a splitting by a symmetry reduction of the environment, thus inducing a lowering of the groundstate energy (Figure 1). While the degeneracy of T groundstates in O_h and T_d ligand fields can be lifted by distortions along a tetragonal or trigonal axis, only tetragonal distortions may split E groundstates (see below). Predictions are possible concerning the extent of the groundstate splitting and the corresponding distortions [2]. If the groundstate is coupled to the ligand environment by strong bonds, the Jahn-Teller theorem works very effectively. Thus the distortion is very large for the σ-antibonding E_g groundstates of octahedrally coordinated d^9, d^4 and low-spin d^7 configurated transition metal ions (Table 1), while the E groundstate in tetrahedral or the T_2 groundstate in octahedral coordination [3] are only π-antibonding or even non-bonding and hence rather stable with respect to distortions and term splittings. On the other hand the T_2

C. D. Flint (ed.), Vibronic Processes in Inorganic Chemistry, 267–282.
© 1989 by Kluwer Academic Publishers.

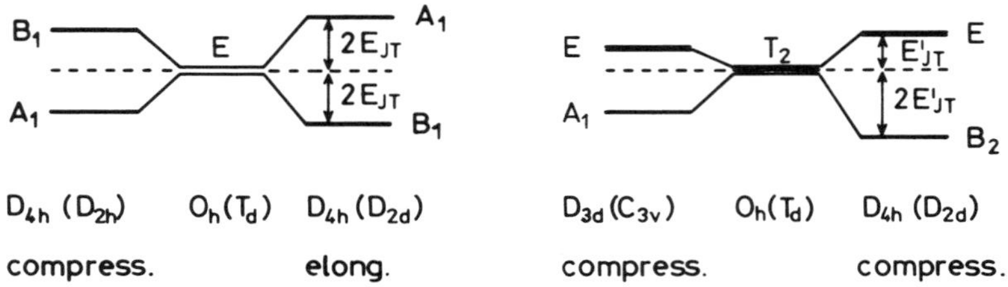

Figure 1. Symmetry splittings and electronic groundstate stabilisations ΔE for E groundstates by tetragonal (O_h: d^9, d^4 ; T_d: d^1, d^6 ; ΔE = 2 E_{JT}) and for T_2 groundstates by tetragonal and trigonal distortions (O_h: d^1, d^6 ; T_d: d^9, d^4 ; ΔE = 2 E'_{JT}).

TABLE 1. Groundstates for d^n cations (low-spin configurations underlined) in octahedral and tetrahedral coordination

octahedral		
E (σ)	d^4, d^9, $\underline{d^7}$	Cr^{2+}, Mn^{3+}; Cu^{2+}, Ag^{2+} Co^{2+}, Ni^{3+}
T_1 (π)	d^2, d^7, $\underline{d^4}$	V^{3+}; Co^{2+}
T_2 (π)	d^1, d^6, $\underline{d^5}$	Ti^{3+}, V^{4+}, Cr^{5+}; Fe^{2+}
tetrahedral		
E (π)	d^1, d^6	
T_1 ($\sigma+\pi$)	d^3, d^8	Ni^{2+}
T_2 ($\sigma+\pi$)	d^4, d^9	Cu^{2+}

groundstate of d^9 and d^4 cations in T_d symmetry is ($\sigma+\pi$)-antibonding and undergoes considerable energetic effects [4]. The specific distortions are induced by the vibronic coupling of the electronic groundstate wave functions with certain normal modes of the considered point group [5], leading to groundstate potential surfaces with energy minima along the distortion pathway of the active vibrations. The conceptional and mathematical background of vibronic interactions of this kind is given in C. Ballhausens article in this volume [6] and will not be treated here.

The ring-like minimum in the groundstate potential surface, resulting from the linear vibronic $E_g \otimes \varepsilon_g$ coupling for octahedral Cu^{2+} (Figure 2a) predicts the same stabilisation energy for any linear coordination of the two Q_Θ and Q_ε components of the ε_g vibration (Figure 2b) with the same radial distortion parameter

$$\rho = \{2(\delta a_x^2 + \delta a_y^2 + \delta a_z^2)\}^{1/2} \quad (1)$$

The δa_i are the deviations of the Cu^{2+}-ligand spacings from those of the octahedral parent geometry. Accounting also for non-linear vibronic interactions generates minima and saddlepoints in those directions of the potential surface, which correspond to D_{4h} symmetries, while all other points of the groundstate potential surface have the lower D_{2h} symmetry. Experiment shows, that the sign of the non-linear interaction terms have to be chosen such, that the minima appear in the directions $\varphi = 0^\circ, 120^\circ, 240^\circ$ of the angular coordinate and saddlepoints at $\varphi = 180^\circ, 300^\circ, 60^\circ$ (Figure 2). The former characterise tetragonally elongated octahedra along z, x, y and the latter correspondingly compressed octahedra, which are slightly less stable (see Figure 1). If the vibrational levels in the potential surface are explicitely introduced in addition, the terminology: "dynamic Jahn-Teller effect" is generally used [6]. If the ratio $E_{JT}/\hbar\omega$ (ω: frequency of the active mode) is much larger than unity, as in most of the treated cases, it is sufficient to

Figure 2. The groundstate potential surface - $E \otimes \varepsilon$ vibronic interaction (a) Linear coupling ("Mexican hat" potential surface) (b) Components of ε normal mode (c) Cross section perpendicular to E axis of (a), but with inclusion of non-linear interaction terms.

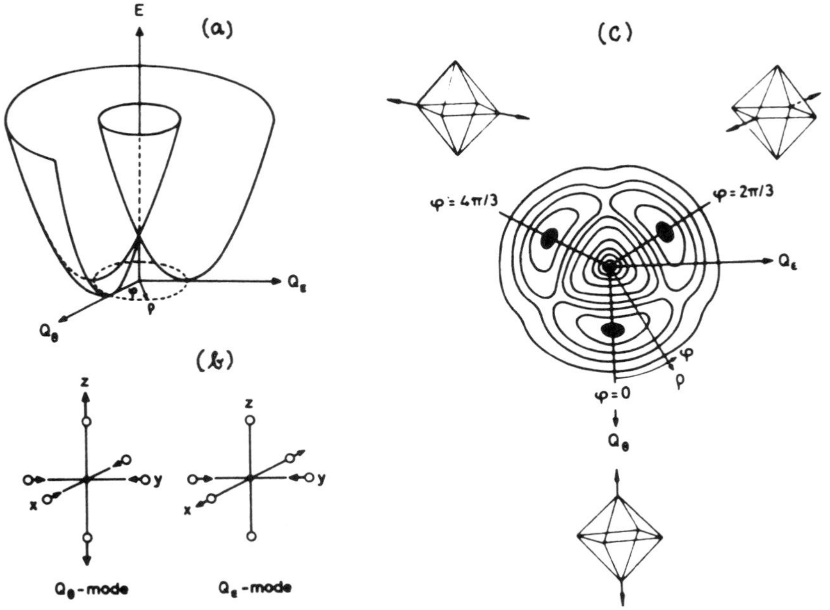

consider the "static" features of the vibronic interaction.
In order to freeze a system into a distorted configuration, one of the minima of Figure 2c must have a lower energy than the other two - otherwise a symmetry break will not occur. Two mechanisms can be discussed:

(a) Random strain induced by lattice defects in a solid compound doped with for example Cu^{2+} ions may be operative in this way, yielding equal probabilities for any Cu^{2+} ion to occupy an elongated octahedral site with the C_4 axis along x, y or z (Figure 2c) [7].
(b) The "cooperative Jahn-Teller effect" considers elastic interactions between different locally distorted polyhedra [5,6], which may eventually orient all long axes of the distorted polyhedra in the same crystallographic direction. Apparently one minimum is stabilised with respect to the other two by this "ferrodistortive" order, which leads to a tetragonal distortion of a cubic unit cell with $c/a > 1$ [8]. If two ferrodistortive sublattices are oriented perpendicular to each other, the order is "antiferrodistortive". In this case two minima are energetically lower than the third and a unit cell distortion with $c/a < 1$ is observed [8]. Various examples for these two extreme types of order or for intermediate situations, which occur in solids with higher concentrations of "Jahn-Teller ions", are known meanwhile [8]. An interesting feature of cooperative Jahn-Teller interactions in structures with widely interconnected polyhedra is, that the extent of distortion (ρ, eq(1)) and the groundstate splitting (Figure 1) of the individual polyhedra may be much larger than in the situation of elastically isolated polyhedra [8,4].

In the next Section experimental evidence for vibronic coupling effects of Cu^{2+} in octahedral coordination will be presented, but it should be stated that completely analogous structural and spectroscopic results are observed for other d^9 cations or cations with d^4 and d^7 low-spin configuration. In Section C Cu^{2+} in five-coordination will be considered, while only a few remarks will be made with respect to Cu^{2+} in tetrahedral coordination.

B. THE $E_g \otimes \varepsilon_g$ COUPLING CASE OF OCTAHEDRAL Cu^{2+}

In Figure 3 the structure of a pseudo-octahedral Cu^{2+} complex with six nitrogen ligator atoms from two tridentate ligands is illustrated [9]. At 298 K a rather small distortion (eq(1): $\rho \simeq 0.14$ Å) is observed, which increases when lowering the temperature (110 K: $\rho \simeq 0.30$ Å). The temperature dependence of the extent of distortion is also nicely reflected by the g-tensor [9], which is experimentally accessible from EPR spectroscopy. Apparently the distortion is maximal at $T \lesssim 120$ K and vanishes at about 340 K. This observation may be readily explained by the occupation of excited vibrational levels in the rather flat minima at $\varphi = 0^\circ$, $2\pi/3$ and $4\pi/3$ in the groundstate potential surface of Figure 2c with increasing temperature. Hence, if kT becomes larger

Figure 3. Static and dynamic distortions of the "CuN_6" polyhedra in $[Cu^{II}(TACN)_2]Cu^{I}(CN)_3$ (TACN: 1,4,7-Triazacyclononane): The geometry of the $Cu(TACN)_2^{2+}$ cation [Cu-N spacings: 2 x 2.32(2.23) Å, 4 x 2.06(2.11) Å at 110(293) K] (left) and the molecular g-values in dependence on temperature (right).

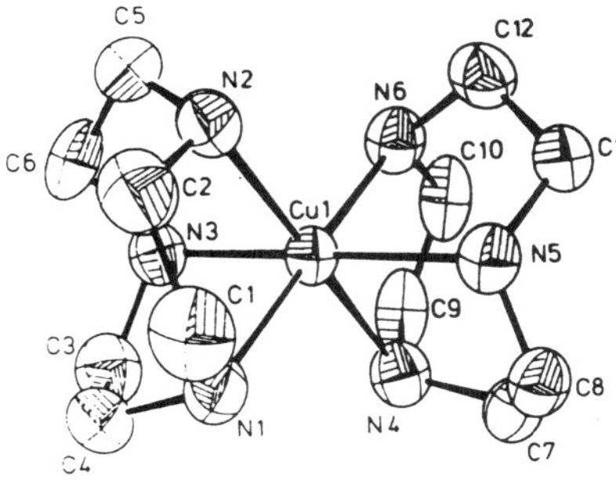

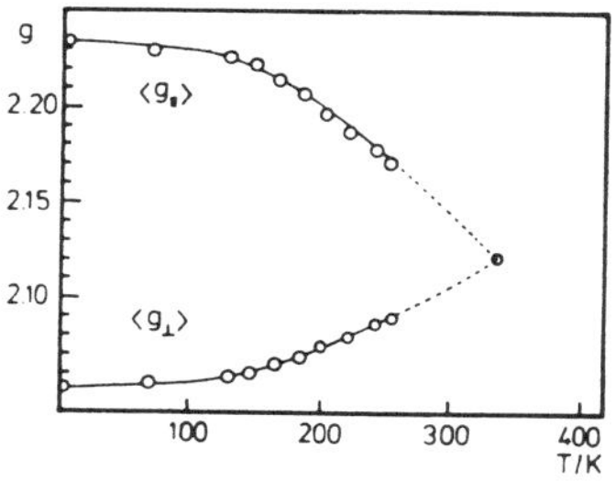

than the warping energy (energy difference between minima and saddle-points), an equilibration between the three octahedral conformations elongated along x, y and z occurs and an undistorted octahedron in time-average results. This process, which is determined by the ratio of the non-linear vibronic interaction energies and the energy of the angular part of the Jahn-Teller active vibration, is called the "dynamic Jahn-Teller effect" by the chemists (see Section A, however), and the author will stick to this nomenclature in this article. It will depend on the time scale of the applied physical method,however, if the instantaneous "distorted" or the time-averaged "undistorted" polyhedron is observed. EPR spectroscopy (time scale $\gtrsim 10^{-9}$ s) - as NMR and Mößbauer spectroscopy - is slow with respect to the angular motion ($\approx 10^{-13}$ s) and sees the dynamical average (motional narrowing). X-ray or neutron diffraction are very fast in the elementary process of the interaction between radiation and electron density or atomic nuclei (10^{-18} s), but - because a series of simultaneous and successive processes constitutes each experimental issue - the result is averaged also. Nevertheless the ellipsoids of thermal motion reflect the instantaneous nuclear positions and allow the quantitative analysis with respect to the underlying statically distorted polyhedra [10,8]. The transition from the static to the dynamic Jahn-Teller effect may be continuous and of second order, extending over a wide temperature range - as in the considered example [9] - or discontinuous and of first order. The latter case is always observed, if the polyhedra are strongly interconnected in the crystal structure [8].

Frequently the host site of a compound, in which Cu^{2+} is isomorphously substituted, deviates from O_h symmetry by the specific packing forces of the chosen crystal structure already. Such a lower symmetry component may distort the potential surface of Figure 2 and shift the minima to other positions. Figure 4 illustrates the cases of the Cu^{2+} doped tetragonal compounds Ba_2ZnF_6 and K_2ZnF_4, in which the ZnF_6 octahedra

Figure 4. Strain influence (tetragonal compression of octahedral site) on the warping structure of the groundstate potential surface (Fig.2c): Angular dependence of the energy (in units of the warping energy β) for increasing strain influence (left) and cross sections perpendicular to the energy axis for cases (a) and (b), corresponding to Cu^{2+} doped Ba_2ZnF_6 and K_2ZnF_4, respectively (right).

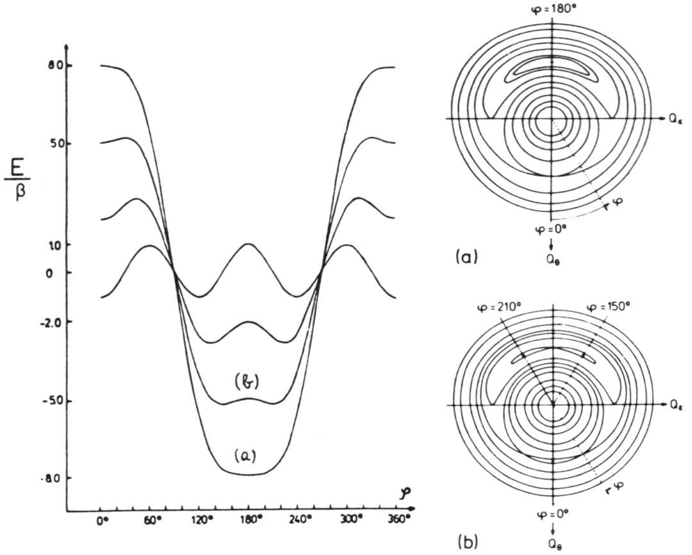

are slightly tetragonally compressed (along z) and hence will relatively stabilise the position of the saddlepoint at $\varphi = 180^\circ$ in the groundstate potential surface of Figure 2c for example [8]. Depending on the ratio between the strain and the warping energy the two minima at $\varphi = 120^\circ$, 240° will be lowered in energy with respect to the third one and move toward $\varphi = 180^\circ$. For a comparatively large strain influence only one minimum at $\varphi = 180^\circ$ remains [11,12].

Cu^{2+} ions can be easily incorporated into NH_4Cl with the CsCl structure. They substitute the NH_4^+ ions of two neighboured unit cells, occupying the centre of the common face. The positions of the two NH_4^+ cations may be taken by OH_2 or NH_3 molecules, creating pseudooctahedral $CuCl_4X_2$ centres I ($X = OH_2$) and II ($X = NH_3$) (Figure 5). The potential surface of Centre II corresponds to case (a) in Figure 4, because the axial nitrogen ligator atoms induce stronger σ-bonds than the axial Cl^- ligands, thus generating a strain corresponding to a tetragonal compression [13]. On the other hand we learn from the temperature dependence of the g-values of Centre I (Figure 5), that the symmetry is decreased from tetragonal to orthorhombic at very low temperatures. This behaviour is due to a groundstate potential surface with two equivalent lower-energy minima, which correspond to o-rhombically distorted $CuCl_4(OH_2)_2$ polyhedra and which are separated by a very flat saddlepoint at $\alpha = 180^\circ$

Figure 5. $CuCl_4X_2$ centres [I: X = OH_2 ; II: X = NH_3] in NH_4Cl: The nature of Centre I (and analogously II) [above, left] and the temperature dependence of the g-values for Centre I (EPR powder data) [above, right]. The angular variation of the g-tensor (Centre I) in the (001) [(010),(100)] plane at 130 and 4 K (EPR, Q band) is given below.

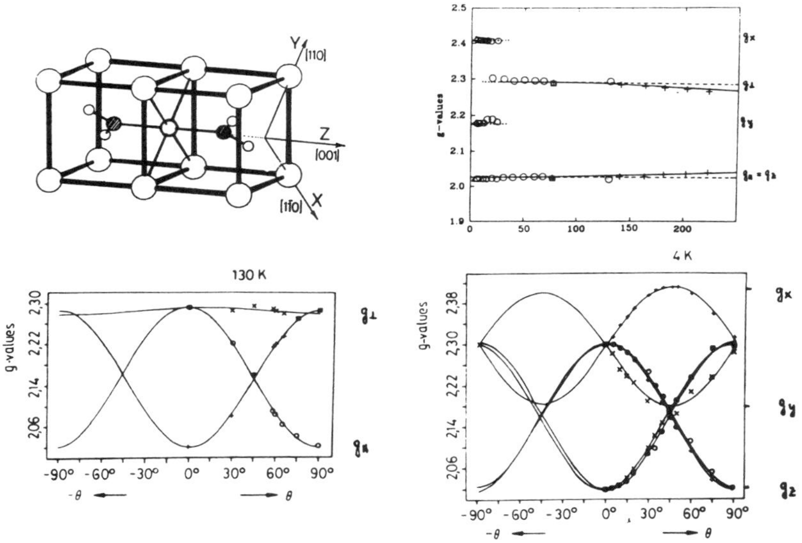

(Figure 6). At very low temperatures the system is frozen into either one of the two minima, which characterize polyhedra with longer Cu-Cl spacings in the x(y) and shorter distances in the y(x) directions (Figure 6). Increasing the temperature induces dynamical averaging between these minima, yielding a tetragonally compressed octahedron in time average. The different behaviour of Centre I with respect to II is due to the comparatively weaker axial ligands (O compared to N ligator atoms), which distort the potential surface of Figure 2c to a smaller extent. The vibrational structure of the groundstate potential surface of Centres I and II as well as the vibronic wavefunctions have been calculated explicitely [13].

Concluding this chapter we may state that the stereochemistry of Cu^{2+} in octahedral coordination is dominated by the linear vibronic Jahn-Teller coupling leading to groundstate splittings of $\gtrsim$ 1 eV and considerable ρ-values (Figure 2a). Though the non-linear interaction terms, which are usually smaller by a factor of 20, favour tetragonally elongated octahedra with D_{4h} symmetry (Figure 2c), strain energies of various kinds comparable to or even larger than the warping energies may shift the minima to other positions of the groundstate potential surface, giving rise to a large variety of distortion geometries [8,14]. In order to

Figure 6. The $CuCl_4(OH_2)_2$ Centre in NH_4Cl: The groundstate potential surface - cross section perpendicular to the energy axis and angular dependence of the energy (with lowest vibronic levels) [above]. The $CuCl_4(OH_2)_2$ geometries, corresponding to the minima A and A' [low-temperature situation], and the time-averaged conformation at higher temperature are given below.

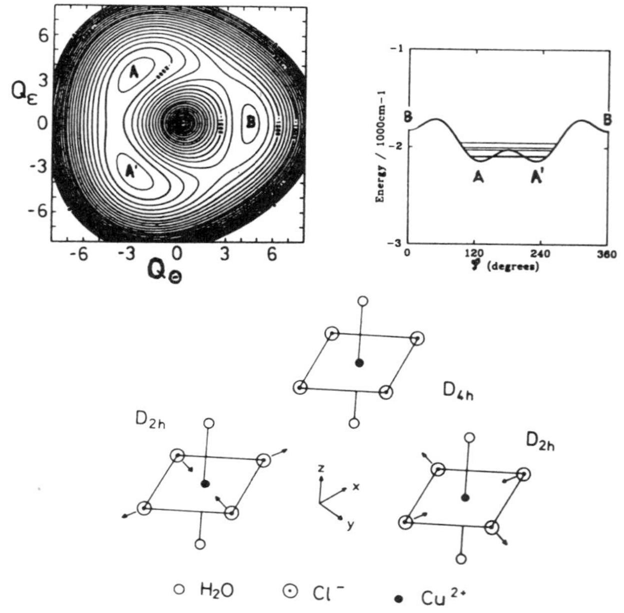

analyse the strain influence and separate it from the geometric implications of the Jahn-Teller coupling, the structural comparison of the Cu^{2+} compound with the corresponding Ni^{2+} or Zn^{2+} compounds is useful, because the latter cations are also from the 3d series, have similar ionic sizes as Cu^{2+}, but possess Jahn-Teller stable $^3A_2$ and $^1A_1$ groundstates, respectively.

C. Cu^{2+} IONS IN FIVE-COORDINATION

Electron pair repulsion calculations show - in agreement with structural data for central ions with noble-gas configuration - that the energetically most favourable coordination geometry is an elongated trigonal bipyramid, which is only slightly preferred with respect to a compressed square pyramid, however [15]. Deviating from this behaviour, five-coordinate Cu^{2+}-complexes have mostly geometries near to elongated square pyramids, which seem to be somewhat more stable than compressed trigonal bipyramids [16]. The $Cu(NCS)_5^{3-}$ polyhedron in the green compound

Figure 7. The trigonal-bipyramidal $CuCl_5^{3-}$ and the square-pyramidal $Cu(NCS)_5^{3-}$ polyhedron in the compounds $[Co(NH_3)_6]CuCl_5$ (298 K) [19] and $(NEt_4)_3[Cu(NCS)_5]$ (200 K) [18], respectively, and temperature-dependent heat capacity data for the former compound.

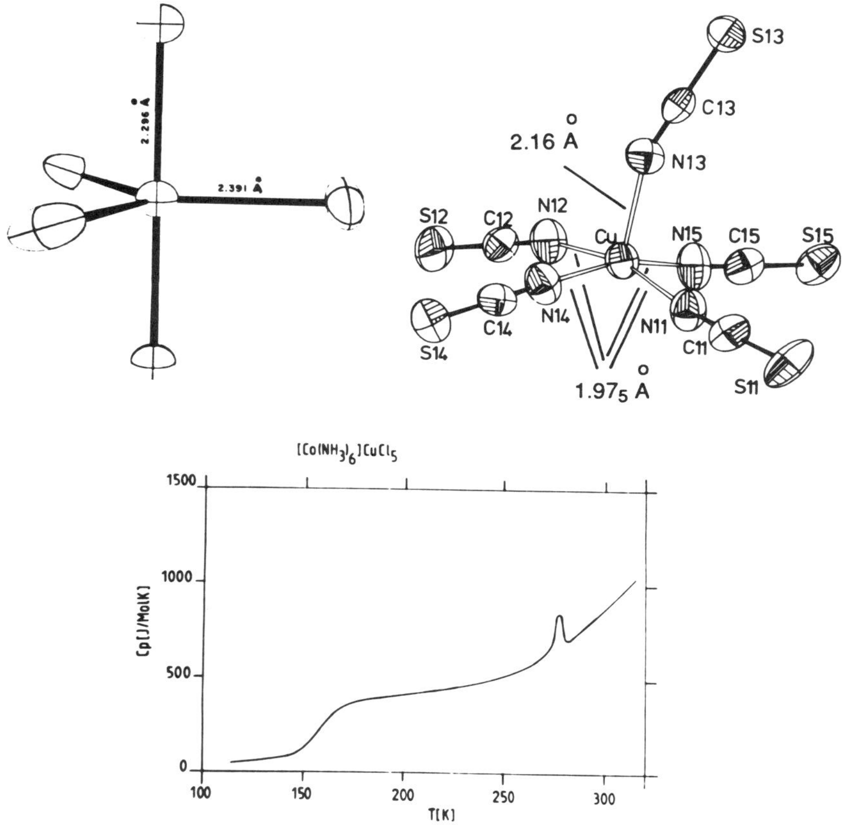

$(NEt_4)_3[Cu(NCS)_5]$, which is square-pyramidal as the $Cu(NH_3)_5^{2+}$ complex [17], is a particularly interesting example (Figure 7), because this compound exhibits a reversible phase transition at about 285 K to a brown compound with square-planar $Cu(NCS)_4^{2-}$ entities [18] (see below). On the other hand the $CuCl_5^{3-}$ polyhedra in $[Co(NH_3)_6]CuCl_5$ are compressed trigonal bipyramids [19] but with ellipsoids of thermal motion of the equatorial Cl^- ligands, which are anomalously enlarged perpendicular to the Cu-Cl bonds and lie approximately in the equatorial plane. This compound shows a first-order transition at 280 K to a low temperature phase, however (Figure 7), which now contains elongated square-pyramids. This could be demonstrated by spectroscopic investigations [16], while

Figure 8. The pseudorotation between three square-pyramidal conformations in the xy plane [above, right], the term diagrams for the $CuCl_5^{3-}$ polyhedron in D_{3h} (trigonal bipyramid ; d_{z^2} groundstate), C_{4v} (square pyramid ; $d_{x^2-y^2}$ groundstate) and "intermediate" C_{2v} geometries (left) [21], and the higher symmetry Q_ε components of the three ε' modes in D_{3h}.

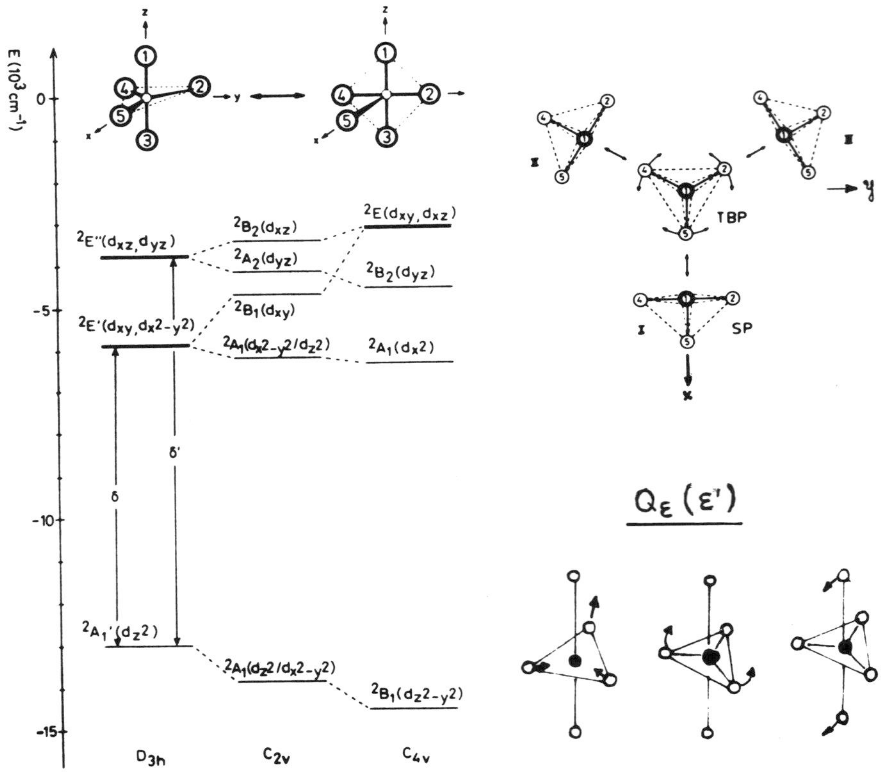

structural low-temperature analyses failed because of extensive crystal twinning. Apparently the trigonal bipyramids are the dynamic averages over differently oriented elongated square pyramids (Figure 8), which freeze in below the transition temperature of 280 K. This dynamical process shows some similarity to the well-known Berry pseudorotation [20], but is restricted essentially to one (the xy) plane. One may explain the described structural results by a vibronic coupling model of the pseudo-Jahn-Teller type, which considers the interaction between the groundstate and a nearlying excited state via certain vibrational modes of the respective point group [21,22].

Figure 8 gives the term diagrams for the alternative trigonal bipyramidal and square-planar geometries. In both cases the groundstate is orbitally non-degenerate and hence stable with respect to first-order Jahn-Teller couplings. The respective groundstates can be lowered, if the trigonal bipyramid is compressed and the square pyramid is elongated - as actually observed. This can be easily deduced from Angular-Overlap considerations [21]. Starting from the D_{3h} geometry one can now demonstrate, that a symmetry reduction to C_{2v} via the ε' vibrations eventually lowers the groundstate further, because in this point-group the groundstate and one split-state of the excited E' state have the same A_1 symmetry and hence may repel each other (Figure 8). Only the (three) ε' modes in D_{3h} (Figure 8) can be active in this way and provide a distortion pathway corresponding to a $A_1' \otimes \varepsilon' \otimes E'$ pseudo Jahn-Teller interaction. One easily recognizes, that the ε' modes additionally give rise to a first-order Jahn-Teller splitting in the excited E' state, which enhances the pseudo-Jahn-Teller coupling [21]. The C_{4v} geometry is verified on the C_{2v} distortion pathway by a particular linear combination of the three higher-symmetry Q_ε components of the stretching and the "in-plane" and "out-of-plane" deformation ε' modes (Figures 8, 9). There are three conformational possibilities for the D_{3h} - $C_{2v}(C_{4v})$ transformation, namely by having the 2-fold (4-fold) axis along the bonds to the ligand 5, 2 or 4.

On the basis of the available structural and spectroscopic data for the $CuCl_5^{3-}$ polyhedron in $[Co(NH_3)_6]CuCl_5$, the groundstate potential surface of Figure 10 is calculated, which resembles the one for the octahedral $E \otimes \varepsilon$ coupling case very much indeed (Figure 2c). The vibronic coupling

Figure 9. The linear combinations of the $Q_\varepsilon(\varepsilon')$ modes in D_{3h} (Figure 8), leading to the square pyramid (+ε) and the "inverse geometry" (-ε), respectively.

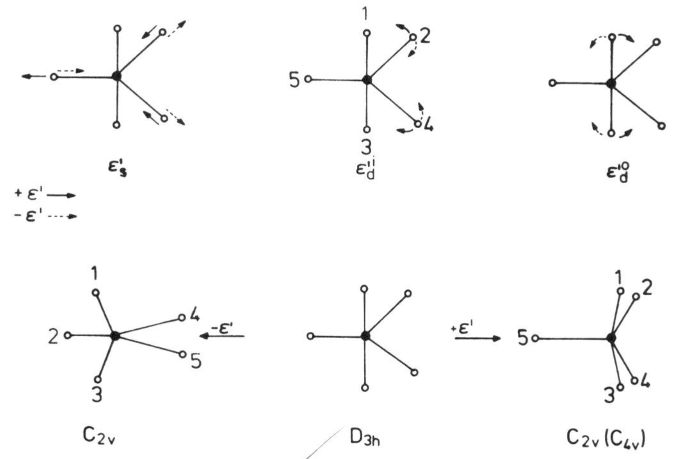

Figure 10. Contour diagram of the groundstate potential surface of the $CuCl_5^{3-}$ polyhedron in $[Co(NH_3)_6]CuCl_5$ (cross-section perpendicular to the energy axis).
Q_ε, Q_ζ : Linear combinations of the three "C_{2v}" and "C_s" components of the ε' modes in D_{3h} - b , d : Vibronic constants, describing the $A_1' \otimes \varepsilon' \otimes E'$ and the $E' \otimes \varepsilon'$ coupling, respectively - k : Force constant of the "combined" ε' modes [21] - δ : Initial $^2A_1' - {}^2E'$ splitting (Figure 8).

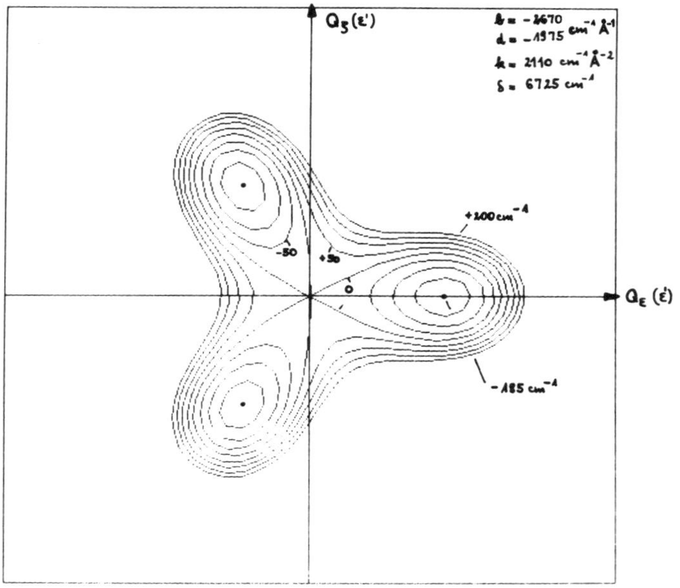

matrix and the calculation procedure is published elsewhere [21]. The three minima correspond to the three square-pyramidal conformations of Figure 8 and are more stable with respect to the D_{3h} symmetry by only 185 cm^{-1}. This value perfectly matches with the phase transition temperature of 280 K and with the anomalous ellipsoids of thermal motion mentioned before (Figure 7), which are due to the very pronounced contribution of the "in-plane" deformation in the pseudorotation process from D_{3h} to C_{2v} (C_{4v}). While values of the angular parameter $\alpha = 0^\circ$, 120°, 240° and 180°, 300°, 60° describe C_{2v} geometries (special points: C_{4v}), all other geometries of the potential surface have the lower C_s symmetry.
Compounds $Cu(NH_3)_5X_2$ [X = Cl^-, Br^-, BF_4^-] also undergo phase transitions at about 280 - 300 K [17,21]. In these cases the square pyramids transform into trigonal bipyramids, the threefold axes of which are not fixed in space, however. This process is the unrestricted Berry rotation, which nevertheless still follows the ε' distortion pathway, but involves the free rotation of the threefold axis [21].
Remembering the small stabilisation energy of the elongated square

Figure 11. The square-pyramidal [X = Cl^-] and "inverse" geometry [X = NCS^- (or Br^-)] of complexes Cu terpy X_2

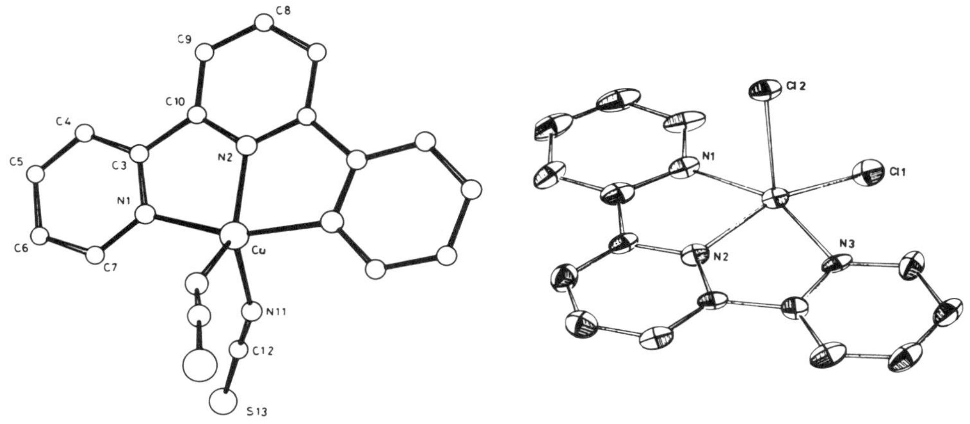

pyramid with respect to the D_{3h} geometry and the saddlepoints (α = 180°, 300°, 60°) - similarly small values are found for NCS^- and NH_3 as ligands [21] - the enormous variety of different geometries for five-coordinate Cu^{2+} complexes observed experimentally [23,24] is not really surprising. Already small strains, due to rigid polydentate ligands and/or to the presence of different ligands with differing σ-bond strength for example, may shift the shallow minima to any point of the groundstate potential surface (Figure 10). Particularly interesting is the tridentate terpyridine ligand ($C_{15}N_3H_{11}$), which imposes a strain on complexes $Cu(terpy)X_2$ [X: Cl^-, Br^-, NCS^- etc] with the symmetry of nuclear ε' motions "inverse" to those leading into the square pyramid (Figure 9). While for X = Cl^- still an (approximate) square-pyramid is observed, for X = Br^-, NCS^- a geometry in the directions of the saddlepoints is stabilised [24] (Figure 11).
Though the groundstate potential surfaces for octahedral (Figure 2c) and five-coordinate Cu^{2+} (Figure 10) are topologically identical, the involved energies are different. While in the latter case the D_{3h} and C_{2v} (C_{4v}) geometries have nearly the same energies, the regular octahedron is energetically much less stable with respect to the D_{4h} polyhedra in case of the octahedral $E \otimes \varepsilon$ vibronic interaction. The five-coordinate species, which are observed in the photochemistry of octahedral Cr^{3+} complexes, seem to be involved into a similar pseudorotation process, as discussed in this section [25].

D. Cu^{2+} IN TETRAHEDRAL COORDINATION

The threefold orbital degeneracy of the electronic T_2 groundstate of

tetrahedral Cu^{2+} can be lifted by vibronic interactions with the ε and τ_2 modes in T_d symmetry. The general topology of the adiabatic $T \otimes (\varepsilon+\tau_2)$ potential surface has recently been analysed in detail [26]. The symmetry reduction to tetragonal (D_{2d}) and trigonal geometries (C_{3v}), introduced by distortions along the four- and threefold axes by the two modes, always corresponds to compressions, because an orbital singlet groundstate in the symmetry-reduced conformation results only under this geometric condition (Figure 1). The structural and spectroscopic evidence [27] for Jahn-Teller distortions in T_d symmetry clearly indicates, that the coupling to the ε mode dominates with respect to the $T_2 \otimes \tau_2$ interaction. This has been confirmed by calculations of the extremum positions of the $T_2 \otimes (\varepsilon+\tau_2)$ groundstate potential surface [4].
We may conclude, that the stereochemistry of four-coordinate Cu^{2+} is determined by flattened tetrahedra of D_{2d} symmetry or in many cases even by a square-planar coordination with D_{4h} symmetry, which may be considered as the extreme of a compression along the $T_2 \otimes \varepsilon$ distortion pathway.

E. CONCLUSIONS

If one analyses the molecular structures and the electronic groundstates of four- and six-coordinate Cu^{2+} entities in isolated complexes or solid compounds with interconnected polyhedra, one finds a very pronounced tendency to avoid orbitally degenerate groundstates. This is well understood in terms of strong vibronic interactions between the respective T_2 and E groundstates and the ε normal modes in T_d and O_h ("Jahn-Teller effect"). In T_d symmetry this coupling leads always to compressed tetrahedra and to the stabilisation of a $d_{x^2-y^2}$ groundstate, with a groundstate splitting of about $\gtrsim 1/2$ eV [27,4]. The splitting is about twice as large in the octahedral case, but, in first order, without fixing the distortion geometry at a specific D_{4h} or D_{2h} geometry (Figure 2a). Nonlinear coupling contributions lead to shallow minima for the elongated octahedron ($d_{x^2-y^2}$ groundstate), which are only a few hundred wave-numbers more stable than the saddlepoints (compressed octahedron, d_{z^2} groundstate) (Figure 2c). This situation introduces interesting dynamic effects and also allows to stabilize various D_{4h} and D_{2h} geometries by strains due to packing forces, rigid ligand effects or the presence of different ligands.
Though the topology of the groundstate potential surface for five-coordinate Cu^{2+} (Figure 10) is equivalent to the one of octahedral Cu^{2+} (Figure 2c), it is generally rather flat and not very steep with respect to the origin as for the six-coordination. The vibronic pseudo-Jahn-Teller interaction stabilizes the (approximate) elongated C_{4v} geometry ($d_{x^2-y^2}$ groundstate) with respect to the trigonal bipyramid (D_{3h} symmetry, d_{z^2} groundstate) (Figure 8) by only a few hundred wave numbers. As for octahedral Cu^{+}, this situation again introduces a high sensitivity with respect to dynamic and strain effects, which is not restricted to the angular coordinate, however, but strongly involves in particular the radial coordinate.

After all the predominant feature in the stereochemistry of Cu^{2+} compounds is the tendency toward a square-planar environment ($d_{x^2-y^2}$ groundstate), which is equally verified in strongly compressed tetrahedra or strongly elongated octahedra and square-pyramids. It has to be emphasized, that Cu^{2+} is only the most prominent example for the presence of vibronic Jahn-Teller interactions. Various other d^n cations (Table 1, Section A) are also subject to large Jahn-Teller distortions, and we should also mention in this connection the stereochemical consequences of the presence of lone electron pairs, which frequently induce vibronic couplings of the pseudo-Jahn-Teller type [28]. One may finally speculate about the importance of vibronic interactions for the superconducting properties of ceramic materials, because all effective "high-temperature superconductors" contain cations with groundstates, which are unstable with respect to Jahn-Teller or pseudo-Jahn-Teller couplings. Concluding it seems interesting to shortly follow the consideration, why the symmetry reduction due to vibronic interactions always occur in such a manner, that the maximal symmetry is still preserved. Examples are the groundstate potential surfaces of Figures 2c and 10, in which the minima are connected with the maximal symmetries D_{4h} and C_{2v} (C_{4v}), respectively. The recently formulated "epikernel principle" comprises these symmetry properties on a general basis and is discussed elsewhere [29].

REFERENCES

1. Jahn, H.A. and Teller, E. (1937), Proc. R. Soc. A 161 , 220.
2. Reinen, D. (1983), Comments on Inorg. Chem. 2, 227.
3. Ameis, R., Kremer, S., and Reinen, D. (1985), Inorg. Chem. 24, 2751.
4a. Reinen, D., Allmann, R., Baum, G., Jakob, B., Kaschuba, U., Massa, W., and Miller, G.J. (1987), Z. anorg. allg. Chem. 548, 7 .
b. Reinen, D., Atanasov, M., Nikolov, G.St., and Steffens, F. (1988), Inorg. Chem. 27, 1678.
5. Bersuker, I.B. (1984) The Jahn-Teller Effect and Vibronic Interactions in Modern Chemistry, Plenum Press, New York.
6. Ballhausen, C.J. (1988), Vibronic Processes in Inorganic Chemistry, NATO ASI Series, this volume.
7. Ham, F.S. (1972) in: Electronic Paramagnetic Resonance, Plenum Press, New York.
8. Reinen, D. and Friebel, C. (1979), Structure and Bonding, Springer-Verlag, 37, 1 - and cited references.
9. Chaudhuri, P., Oder, K., Wieghardt, K., Weiss, J., Reedijk, J., Hinrichs, W., Wood, J., Ozarowski, A., Stratemeier, H., and Reinen, D. (1986), Inorg. Chem. 25, 2951.
10. Ammeter, J.H., Bürgi, H.B., Gamp, E., Meyer-Sandrin, V., Jensen, W.P. (1979), Inorg. Chem. 18, 733.
11a. Reinen, D. and Krause, S. (1981), Inorg. Chem. 20, 2750.
b. Riley, M.J., Hitchman, M.A., and Reinen, D. (1986), Chem. Phys. 102, 11.

12. Steffen, G., Reinen, D., Stratemeier, H., Riley, M.J., Hitchman, M.A., Matthies, H.E., Recker, K., Wallrafen, F., and Niklas, J.R., Inorg. Chem., to be published.
13. Riley, M.J., Hitchman, M.A., Reinen, D., and Steffen, G. (1988), Inorg. Chem. 27, 1924 - and cited references.
14. Hathaway, B.J. (1984), Structure and Bonding, Springer-Verlag, 57, 55.
15. Kepert, D.L. (1982) Inorganic Stereochemistry, Springer-Verlag.
16. Reinen, D. and Friebel, C. (1984), Inorg. Chem. 23, 791.
17a. Duggan, M., Ray, N., and Hathaway, B.J. (1980), J. Chem. Soc. Dalton Trans. 1342.
b. Tomlinson, A.A.G. and Hathaway, B.J. (1968), J. Chem. Soc.(a), 1905.
18. Wilk, A., Reinen, D., and Friebel, C., Inorg. Chem., to be published.
19a. Raymond, K.N., Meek, D.W., and Ibers, J.A. (1968), Inorg. Chem. 7, 1111.
b. Bernal, I., Korp, J.D., Schlemper, E.O., and Hussain, M.S. (1982) Polyhedron 1, 365.
20. Berry, R.S. (1960), J. Chem. Phys. 32, 933.
21a. Craubner, H., Friebel, C., Henke, W., Reinen, D., and Stratemeier, H. (1985), Jahn-Teller Conference, Marburg.
b. Reinen, D. and Atanasov, M. (1989), Chem. Phys., in press.
22. Bacci, M. (1986), Chem. Phys. 104, 191.
23a. Foley, J., Tyagi, S., and Hathaway, B.J. (1984), J. Chem. Soc. Dalton Trans. 1.
b. Foley, J., Dennefick, D., Phelan, D., Tyagi, S., and Hathaway, B.J. ibid. (1983), 2333.
24a. Henke, W., Kremer, S., and Reinen, D. (1983), Inorg. Chem. 22, 2858.
b. Arriortua, M.I., Mesa, J.L., Rojo, T., Debaerdemaeker, T., Beltrán-Porter, D., Stratemeier, H., and Reinen, D. (1988), Inorg. Chem. 27, 2976.
25. Ceulemans, A. (1988) Vibronic Processes in Inorganic Chemistry, NATO ASI Series, this volume.
26. Ceulemans, A., Beyens, D., and Vanquickenborne, L.G. (1984), J. Am. Chem. Soc. 106, 5824.
27. Smith, D.W. (1976), Coord. Chem. Rev. 21, 93.
28. Pearson, R.G. (1976) Symmetry rules for chemical reactions, Wiley-Interscience.
29. Ceulemans, A. and Vanquickenborne, L.G. (1989), Structure and Bonding, Springer-Verlag, 71, in press.

COOPERATIVE VIBRONIC EFFECTS:
ELECTRON TRANSFER IN SOLUTION AND THE SOLID STATE

P. Day
Institut Laue-Langevin
156X
38042 Grenoble
France

ABSTRACT. Vibronic interactions are discussed for systems comprising more than a single metal ion surrounded by ligands. The cases of two sites and one transferable electron, and two sites and two electrons are considered in detail. The case of N sites and N electrons illustrates the distinction between spin density wave and charge density wave ground states. Brief mention is made of superconductivity as a vibronic phenomenon.

1. INTRODUCTION

This chapter concerns vibronic interactions involving more than one metal centre and its surrounding ligands. Such interactions are most important when an electron is transferred from one metal centre to another. The simplest case is a dimer which may be a transient collision complex, as in oxidation-reduction in solution, or a discrete mixed valency molecule like the Creutz-Taube complex. The starting point is the semi-classical models of adiabatic (Marcus-Hush) and non-adiabatic (Levich) electron transfer. The older static Robin-Day (RD) model of mixed valency is quite inadequate and the vibronic model of Piepho, Krausz and Schatz (PKS) will be described in its adiabatic and full vibronic forms. The PKS approach is an instance of the two-(deformable) site, one-electron problem but is capable of extension to the two-site two-electron case which introduces the question of single-site electron repulsion. Further extension of the latter to N-sites and N-electrons widens its applicability into the solid state and leads to a distinction between charge density wave (CDW) and spin density wave (SDW) ground states. Finally, removal of the Born-Oppenheimer approximation in this model leads to superconductivity, some salient features of which are briefly mentioned, especially for the new high temperature oxide superconductors.

2. ADIABATIC ELECTRON TRANSFER IN SOLUTION

The first approach to the analysis of outer sphere electron transfer is associated with the names of Marcus and Hush, who adopted an adiabatic

C. D. Flint (ed.), Vibronic Processes in Inorganic Chemistry, 283–300.
© 1989 by Kluwer Academic Publishers.

viewpoint in which the reaction takes place along a single potential energy surface. They wrote the electron transfer rate constant k as

$$k = Z. \exp(-G^{\neq}/kT) \tag{1}$$

where Z_0 is the effective collision number and $\Delta G^{\neq}$ is the activation free energy. The latter contains contributions from the interionic electron transfer itself and the resultant polarisation energy change in the solvent. Thus

$$k = Z. \exp[-(\Delta G_0 + E_p)^2 / 4E_p kT] \tag{2}$$

where ΔG_0 is the free energy of reaction and E_p is the solvent polarization energy given by

$$E_p = (\Delta e)^2 \left(\frac{1}{2a_1} + \frac{1}{2a_2} - \frac{1}{R}\right)\left(\frac{1}{\varepsilon_{op}} - \frac{1}{\varepsilon_s}\right) \tag{3}$$

where Δe is the electronic charge transferred, a_1 and a_2 are the radii of the solvated oxidant and reductant, R is the separation between their centres and ε_{op} and ε_s are the optical and static dielectric constants of the solvent.

A more recent alternative description of outer sphere electron transfer is based on time-dependent first order perturbation theory, making use of the Fermi golden rule

$$W_{fi} = \frac{4\pi^2}{h} \Sigma \, | \langle f|V|i \rangle |^2 \, \delta\, (E_i - E_f) \tag{4}$$

where W_{fi} is the probability per unit time of transition from an initial state i of energy E_i to a set of closely spaced final states f with energies E_f after turning on a perturbation V.

One assumes that the wave functions of the oxidant and reductant vibronic states can be written as products of electronic and vibrational wave functions, and gets the overall rate constant by making a thermal average over all states i

$$W = \sum_i W_{fi} \exp(-E_i/kt) \Big/ \sum_i \exp(-E_i/kT) \tag{5}$$

In the high temperature limit the expression becomes

$$W = \left(\frac{2\pi}{h}\right) |V_{if}|^2 \left(\frac{\pi}{E_p kT}\right)^{1/2} \exp\left\{-\frac{(\Delta J + E_p)^2}{4E_p kT}\right\} \tag{6}$$

where V_{if} is the electronic coupling integral from eq. (4), E_p is the polarization energy of the solvent and ΔJ is the change in the electronic and solvation energy from i to f.

A method of estimating the magnitude of V_{if} is offered by the static RD model which considers the perturbation mixing two valence bond configurational wave functions, $\Psi_o = \psi_n^A \psi_{n+1}^B$ and $\Psi_1 = \psi_{n+1}^A \psi_n^B$ where A and B are the metal ion states and n, n+1 represent the oxidation states.

The new ground state wave function after interaction with the intervalence charge transfer state function is

$$\Psi_G = (1-\alpha^2)^{1/2} \Psi_0 + \alpha\Psi_1 \tag{7}$$

whilst the intervalence charge transfer state function after interaction with the ground state function is

$$\Psi_E = \beta\Psi_0 + (1-\beta^2)^{1/2} \Psi_1 \tag{8}$$

Clearly an optical or adiabatic transition can only occur if the resonance integral or transfer integral $V_{01} = \langle\Psi_0|\mathbf{H}|\Psi_1\rangle$ is not zero.

If the atomic orbitals of A and B are well separated, as in some outer sphere electron transfer reactions, $V_{01} \sim \langle \psi^A |\mathbf{H}|\psi^B \rangle \sim 0$, and one has to use filled and empty orbitals ϕ_L and ϕ_L^* of the intervening ligands to construct 'local' charge transfer configurations based on $\phi_L \rightarrow \psi_{n+1}$ and $\psi_n \rightarrow \phi_L^*$ yielding further functions Ψ_2, Ψ_3 that can interact with Ψ_0 and Ψ_1.

Second order perturbation using Ψ_2 and Ψ_3 leads to expressions for the (RD) valence delocalization coefficients α and β as follows

$$\alpha = \sum_{n=2,3} (V_{on}V_{1n}) / (E_1 - E_o)(E_n - E_o) \tag{9a}$$

$$\beta = \sum_{n=2,3} (V_{on}V_{1n}) / (E_1 - E_o)(E_n - E_o) \tag{9b}$$

For conjugated ligands with many orbitals, summation is taken over all $\psi^A \rightarrow \phi_{Lj}^*$ and $\phi_{Lk} \rightarrow \psi^B$ with energies $E_{j,k}$. Most commonly, one can get experimental values of $E_{j,k}$ from the spectra of mixed and single valency compounds and the V estimates from calculated metal-ligand overlap integrals or observed intensities of metal-ligand charge transfer transitions. To validate the model one can compare the intensity of the intervalence charge transfer absorption band with that calculated from the expression

$$M_{GE} = \langle\Psi_G|er|\Psi_E\rangle = \frac{1}{2} e (\alpha + \beta) R \tag{10}$$

Some examples of V_{01} calculated in this way are given in Table 1.

3. THE VIBRONIC (PKS) MODEL OF MIXED VALENCY

In the static RD model, when $E_0 = E_1$, the 'extra' electron would have to be fully delocalized between the two metal ions A and B. However, to explain the possibility that a mixed valency dimer with identical ligands around the two metal ions (i.e. with $E_0 = E_1$ in Figure 1) could be capable of being asymmetrical, it is absolutely essential to take into account the interaction of electronic and vibrational motion. This can be done at two levels of approximation, namely with or without making the Born-Oppenheimer separation. Both approaches were first considered in detail by Piepho, Krausz and Schatz, whose arguments will now be summarized.

TABLE 1. Intensities of intervalence absorption bands and transfer integrals of $[(NH_3)_5Ru^{II}(L)Ru^{III}(NH_3)_5]^{5+}$

	R_{MM}	Intensity of MV absorption band		Transfer integral V_{01}
		obs	calc	
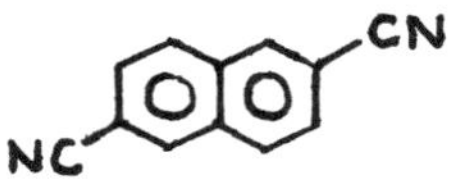	14 Å	0.16	0.09 eÅ	148 cm^{-1}
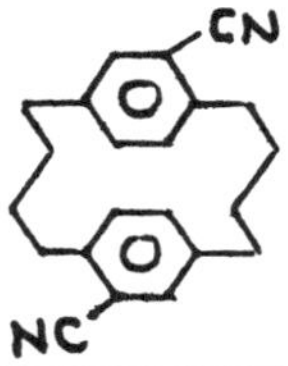	12 Å	0.045	0.007 eÅ	48 cm^{-1}

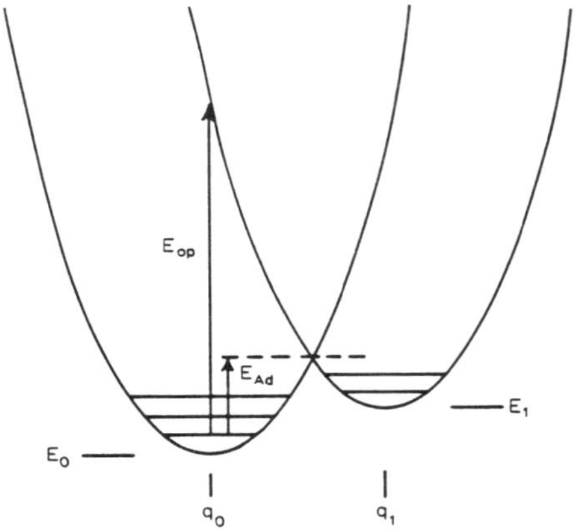

Figure 1. Potential energy surfaces of Ψ_0 and Ψ_1 (defined in the text). E_{op} and E_{Ad} are the optical (Franck-Condon) and adiabatic intervalence electron transfer energies.

3.1. The Adiabatic Limit

For two subunits A and B with metal ion oxidation states (orbital occupancies) M and N, the zeroth order electronic wave functions are

$$\psi_a = \psi_M^A \psi_N^B , \quad \psi_b = \psi_N^A \psi_M^B \tag{11}$$

For the present we only consider the case for which N=M±1 (e.g. Ru(II,III)). Assuming that each subunit has a single symmetric normal coordinate labelled respectively Q_A and Q_B, the vibrational potential energy of A in oxidation state M can be written

$$W_M^A = W_M^o + l_M^A Q_A + \frac{1}{2} k_M^A Q_A^2 \tag{12}$$

with corresponding expressions for W_N^A, W_M^B, W_N^B, and defining $W_M^o \equiv W_M^A(Q = 0) = W_M^B(Q_B = 0)$.

The Hamiltonian for subunit A

$$H^A = H_{el}^A + T_n^A \tag{13}$$

consists of an electronic and a nuclear kinetic energy part. For the dimer with subunit A in oxidation state M and subunit B in oxidation state N we have

$$(H_{el}^A + H_{el}^B)\ \psi_a = (W_M^A + W_N^B)\ \psi_a = W_a \psi_a \tag{14}$$

For a dimeric complex with identical ligands around sites A and B we can put $l_M^A = l_M^B \equiv l$, $k_M^A = k_M^B \equiv k_M$, $k_N^A = k_N^B \equiv k_N$, and hence

$$\begin{aligned} W_a &= lQ_A + \frac{1}{2} k_M Q_A^2 + \frac{1}{2} k_N Q_B^2 \\ W_b &= lQ_B + \frac{1}{2} k_M Q_B^2 + \frac{1}{2} k_N Q_A^2 \end{aligned} \tag{15}$$

To take advantage of the interchange symmetry, new vibrational coordinates were defined as follows

$$Q\pm = \frac{1}{\sqrt{2}} (Q_A \pm Q_B) \tag{16}$$

so that eq (15) is rewritten as

$$W_a = \left(\frac{1}{\sqrt{2}}\right) lQ_- + \frac{1}{4}(k_N + k_M) Q_-^2 +$$

$$+ \frac{1}{2}(k_M - k_N)Q_+Q_- + \left(\frac{1}{\sqrt{2}}\right)lQ_+ + \frac{1}{4}(k_N + k_M)Q_+^2 \tag{17}$$

W_b is obtained if $(-Q_-)$ is substituted for Q_-. To separate Q_+ and Q_- we have to make the apparently drastic approximation that $k_M = k_N \equiv k$. However, in practice k_M and k_N will usually only differ by about 10%, e.g. k(Fe-C) in $Fe(CN)_6^{3-}$ and $Fe(CN)_6^{4-}$. With this approximation

$$\begin{aligned} W_a &= \left(\frac{1}{\sqrt{2}}\right) lQ_- + \frac{1}{2} kQ_-^2 + \left(\frac{1}{\sqrt{2}}\right) lQ_+ + \frac{1}{2} kQ_+^2 \\ W_b &= \left(-\frac{1}{\sqrt{2}}\right) lQ_- + \frac{1}{2} kQ_-^2 + \left(\frac{1}{\sqrt{2}}\right) lQ_+ + \frac{1}{2} kQ_+^2 \end{aligned} \tag{18}$$

The cross term in Q_+Q_- has vanished and so the problem is separable with respect to Q_+ and Q_-. In Q_+ space, W_a and W_b give identical potential energy surfaces with minima at $Q_+ = (-l/k\sqrt{2})$. PKS further define the dimensionless variables

$$\begin{aligned} q &= 2\pi(\nu_-/h)^{1/2}\, Q_- \\ \lambda &= (8\pi^2 h\nu_-^3)^{-1/2}\, l \end{aligned} \tag{19}$$

where $\nu_- = (2\pi)^{-1}\sqrt{k}$ is the fundamental vibrational frequency associated with normal coordinate Q_- (or q). Then eq. (18) is simply

$$\begin{aligned} W_a/h\nu_- &= \lambda q + \frac{1}{2} q^2 \\ W_b/h\nu_- &= -\lambda q + \frac{1}{2} q^2 \end{aligned} \tag{20}$$

Electron transfer can only take place between the two sites if there is a non-zero off-diagonal matrix element connecting Ψ_a and Ψ_b, analogous to V_{if} of eq. (6). Defining

$$V_{ij}^o \equiv \langle\Psi_i|V^{AB}|\Psi_j\rangle^o \;;\; i,j = a,b \tag{21}$$

as the value of V_{ij} at q=0, one further defines a dimensionless electronic coupling parameter ε by $(h\nu_-)\varepsilon \equiv V_{ab}^o$.

Using these definitions and Eq. (20) the secular determinant in the substitution approximation becomes

$$\begin{vmatrix} \lambda q + \frac{1}{2} q^2 - W_k & \varepsilon \\ \varepsilon & -\lambda q + \frac{1}{2} q^2 - W_k \end{vmatrix} = 0 \tag{22}$$

with W_k in units of $h\nu_-$. The roots of Eq. (22) are

$$W_{2,1} = \frac{q^2}{2} \pm \sqrt{\varepsilon^2 + \lambda^2 q^2} \tag{23}$$

and the corresponding electronic wave functions

$$\begin{aligned}\psi_1 &= -(1/N\surd 2)\,[(\varepsilon + \lambda q - \sqrt{\varepsilon^2 + \lambda^2 q^2})\,\psi_a \;+ \\ &\quad + (\varepsilon - \lambda q - \sqrt{\varepsilon^2 + \lambda^2 q^2})\,\psi_b] \\ \psi_2 &= (1/N\surd 2)\,[-(\varepsilon - \lambda q - \sqrt{\varepsilon^2 + \lambda^2 q^2})\,\psi_a \;+ \\ &\quad + (\varepsilon + \lambda q - \sqrt{\varepsilon^2 + \lambda^2 q^2})\,\psi_b\,]\end{aligned} \tag{24}$$

It can be seen that ε has caused a mixing between ψ_a and ψ_b which is dependent on q. Accordingly, as $\varepsilon \gtrless \lambda^2$ the electronic ground state potential energy surface W_1 will have either one or two minima corresponding to equal or unequal equilibrium bond lengths in the two subunits. The older RD model simply had $\lambda^2 = 0$.

3.2. Relation between the PKS and RD Models

To extend the results of section 3.1. to this situation where the two metal ion sites are surrounded by different ligands (or different numbers of the same ligands) we note that $W_N^A \neq W_N^B$: $W_M^A \neq W_M^B$. Defining $W = E_0 - E_1$, a development analogous to that of section (3.1) for the symmetrical case leads to

$$W_{1,0} = (1/2)q^2 \pm [\varepsilon^2 + (\lambda q + W)^2]^{1/2} \tag{25}$$

The corresponding wave functions can be cast into a form similar to that used by RD in the following way:

$$\begin{aligned}\psi_G^{PKS} &= -(1/N\surd 2)[(\varepsilon - K + \sigma)\psi_0 + (\varepsilon - K - \sigma)\psi_1] \\ \psi_E^{PKS} &= (1/N\surd 2)[-(\varepsilon - K + \sigma)\psi_0 + (\varepsilon - K - \sigma)\psi_1]\end{aligned} \tag{26}$$

where $N = [\sigma^2 + (\varepsilon - K)^2]^{1/2}$, $K = [\varepsilon^2 + \sigma^2]^{1/2}$, $\sigma = (\lambda q + W)$.

At the point on the lower potential energy surface where q is equal to λ, there is then a direct relation between the valence delocalisation coefficient α of the RD model and the parameters of the PKS model, as follows

$$\alpha^2 = \tfrac{1}{2}\left[1 - \left\{\frac{(\lambda^2 + W)^2}{\varepsilon^2 + (\lambda^2 + W)^2}\right\}^{1/2}\right] \tag{27}$$

The three RD classes of mixed valency behaviour therefore correspond to the inequalities

$$\text{Class I} : |\varepsilon| \ll (\lambda^2 + W),$$
$$\text{Class II} : |\varepsilon| \lesssim (\lambda^2 + W)$$
$$\text{Class III} : |\varepsilon| > (\lambda^2 + W)$$

3.3 The full vibronic treatment

The vibronic treatment so far has remained within the framework of the Born-Oppenheimer approximation, which may be sufficient for many purposes, but to go beyond the adiabatic limit one should write a general vibronic wave function

$$\Psi_k(r,q) = \psi_1(r,q)\chi_{1,k}(q) + \psi_2(r,q)\chi_{2,k}(q) \tag{28}$$

with the same Hamiltonian as eq. (13). Now, by definition, H_{el} is diagonal in ψ_1 and ψ_2, and so the secular equations arising from the wave functions of eq. (28) are

$$\sum_{j=1,2} [(W_i - E_k)\delta_{ij} + \langle\psi_i|T_n(q)|\psi_j\rangle]\chi_{j,k} = 0 \tag{29}$$

$$i = 1,2; \quad k = 1, \ldots, \infty$$

where $\langle\psi_i|T_n(q)|\psi_j\rangle\chi_{j,k} \equiv \int \psi_i^*(r,q)T_n(q)\psi_j(r,q)\chi_{j,k}(q)d\tau_{el}$, and E_k are the vibronic eigenvalues. Physically, this means that the nuclear motion is no longer confined to one potential energy surface. For the electron wave functions which have no explicit q dependence, we still use ψ_a and ψ_b but for the symmetrical case put

$$\psi_\pm = \frac{1}{\sqrt{2}}(\psi_a \pm \psi_b) \tag{30}$$

Then our general vibronic function is

$$\Phi_\nu(r,q) = \psi_+(r)\chi_{+,\nu}(q) + \psi_-(r)\chi_{-,\nu}(q) \tag{31}$$

where the χ_+ and χ_- are (q-dependent) linear combinations of the χ_1 and χ_2, and the corresponding secular equations are

$$\sum_{j=+,-} [\langle\psi_i|H_{el} + T_n|\psi_j\rangle - \langle\psi_i|\psi_j\rangle E_\nu]\chi_{j,\nu} = 0 \tag{32}$$

$$i = +,-; \quad \nu = 1,2 \ldots,\infty$$

From the definition of $\psi_\pm$, $\langle\psi_i|\psi_j\rangle = \delta_{ij}$, but $\langle\psi_i|H_{el}|\psi_j\rangle$ is not diagonal, so it is necessary to assume that ψ_a and ψ_b are only slowly varying functions of q and hence that $T_n(q)$ commutes with ψ_a and ψ_b. Then it follows that $\langle\psi_i|T_n|\psi_j\rangle\chi_{j,\nu} = \delta_{ij}T_n\chi_{j,\nu}$ and eq. (32) becomes

$$(T_n(q) + \tfrac{1}{2}q^2 + \varepsilon - E_\nu)\ \chi_{+,\nu}(q) + \lambda q \chi_{-,\nu}(q) = 0$$

$$\lambda q \chi_{+,\nu}(q) + (T_n(q) + \tfrac{1}{2}q^2 - \varepsilon - E_\nu)\ \chi_{-,\nu}(q) = 0 \tag{33}$$

$$\nu = 1, 2, \ldots, \infty$$

with E_ν in units of $h\nu_-$.

Harmonic oscillator functions in coordinate q provide the vibrational basis, i.e.

$$\chi_{+,\nu} = \sum_{n=0}^{\infty} c_{\nu,n}\chi_n \ ; \quad \chi_{-,\nu} = \sum_{n=0}^{\infty} c'_{\nu,n}\ \chi_n \tag{34}$$

so when the vibronic functions Φ_ν are classified according to their interchange behaviour, eq. (31) becomes

$$\Phi_\nu^+ = \psi_+ \sum_{n=0,2,4,\ldots}^{\infty} r_{\nu n}\chi_n + \psi_- \sum_{n=1,3,5,\ldots}^{\infty} r_{\nu n}\chi_n$$

$$\Phi_\nu^- = \psi_+ \sum_{n=1,3,5,\ldots}^{\infty} s_{\nu n}\chi_n + \psi_- \sum_{n=0,2,4,\ldots}^{\infty} s_{\nu n}\chi_n \tag{35}$$

where the summations encompass either all even or all odd χ so as to match the interchange symmetry of $\psi_\pm$ to give $\Phi_\nu^\pm$.

Solution of the vibronic problem demands calculations of the coefficients $r_{\nu n}$ and $s_{\nu n}$ for each value of ν. The χ_n are eigenfunctions of $(T_n + \tfrac{1}{2}q^2)$ with eigenvalues $(n + \tfrac{1}{2})$ in units of $h\nu_-$ and furthermore

$$\langle\chi_n|q|\chi_{n+1}\rangle = \langle\chi_{n+1}|q|\chi_n\rangle = \left(\frac{(n+1)}{2}\right)^{1/2}$$

The harmonic oscillation wavefunctions of eq. (51) are next substituted into eq. (32) to give the two following equations that are analogous to eq. (33):

$$(T_n(q) + \tfrac{1}{2}q^2 + \varepsilon - E_\nu^+) \sum_{n,\mathrm{even}} r_{\nu n}\chi_n + \lambda q \sum_{n,\mathrm{odd}} r_{\nu n}\chi_n = 0$$

$$\lambda q \sum_{n,\mathrm{even}} r_{\nu n}\chi_n + (T_n(q) + \tfrac{1}{2}q^2 - \varepsilon - E_\nu^+) \sum_{n,\mathrm{odd}} r_{\nu n}\chi_n = 0 \tag{36}$$

Finally, the latter are multiplied by each χ_n in turn $(n=0,1,2,\ldots\infty)$ and integrated over q. The following set of secular equations is obtained:

$$\sum_{n=0}^{\infty} r_{\nu n}(H_{mn} - \delta_{mn}E_\nu^+) = 0; \qquad \begin{matrix} m = 0,1,2,\ldots \\ \nu = 0,1,2,\ldots \end{matrix} \tag{37}$$

with

$$H_{mn} = \lambda \left[\left(\frac{m}{2}\right)^{1/2} \delta_{m,n+1} + \left(\frac{(m+1)}{2}\right)^{1/2} \delta_{m,n-1} \right] + \qquad (38)$$

$$+ \left(m + 1/2 + (-1)^m \varepsilon\right) \delta_{m,n}$$

The corresponding secular determinant yields the eigenvalues, E_ν^{+}. Computations for particular cases of interest involve choosing sets of values for ε, λ (and W in the unsymmetrical case that has not been described here - see the review of Wong and Schatz for details). The summation over n can be computed for as many quanta as may be necessary to give an adequate convergence.

Armed with $\Phi_\nu^{\pm}$ and $E_\nu^{\pm}$ one can then calculate physical properties of interest, principal among which is the intervalence absorption band profile given by

$$D^e\ (\nu' \to \nu) = \sum_{\gamma=x,y,z} \frac{(N_{\nu'} - N_\nu)}{N} \left| \langle \Phi_{\nu'}^{+} | m_\gamma^e | \Phi_\nu^{-} \rangle \right|^2 \qquad (39)$$

where $N_{\nu'}/N$ is the fractional population in state ν'. $N_{\nu'} = \exp(-E_{\nu'}/kT)$ and $N = \Sigma_{\nu'}/N_{\nu'}$. m_γ is the γth component of the molecule-fixed electric dipole operator, $\mathbf{m} = \Sigma_i e_i r_i$. Putting in the $\Psi_{N,M}^{A,B}$ basis and defining the origin for the components of m_γ^e (m_x, m_y, $m_z = \sum_i e_i z_i$ etc) one gets for D^e:

$$D^e(\nu' \to \nu) = \left(\frac{N_{\nu'} - N_\nu}{N}\right) S_{\nu'\nu}^2 \left| \langle \psi_+ | m_z^e | \psi_- \rangle \right|^2 \qquad (40)$$

where $\left| \langle \psi_+ | m_z^e \ | \psi_- \rangle^{o} \right|^2 = e^2 R^2/4$, and

$$S_{\nu'\nu} = \sum_n r_{\nu' n} S_{\nu n} \qquad (41)$$

Note that the intervalence transition only has electric dipole strength for polarisation along the line joining the two centres.

4. VIBRONIC COUPLING IN TWO-ELECTRON SYSTEMS

In the later 4d and 5d elements, as well as in the post transition elements, there are many examples of mixed valency systems in which the difference in formal oxidation states at the two sites is two rather than one. Instances are Pt(II,IV), Au(I,III) and Sb(III,V). To treat these cases we have to introduce a third vibrational potential energy surface and also take explicit account of electron repulsion.

4.1 Theory

Suppose we have two subunits A and B associated with formal oxidation states N and N-2 respectively. As before, the electronic Hamiltonian operators associated with them are $H_{el}{}^A$ and $H_{el}{}^B$. Then

$$H_{el}{}^{A}\psi_{N}{}^{A} = W_{N}{}^{A}\psi_{N}{}^{A} \tag{42}$$

$$H_{el}{}^{B}\psi_{N-2}{}^{B} = W_{N-2}{}^{B}\psi_{N-2}{}^{B}$$

If there is no interaction between the subunits, the electronic Schrödinger equation for the system is

$$(H_{el}{}^{A} + H_{el}{}^{B})\psi_{N}{}^{A}\psi_{N-2}{}^{B} \equiv (H_{el}{}^{A} + H_{el}{}^{B})\psi_a =$$

$$= (W_{N}{}^{A} + W_{N-2}{}^{B})\psi_a \equiv W_a\psi_a \tag{43}$$

If one or two electrons are transferred from B to A, the zeroth-order electronic wave functions of the new states are respectively

$$\psi_b = \psi_{N-1}{}^{A}\psi_{N-1}{}^{B}$$

$$\psi_c = \psi_{N-2}{}^{A}\psi_{N}{}^{B} \tag{44}$$

Then equations exactly analogous to (43) apply.

Just as in section 3.1. the vibrational potential energy of subunit j in oxidation state i can be written

$$W_i^j = W_i^{oj} + l_i^j Q_j + \tfrac{1}{2}k_i^j Q_j^2, \quad j = A, B; \quad i = N, N-1, N-2 \tag{45}$$

where $W_i^{oj} \equiv W_i^j(Q_j = 0)$.

Assuming that A and B are equivalent, we define the zero of energy by $W_N{}^{OA} + W_{N-2}{}^{OB} = W_{N-2}{}^{OA} + W_N{}^{OB} = 0$ and define $2\bar{W} \equiv (W_{N-1}{}^{OA} + W_{N-1}{}^{OB}) - (W_N{}^{OA} + W_{N-2}{}^{OB})$. We introduce the sum and difference coordinates of eq. (16) and make similar bold assumptions as in section 3.1. (which also apply for A → B) that

$$k_N{}^{A} = k_{N-1}{}^{A} = k_{N-2}{}^{A} \equiv k$$

$$l_{N-1}{}^{A} = \tfrac{1}{2}(l_N{}^{A} + l_{N-2}{}^{A}) \equiv l \tag{46}$$

so that the Q_+ coordinate decouples from the problem. We choose origins by setting Q_A and Q_B to zero at the minima of $W_{N-2}{}^{A}$ and $W_{N-2}{}^{B}$ respectively. Introducing the dimensionless variables

$$q = 2\pi(\nu_-/h)^{1/2}Q_-$$

$$\lambda = (8\pi^2 h\nu_-{}^3)^{-1/2}l \tag{47}$$

$$2W = 2\bar{W}/h\nu_- + \lambda^2/2$$

where $\nu_- = (2\pi)^{-1}k^{1/2}$ is the vibrational frequency associated with Q_-, we obtain (after changing the energy zero by $\lambda^2/2$)

$$W_a/h\nu_- = \tfrac{1}{2}(q + \lambda)^2$$
$$W_b/h\nu_- = 2W + \tfrac{1}{2}q^2 \qquad (48)$$
$$W_c/h\nu_- = \tfrac{1}{2}(q - \lambda)^2$$

To allow the centres to interact we hold the nuclei at $q = 0$ and define

$$V_{ij}^0 \equiv \langle\psi_i|V^{AB}|\psi_i\rangle^0 \; ; \quad i,j = a,b,c \qquad (49)$$

In the symmetrical case, $V_{aa}^0 = V_{cc}^0$, but in contrast to the 2-site, 1-electron case, we now have two electronic interaction parameters ε and ε' defined by

$$\varepsilon \equiv V_{ab}^0/h\nu_-$$
$$\varepsilon' \equiv V_{ac}^0/h\nu_- \qquad (50)$$

Changing the zero of energy by $V_{aa}^0/h\nu_-$ and substituting the above definitions and eq. (48) into eq. (43) we obtain the vibronic matrix in the ψ_a, ψ_b, ψ_c electronic basis:

$$\begin{pmatrix} \tfrac{1}{2}(q + \lambda)^2 & \varepsilon & \varepsilon' \\ \varepsilon & \tfrac{1}{2}q^2 + 2W & \varepsilon \\ \varepsilon' & \varepsilon & \tfrac{1}{2}(q-\lambda)^2 \end{pmatrix} \qquad (51)$$

The constant term $(V_{bb}^0 - V_{aa}^0)/h\nu_-$ has been absorbed into 2W.

It is important to notice that the definition of W in eq. (47) reflects competition between electron-vibrational coupling symbolised by $\lambda^2/2$ and electron repulsion, by W. The latter measures the energy difference between the spin singlet arising from two electrons in an orbital on the same site $|\phi_A\alpha\phi_B\beta\rangle$ and those from electrons on two sites $|\phi_A\alpha\phi_B\beta\rangle - |\phi_A\beta\phi_B\alpha\rangle$.

4.2. An example

A convenient example of a mixed valency system with an oxidation state difference of two, that has been thoroughly studied experimentally, is that of SbX_6^- + SbX_6^{3-} (X = Cl,Br). These two ions may be isolated by dilution in a high symmetry host lattice such as A_2SnX_6, where A = Cs gives a cubic crystal while A = CH_3NH_3 forms flat plates suitable for optical transmission spectroscopy. The crystal structures of the parent compounds $M^I_4(Sb^{III}X_6)(Sb^VX_6)$ (M = Rb, Cs) have been determined by high resolution powder neutron diffraction so the (small) differences in bond lengths and angles around the two Sb sites are known: both are octahedral and, for example, $[Sb^{III}\text{-}Cl] - [Sb^V\text{-}Cl]$ is 0.28 Å. Thus the origins of the $Q_{N,N-2}$ potential energy surfaces are fixed.

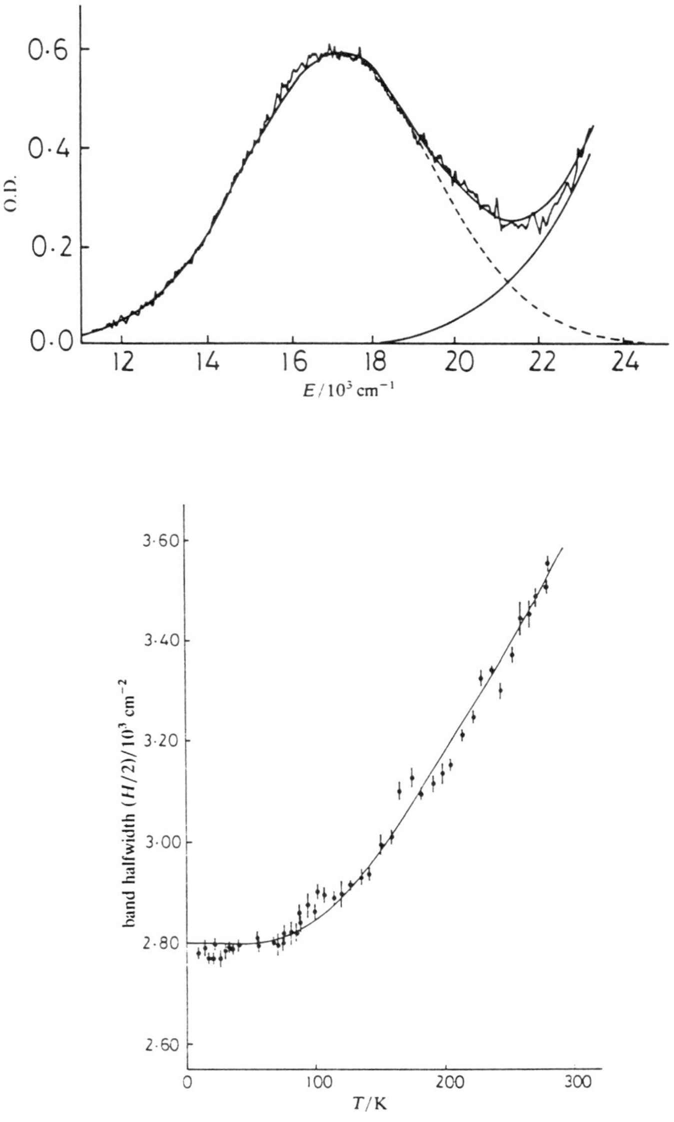

Figure 2. (a) The intervalence band in $(CH_3NH_3)_2Sb_xSn_{1-x}Cl_6$ at 55K. The full and dashed lines are Gaussians. (b) Variation with temperature of the halfwidth of the band. The full line is calculated from eq. (52) with $\nu_- = 290$ cm^{-1} and $\lambda = 5.8$.

In the limit of weak coupling ($\varepsilon \ll \lambda^2$) PKS show that the inter valence optical transition should be Gaussian, with a halfwidth H given, in the high temperature limit, by

$$H^2 = (16/\ln 2)\lambda^2(h\nu_-)^2 \coth(h\nu_-/2kT) \qquad (52)$$

In Figure 2(a) we illustrate that the intervalence band in $(CH_3NH_3)_2Sb_xSn_{1-x}Cl_6$ is indeed Gaussian and in Figure 2(b) that temperature variation of its halfwidth closely obeys eq. (52), yielding values of ν_- and λ. With these parameters, and taking into account the observed frequency of the intervalence band maximum, one arrives at the 'experimental' potential energy surfaces shown in Figure 3.

It is important to note that the p.e. surfaces of Figure 3 are constructed with respect to the single vibrational coordinate q of eq. (47) and that they assume the adiabatic approximation. The assumption is made that only the totally symmetric Sb-Cl local stretching modes of $SbCl_6^{3-}$ contribute to q. However, in an infinite lattice, not only the short-range elastic energy, but also the long-range Coulomb energy, contribute to the change in the total internal energy of the crystal when it is transformed from the equilibrium bond lengths in the observed structure of, say, $Cs_4[Sb^{III}Cl_6][Sb^{V}Cl_6]$ to the situation where all the Sb-Cl bond lengths are equal. For the latter there are two alternative electron configurations which we could symbolise as $Cs_4[Sb^{IV}Cl_6][Sb^{IV}Cl_6]$ and $Cs_4[Sb^{III}Cl_6][Sb^{V}Cl_6]^*$. In terms of Sb atomic orbital percentage, these correspond respectively to $(5s^1)^A(5s^1)^B$ and $(5s^2)^A(5s^0)^B$, the energy difference between which is given by the single-centre electron repulsion integral, normally called U.

In Figure 4 there are, therefore, three energy quantities of interest, shown as E_A, E_B, E_C. We can determine E_A if we know all the force constants, not just of the totally symmetric Sb-Cl stretching mode but of the lattice modes too. The best available method for determining these constants is inelastic neutron scattering, since one is then not bound by symmetry selection rules and, furthermore, not confined to excitations having $\Delta q = 0$, i.e. at the Brillouin zone centre. Unfortunately, large single crystals of the Sb(III,V) compounds are not available but the phonon density-of-states can be measured on powder samples by the technique of incoherent inelastic neutron scattering. Such experiments have been done for $Cs_4[Sb^{III}Cl_6][Sb^{V}Cl_6]$ and the force constants derived. The short-range energy contribution to E_A, $\Delta\Psi^{SR}$ is calculated from these as 2.1 eV. The long-range Coulomb contribution $\Delta\Psi^{C}$ is found from the stress matrix evaluated as part of the lattice dynamics calculated to ensure the stability of the crystal against external strain. Since it is known that the Sb(III,V) ground state is stable with respect to Sb(IV,IV), we have

$$-U_{eff} + \Delta\Psi^{SR} + \Delta\Psi^{C} > 0$$

so that $U_{eff} < 5.9$ eV. The endothermicity of the process Sb(III,V) → Sb(IV,IV) can be estimated from the intervalence optical absorption band energy, though the energy E_C in Figure 4 is not represented by the energy of the absorption maximum (~ 2.1 eV) but by the onset of the band at ~ 1.5 eV. Thus, $U_{eff} < 4.4$ eV, which should be compared with the

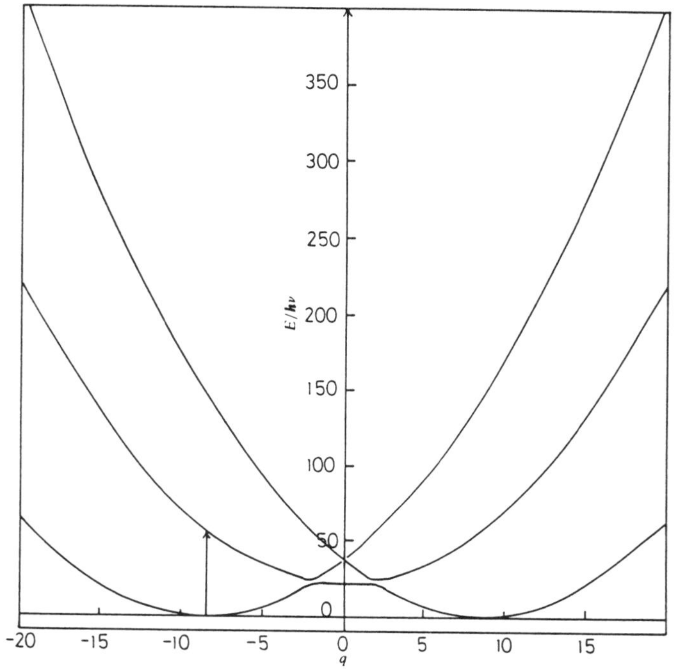

Figure 3. Potential energy surfaces for Sb(III,V), Sb(IV,IV) and Sb(V,III) derived from experiment (λ = 4.3, W = 10.5, ε = -1.8, ν_- = 300 cm^{-1}).

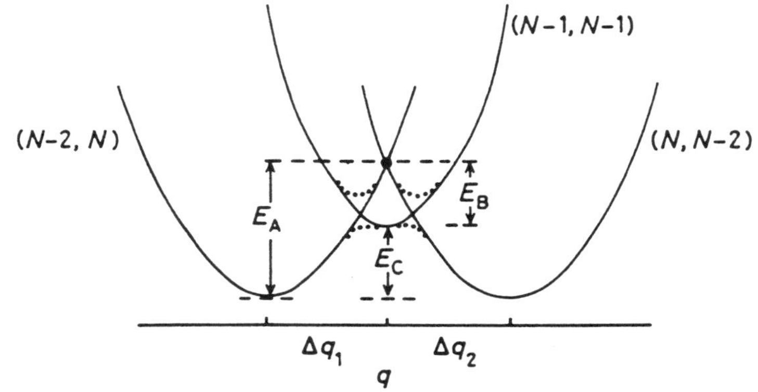

Figure 4. Schematic potential energy surfaces for (N-2,N), (N-1,N-1) and (N,N-2) oxidation states. E_A etc. are defined in the text. The dotted lines indicate the major interactions between the surfaces.

value for the gaseous ions $Sb^{3+,4+,5+}$ of 11.9 eV, obtained from the respective ionization potentials. The diminution of U_{eff} from the gas phase to the solid represents the screening of the 5s electrons from the Sb nucleus by electrons of the ligands, or by metal-ligand covalency.

5. N-SITE, N-ELECTRON SYSTEMS: PHASE DIAGRAMS & SUPERCONDUCTIVITY

Generalising the 2-site, 2-electron problem of section 4.1. to N sites and N electrons, i.e. to an infinite lattice, we can see at once that there are two limiting cases: either sites are alternately occupied by two and zero electrons, or every site is occupied by one electron. In the latter case, if we consider the 'singlet' ground state (spins alternately up and down on neighbouring sites), one would call it anti-ferromagnetically ordered or, in the general case, a spin density wave (SDW). The former, which has an alternation of electron density, is called a charge density wave (CDW). The three variables that determine the state of lowest energy are those defined by eqs. (47) and (50), i.e. λ, $\bar{W}$ (or U) and ε. Consequently, one can represent the regions of stability of SDW and CDW ground states on a triangular phase diagram. Figure 5(a) shows such a diagram worked out by Nasu and Toyazawa for the one-dimensional N-site, N-electron case. In this Figure the ground state potential energy surfaces corresponding to each state are also drawn. Note that those of Figures 3 and 4 represent the CDW (SDW) region in Figure 5(a).

Another way of looking at the CDW state is to say that pairs of electrons have apparently been attracted together, i.e. that W in eq. (47), which should be positive, is in fact negative. Physicists have

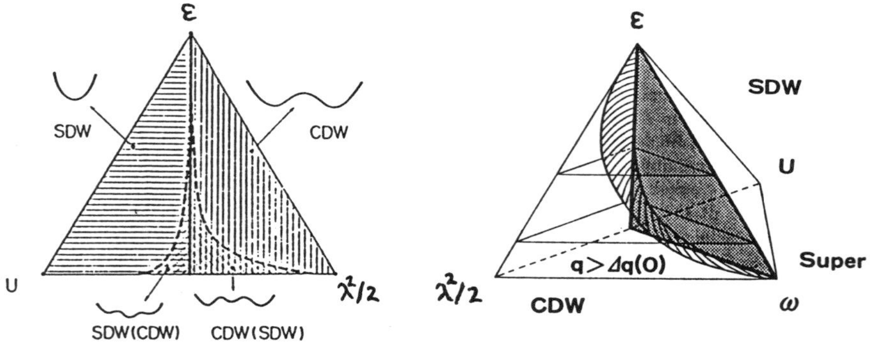

Figure 5. Phase diagrams of the N-site, N-electron system (Nasu 1985). (a) The adiabatic case. (b) The non-adiabatic case.

independently called this situation 'negative U', but it is clear that electrons do not really attract one another and that, in reality, it is the vibronic coupling term $\lambda^2/2$ that overcomes $\bar{W}$ or U. Now, of great interest is the fact that just such an apparent electron attraction is the origin for the formation of current-carrying entities, the Cooper pairs, in superconductors. In conventional superconductors it is likewise the vibronic coupling that overwhelms the electron repulsion, so forming a cloud of Bosons, in contrast to the Fermions that carry the current in normal metals. Therefore, in a sense, superconductivity is the most spectacular consequence of vibronic coupling!

However, it turns out that in the adiabatic limit of section 4.1, only CDW or SDW ground states are obtained; the CDW cannot fluctuate and there is no superconductivity. Only by going beyond the Born-Oppenheimer approximation to solve the full vibronic problem can the collective ground state be superconducting. The validity of the adiabatic limit is determined by the ratio ν/ε. When ν/ε is small, one can say that the mobile electron 'follows' the lattice but, conversely, that the lattice is too heavy to 'follow' the migration of the electron. Under these circumstances one has the phase diagram of Figure 5(a). However, as ν becomes larger relative to ε, the 'bare' mobile electron becomes a polaron (an electron surrounded by lattice polarisation). At the extreme, where ε dominates λ and U($\bar{W}$) (i.e. the top vertex of Figure 5(a)), in the adiabatic limit one would have a normal metallic ground state but in the non-adiabatic case or (in the limit) the anti-adiabatic case, where $\nu \gg \varepsilon$, the normal metallic state is replaced by a superconducting one. To take into account the role of the vibrational frequency, Figure 5(a) has to be extended to make it tetrahedral (Figure 5(b)).

In conventional superconductors, the vibrational modes responsible for the Cooper pair formation are low frequency lattice modes. This is because the electronic wavefunctions are not so far removed from plane waves, so that the 'coherence length' (the spacial extension of the Cooper pair) is quite long (typically 100-500 Å). By contrast, the coherence lengths measured in the new oxide high T_c superconductors are only a very few multiples of the unit cell. Another way of expressing it is to say that the electron pairing takes place in real space as opposed to momentum space. Thus, more localised vibrational modes should also be important. Superconductivity is not only a vibronic phenomenon but one to which chemical concepts, like those outlined in the last sections, may well prove relevant.

REFERENCES

Since this chapter presents a broad overview of a large topic, no detailed references are given. Rather, we list a number of books, reviews and general articles that provide a gateway into the extensive primary literature.

Cannon, R. D. (1977) Electron Transfer Reactions, Butterworth (overview of facts and theories about solution reactions)

Robin, M. B. and Day, P. (1967) Adv. Inorg. Chem. Radiochem. 10, 247 (static model of mixed valency and survey of Periodic Table)

Allen, G. C. and Hush, N. S. (1967) Prog. Inorg. Chem. 8, 357
(mixed valency theory)
Brown, D. B. (1980) Mixed Valency Compounds, D. Reidel Publishing Co.
(NATO ASI Proceedings, many aspects covered)
Day, P. (1981) Int. Rev. Phys. Chem. 1, 149
(review of mixed valency including vibronic aspects)
Wong, K. Y. and Schatz, P. N. (1981) Prog. Inorg. Chem. 28, 369
(thorough review of the PKS vibronic model)
Prassides, K. and Day, P. (1984) J. Chem. Soc., Faraday Trans. 2, 80, 85
(PKS model for 2-site, 2-electron systems)
Prassides, K., Schatz, P. N., Wong, K. Y. and Day, P. (1986) J. Phys. Chem. 90, 5588
(vibronic coupling for 2-sites and 2-electrons)
Day, P. (1987) Ch. in 'Organic and Inorganic Low-Dimensional Crystalline Materials', ed. P. Delhaes & M. Drillon, Plenum, p.33
(N-site, N-electron chains with vibronic coupling)
Nasu, K., (1985) J. Phys. Soc. Japan, 54, 1933
(electron-electron vs. electron-phonon coupling giving CDW, SDW and superconducting ground states in N-site, N-electron chains)

RAMAN, RESONANCE RAMAN AND ELECTRONIC RAMAN SPECTROSCOPY

Robin J.H. CLARK
Christopher Ingold Laboratories
University College London
20 Gordon Street
London WC1H 0AJ

ABSTRACT. The theory and practice of Raman, resonance Raman, and electronic Raman spectroscopy are outlined. Resonance Raman spectra are usually characterized by the development of intense overtone and combination tone progressions (A-term Scattering). This form of spectroscopy yields detailed information regarding the resonant excited state since only vibrational modes associated with the chromophore are resonance enhanced. Examples are given of the application of resonance Raman spectroscopy to the detection of species at very low concentrations, and of matrix-isolated species, and to the qualitative and quantitative calculation of excited state geometries. The importance of the technique in understanding the nature of mixed-valence linear-chain complexes is also outlined, together with a brief account of electronic Raman spectroscopy.

1. INTRODUCTION

Raman spectroscopy is concerned with the phenomenon of the change in frequency of a monochromatic beam of light when it is scattered by molecules. The name given to the subject honours C.V. Raman who announced the discovery of the effect in March 1928 during the course of extensive research on molecular light scattering. Almost simultaneous with Raman's report was that of Landsberg and Mandelstam on light scattering by quartz. The basis of the effect can be understood as follows.

Consider a clear substance (solid, liquid or gas) irradiated by a monochromatic beam of light (usually in the visible region of the spectrum) whose frequency, ν_0, is chosen so that it does not coincide with that of any absorption band of the sample. Most of the light beam will pass through the sample essentially unaffected, but a small fraction will be scattered in directions different from that of the incident beam. When this scattered light is analysed spectroscopically, it is found that most of it by far has the same frequency as that of the incident beam. The existence of this type of scattering constitutes the phenomenon of Rayleigh scattering, the intensity of

C. D. Flint (ed.), Vibronic Processes in Inorganic Chemistry, 301–325.
© 1989 by Kluwer Academic Publishers.

which is (among other factors) proportional to ν_0^4 and α_0^2, where α_0 denotes the equilibrium value of the polarisability of the molecule.

The key discovery by Raman was that, in addition to Rayleigh scattering at frequency ν_0, the scattered radiation contained a series of lines placed symmetrically on either side of the Rayleigh line, at frequencies $\nu_0 \pm \nu_i$. The frequency shifts, ν_i, referred to as the Raman frequencies, are independent of ν_0 and relate to changes in the rotational, vibrational or electronic energies of the molecule; they thus provide key information regarding the scattering species. The intensities of vibrational Raman bands are, among other factors, proportional to $(\nu_0 \pm \nu_i)^4$ and $(\partial\alpha/\partial Q_i)_0^2$, where $(\partial\alpha/\partial Q_i)_0$ denotes the equilibrium value of the rate of change of α with the normal coordinate Q_i. Such vibrational frequencies, ν_i, when observed directly by absorption techniques, lie in the infrared spectral region. However, in Raman spectroscopy, they can be observed in a region (the visible) which is much more convenient in terms of source output, detector sensitivity, etc., than is the infrared.

Raman spectroscopy is not simply an alternative to infrared spectroscopy because the two differ fundamentally in mechanism and therefore in the information they provide; the intensity of the former depends classically upon $(\partial\alpha/\partial Q_i)_0^2$ while that of the latter depends upon $(\partial\mu/dQ_i)_0^2$, where μ denotes dipole moment of the scattering molecule. The selection rules are different for each form of spectroscopy; indeed a fundamental, depending on its type and on the point group to which the molecule belongs, may be Raman-active only (e.g. g-modes of a centrosymmetric molecule), infrared-active only (e.g. u-modes of a centrosymmetric molecule), both Raman and infrared active (giving rise to a so-called coincidence) or inactive. The combination of infrared and Raman spectroscopy is thus particularly powerful for molecular structure determination.

The Raman effect is concerned with the dipole moment induced in a molecule by the electric field of the incident light beam, which may thereby induce a transition in the molecule to a higher or lower rotational or vibrational state. In consequence, radiation of frequency different from that of the incident beam is scattered. Raman bands on the low frequency side of ν_0 are, for historical reasons, referred to as Stokes bands, whereas those on the high frequency side are referred to as anti-Stokes bands. The former are the more intense since at temperature equilibrium the population of the lower state (from which, in the absence of "hot" bands, Stokes bands originate) is always greater than that of any upper states. In consequence, Raman spectra are usually recorded only for Stokes bands. The ratio of Stokes to anti-Stokes band intensities is related in normal Raman scattering directly to the temperature of the scattering species.

The Raman effect is very weak, and it is essential that Raman spectrometers be able to discriminate very effectively against the much more intense Rayleigh line; highly efficient collection optics, coupled to a spectrometer which can discriminate effectively against stray light, are therefore of the utmost importance.

Raman spectroscopy celebrated its 60th anniversary this year (1988) in London in the form of the Eleventh International Conference on Raman spectroscopy (ICORS XI). That the 650 participants remained the full week in close and enthusiastic attendance, despite the many counter-claims of London for their attention, speaks for the very great and growing interest in the subject. The proceedings, which contain the texts of the plenary lectures and of the nearly 500 poster presentations are available[1], as is a key background text[2]. A review series[3], running now to seventeen volumes, contains up-to-date chapters on all aspects of the theory and practice of Raman spectroscopy.

The following outline of the theory of the Raman effect and of its intimate links with vibronic coupling follows that developed by Clark and Dines[4] to whose review the reader if referred for more information on the extensive background literature on the subject.

2. THEORETICAL BACKGROUND

2.1 Normal Raman Scattering

For a Raman transition between two states $|i\rangle$ and $|f\rangle$ of a scattering system, the intensity of light scattered at 90° to the direction of irradiation is given by Equation (1).

$$I_{fi}(\pi/2) = \frac{\pi^2}{\varepsilon_0^2}(\tilde{\nu}_0 \pm \tilde{\nu}_{fi})^4 \mathscr{I}_0 \sum_{\rho,\sigma} [\alpha_{\rho\sigma}]_{fi} [\alpha_{\rho\sigma}]^*_{fi} \tag{1}$$

$\mathscr{I}_0$ is the irradiance of the incident radiation, $\tilde{\nu}_0$ and $\tilde{\nu}_{fi}$ are the wavenumbers of the exciting line and of the Raman transition $|f\rangle \leftarrow |i\rangle$, respectively, ε_0 is the permittivity of free space, and $[\alpha_{\rho\sigma}]_{fi}$, which is equivalent classically to $(\partial\alpha/\partial Q)_0$, is the $\rho\sigma$th element of the transition polarizability tensor given by the expression

$$[\alpha_{\rho\sigma}]_{fi} = \frac{1}{hc}\sum_r \left(\frac{[\mu_\rho]_{fr}[\mu_\sigma]_{ri}}{\tilde{\nu}_{ri} - \tilde{\nu}_0 + i\Gamma_r} + \frac{[\mu_\sigma]_{fr}[\mu_\rho]_{ri}}{\tilde{\nu}_{rf} + \tilde{\nu}_0 + i\Gamma_r} \right) \tag{2}.$$

In this equation $[\mu_\rho]_{fr}$ is the ρth component of the transition dipole moment associated with the transition $|f\rangle \leftarrow |r\rangle$ and $i\Gamma_r$ is a damping factor, which is related to the lifetime of the state $|r\rangle$. Strictly, the summation is over all states $|r\rangle$ of the system, including $|i\rangle$ and $|f\rangle$, but it has been shown that, for Raman scattering, $|i\rangle$ and $|f\rangle$ may be excluded from the sum. (In the SI system, I is in W, $\mathscr{I}_0$ in W m^{-2}, ε_0 in F m^{-1}, $[\alpha_{\rho\sigma}]_{fi}$ in C V^{-1} m^2 and $[\mu_\rho]_{fr}$ in C m. The use of the unit cm^{-1} for wavenumber requires that the right-hand side of Equation (2) be multiplied by 10^8.) For vibrational Raman scattering excited far from resonance, the tensor is symmetric, i.e. $\alpha_{\rho\sigma} = \alpha_{\sigma\rho}$, but it may become asymmetric for electronic Raman (ER) and RR scattering.

The nature of the Raman effect is determined by the initial and final eigenstates $|i\rangle$ and $|f\rangle$, and by the proximity of the wavenumber of the exciting radiation to that of any electronic transition of the system. Clearly, the RR effect occurs when $\tilde{\nu}_0 \approx \tilde{\nu}_{ri}$, resulting in an increase of $[\alpha_{\rho\sigma}]_{fi}$ and consequently an enhancement of the Raman intensity. In order to answer such questions as which Raman bands may undergo enhancement and why overtone progressions are observed in RR spectra it is necessary to probe the dependence of $[\alpha_{\rho\sigma}]_{fi}$ on the properties of excited states by looking closely at Equation (2).

The adiabatic Born-Oppenheimer approximation is invoked, in which the vibronic states $|i\rangle$, $|f\rangle$, and $|r\rangle$ are formed by the products of the pure vibrational and the pure electronic states. The latter are referred to some fixed positions of the nuclei and are parametrically dependent on the normal coordinates, Q_k, of the molecule.

Assuming that the system is initially and finally in the ground electronic state $|g\rangle$ (i.e. a vibrational Raman transition, for instance, is under consideration), we may write:

$$|i\rangle = |gm\rangle = |g\rangle|m\rangle, \quad |f\rangle = |gn\rangle = |g\rangle|n\rangle,$$
$$|r\rangle = |ev\rangle = |e\rangle|v\rangle \tag{3}$$

where $|e\rangle$ denotes an excited electronic state and $|m\rangle$, $|n\rangle$, and $|v\rangle$ denote vibrational states of the scattering species. This enables the transition polarizability to be rewritten in the form

$$[\alpha_{\rho\sigma}]_{gn,gm} = \frac{1}{hc} \sum_{ev} \left(\frac{\langle n|[\mu_\rho]_{ge}|v\rangle\langle v|[\mu_\sigma]_{eg}|m\rangle}{\tilde{\nu}_{ev,gm}-\tilde{\nu}_0+i\Gamma_{ev}} + \frac{\langle n|[\mu_\sigma]_{ge}|v\rangle\langle v|[\mu_\rho]_{eg}|m\rangle}{\tilde{\nu}_{ev,gn}+\tilde{\nu}_0+i\Gamma_{ev}} \right) \tag{4}$$

where $[\mu_\rho]_{ge}$ is the pure electronic transition moment associated with the transition $|g\rangle\leftarrow|e\rangle$. Under the conditions for which the Born-Oppenheimer condition is valid, the dependence of such an electronic transition moment on the k normal coordinates of the system, Q_k, is small. Thus, $[\mu_\rho]_{ge}$ may be expressed as a rapidly converging Taylor series expanded around the equilibrium position:

$$[\mu_\rho]_{ge} = [\mu_\rho]^0_{ge} + \sum_k [\mu_\rho]'_{ge} Q_k + \ldots \tag{5}$$

where $[\mu_\rho]'_{ge}=\partial[\mu_\rho]_{ge}/\partial Q_k$. Higher order terms in this series are, in most cases, negligibly small. A second simplification arises from the Born-Oppenheimer approximation in that the electronic and vibrational parts of the integrals in Equation (4) may be separated, such that, for example,

$$\langle n|[\mu_\rho]_{ge}|v\rangle = [\mu_\rho]^0_{ge}\langle n|v\rangle + \sum_k [\mu_\rho]'_{ge}\langle n|Q_k|v\rangle \tag{6}.$$

The following approximations may also be made in the ideal limit of excitation far from any region of electronic absorption:

(a) The denominators in the transition polarizability are large and insensitive to the vibrational quantum numbers m, n, and v, such that differences between the various $\tilde{\nu}_{ev,gm}$ and $\tilde{\nu}_{ev,gn}$ may be neglected;

(b) The states $|v\rangle$ represent a complete orthonormal set and the sums over them may be evaluated by invoking the closure theorem, which states that $\sum_v |v\rangle\langle v| = 1$;

(c) Since $(\tilde{\nu}_{ev,gm}-\tilde{\nu}_0)$ and $(\tilde{\nu}_{eg,gm}+\tilde{\nu}_0)$ are much larger than the damping factors, $i\Gamma_{ev}$, the latter may be neglected.

With these approximations, and noting that

$$(\tilde{\nu}_e-\tilde{\nu}_0)^{-1}+(\tilde{\nu}_e+\tilde{\nu}_0)^{-1} = 2\tilde{\nu}_e(\tilde{\nu}_e^2-\tilde{\nu}_0^2)^{-1}$$

the transition polarizability may be recast in the following form:

$$[\alpha_{\rho\sigma}]_{gn,gm} = \frac{1}{hc}\sum_e \frac{2\tilde{\nu}_e}{(\tilde{\nu}_e^2-\tilde{\nu}_0^2)}[\mu_\rho]_{ge}^0[\mu_\sigma]_{eg}^0\langle n|m\rangle +$$
$$+ \frac{1}{hc}\sum_e\sum_k \frac{2\tilde{\nu}_e}{(\tilde{\nu}_e^2-\tilde{\nu}_0^2)}\{[\mu_\rho]'_{ge}[\mu_\sigma]_{eg}^0+[\mu_\rho]_{ge}^0[\mu_\sigma]'_{eg}\}\langle n|Q_k|m\rangle$$
$$+ \frac{1}{hc}\sum_e\sum_{k,k'} \frac{2\tilde{\nu}_e}{(\tilde{\nu}_e^2-\tilde{\nu}_0^2)}[\mu_\rho]'_{ge}[\mu_\sigma]'_{eg}\langle n|Q_kQ_{k'}|m\rangle \qquad (7)$$

Because n and m are both vibrational quantum numbers of the ground electronic state, the overlap integral $\langle n|m\rangle=\delta_{nm}$. Thus, the first term in Equation (7) contributes only to Rayleigh scattering. The integral $\langle n|Q_k|m\rangle$ is non-zero if $n=m\pm1$ and has the value $[h(m+1)/8\pi^2c\tilde{\nu}_k]^{\frac{1}{2}}$ for $n=m+1$ and $[hm/8\pi^2c\tilde{\nu}_k]^{\frac{1}{2}}$ for $n=m-1$. It follows that the second term in Equation (7), the vibronic coupling term, is responsible for both the Stokes and the anti-Stokes Raman scattering for vibrational fundamentals. The third term gives rise to first overtones ($k=k'$) and binary combination tones ($k\neq k'$), but is usually so small that bands so attributed are very weak. It is obvious that the magnitude of $[\alpha_{\rho\sigma}]_{gn,gm}$ is unaffected by transposition of the coordinate suffixes ρ and σ. Hence the transition polarizability tensor is symmetric about the leading diagonal and has only six independent components. It may be regarded essentially as a ground state property of the system.

2.2 Resonance Raman Scattering

In the usual treatment of RR scattering the adiabatic Born-Oppenheimer approximation is assumed to be valid, but the other assumptions associated with normal Raman scattering are assumed no longer to apply. That is to say:

(a) The v-dependence of the denominators in Equation (4) cannot be neglected, so that the closure theorem may not be invoked;

(b) the wavenumber difference ($\tilde{\nu}_{ev,gm}-\tilde{\nu}_0$) is comparable with the damping factor, $i\Gamma_{ev}$, so that the latter may not be neglected;

(c) the first-order term $\partial[\mu_\rho]_{ge}/\partial Q_k = [\mu_\rho]'_{ge}$ in the Taylor expansion of Equation (5) needs to be defined explicitly.

In the Herzberg-Teller perturbation description of vibronic coupling the derivative of the transition moment arises because the variation in the Hamiltonian with respect to the normal coordinate Q_k can mix the state $|e>$ with other states $|s>$ of the appropriate symmetry, viz.

$$[\mu_\rho]_{ge} = [\mu_\rho]^0_{ge} + \sum_s \sum_k [\mu_\rho]^0_{gs} \frac{h^k_{es}}{\Delta\tilde{\nu}_{es}} Q_k + \ldots \tag{8}$$

$$h^k_{es} = <e|\partial H/\partial Q_k|s>_{Q_{k=0}}$$

where $|s>$ represents another excited state. As before, the higher order terms are generally negligible.

The term $h^k_{es}/\Delta\tilde{\nu}_{es}$ is a measure of the strength of vibronic coupling of the states $|e>$ and $|s>$ via the normal coordinate Q_k. The Herzberg-Teller expansion is only valid for weak vibronic coupling, i.e. within the framework of the adiabatic Born-Oppenheimer approximation. It is not applicable to cases in which there is strong vibronic coupling, i.e. when $\Delta\tilde{\nu}_{es}$ is of the order of vibrational wavenumbers (in which case non-adiabatic coupling is important) or zero (such as for a degenerate state, in which Jahn-Teller (JT) coupling occurs).

Considerable simplification of the transition polarizability is possible for excitation under resonance conditions. This is because, as $\tilde{\nu}_0$ approaches some particular transition wavenumber $\tilde{\nu}_{ev,gm}$, the relevant excited state will dominate the sum over states. It is therefore generally sufficient to consider only one, or at most, two electronic manifolds. Also, the non-resonant part of the transition polarizability may be neglected at resonance. Applying these considerations to Equation (4) it is evident that there are now four contributions (the so-called A, B, C, and D-terms) to the transition polarizability, viz.

$$[\alpha_{\rho\sigma}]_{gn,gm} = A + B + C + D \tag{9}$$

where

$$A = \frac{1}{hc}[\mu_\rho]^0_{ge}[\mu_\sigma]^0_{eg} \sum_v \frac{<n_g|v_e><v_e|m_g>}{\tilde{\nu}_{ev,gm}-\tilde{\nu}_0+i\Gamma_{ev}} \tag{10}$$

$$B = \frac{1}{h^2c^2} \sum_{s\neq e} [\mu_\rho]^0_{gs}[\mu_\sigma]^0_{ge} \frac{h^k_{se}}{\Delta\tilde{\nu}_{se}} \sum_v \frac{<n_g|Q_k|v_e><v_e|m_g>}{\tilde{\nu}_{ev,gm}-\nu_0+i\Gamma_{ev}} +$$

$$+ \frac{1}{h^2c^2} \sum_{s\neq e} [\mu_\rho]^0_{ge}[\mu_\sigma]^0_{sg} \frac{h^k_{es}}{\Delta\tilde{\nu}_{es}} \sum_v \frac{<n_g|v_e><v_e|Q_k|m_g>}{\tilde{\nu}_{eg,gm}-\tilde{\nu}_0+i\Gamma_{ev}} \tag{11}$$

$$C = \frac{1}{h^2c^2} \sum_{s \neq g} [\mu_\rho]^0_{se} [\mu_\sigma]^0_{eg} \frac{h^k_{gs}}{\Delta\tilde{\nu}_{gs}} \sum_v \frac{\langle n_g|Q_k|v_e\rangle\langle v_e|m_g\rangle}{\tilde{\nu}_{ev,gm} - \tilde{\nu}_0 + i\Gamma_{ev}} +$$

$$+ \frac{1}{h^2c^2} \sum_{s \neq g} [\mu_\rho]^0_{ge} [\mu_\sigma]^0_{es} \frac{h^k_{sg}}{\Delta\tilde{\nu}_{sg}} \sum_v \frac{\langle n_g|v_e\rangle\langle v_e|Q_k|m_g\rangle}{\tilde{\nu}_{ev,gm} - \tilde{\nu}_0 + i\Gamma_{ev}} \qquad (12)$$

$$D = \frac{1}{h^3c^3} \sum_{s,s' \neq e} [\mu_\rho]^0_{gs} [\mu_\sigma]^0_{s'g} \frac{h^k_{es} h^{k'}_{es'}}{\Delta\tilde{\nu}_{es}\Delta\tilde{\nu}_{es'}} \sum_v \frac{\langle n_g|Q_k|v_e\rangle\langle v_e|Q_{k'}|m_g\rangle}{\tilde{\nu}_{ev,gm} - \tilde{\nu}_0 + i\Gamma_{ev}} \qquad (13)$$

In these equations the vibrational quantum numbers n, v and m have been subscripted to indicate the electronic state to which they refer, all other terms having been defined previously. We now look closely at the A and B contributions, which are the largest.

2.3 A-term Resonance Raman Scattering

For the A-term contribution to the transition polarizability to be non-zero it is necessary that the following two conditions are fulfilled:
(a) The transition dipole moments $[\mu_\rho]^0_{ge}$ and $[\mu_\sigma]^0_{eg}$ must be non-zero;
(b) The products of vibrational overlap integrals (Franck-Condon factors), $\langle n_g|v_e\rangle\langle v_e|m_g\rangle$, must be non-zero for at least some values of v.

The first of these two conditions dictates that the resonant electronic transition should be electric-dipole-allowed. Excitation within the contour of a weak band, e.g. one resulting from a ligand-field or spin-forbidden transition, would thus not be expected to produce a significant A-term. The restrictions imposed by the second condition are less obvious.

Vibrational overlap integrals of the type $\langle n_g|v_e\rangle$ are finite only if the vibrational wave functions χ_n and χ_v are non-orthogonal. For any one vibrational mode ν_k of a molecule non-orthogonality of these wave functions will subsist if, between the ground state and excited states, there is either (a) a difference of vibrational wavenumber ($\tilde{\nu}^e_k \neq \tilde{\nu}^g_k$), i.e. a change in shape of the potential energy surface, or (b) a displacement of the potential energy minimum along the normal coordinate, ΔQ_k. Symmetry arguments dictate that such a displacement may only occur for totally symmetric modes unless the molecular symmetry is altered in the excited state. However, if a change in molecular symmetry accompanies the electronic transition, this restriction no longer applies. The term totally symmetric then applies not to the ground state point group but to the subgroup formed by the operations that are common to both the ground- and excited-state point groups (the common group). In practice it is found that a significant change of vibrational wavenumber usually occurs only where there is a displacement of the potential energy minimum. Dependent on the magnitude of ΔQ, the Franck-Condon factors, $<n_g|v_e><v_e|m_g>$, may have appreciable size for up to several quanta of n. Thus the A-term can give rise to overtones of intensity comparable to that of the fundamental.

2.4 B-Term Resonance Raman Scattering

The B-term contribution to the transition polarizability arises from the vibronic coupling of the resonant state to other excited states. Its magnitude is dependent on the vibronic coupling integral h^k_{es}, on the products of vibrational transition integrals and overlap integrals, e.g. $<n_g|Q_k|v_e><v_e|m_g>$, on the transition dipole moments $[\mu_\rho]^0_{ge}$ and $[\mu_\sigma]^0_{sg}$, and on the energy difference between the coupled states, $hc(\tilde{\nu}_s-\tilde{\nu}_e)$. For the B-term to be non-zero it is a requirement that both the resonant transition and the transition $|s>\leftarrow|g>$ should be electric-dipole-allowed. Because of the additional factor $h^k_{es}(\nu_s-\nu_e)^{-1}$ in the equation, the B-term is usually very much smaller than the A-term. However, if the resonant transition gives rise to zero or very small displacements ΔQ_k, then only the diagonal overlap integrals (i.e. n = v, v = m) are non-vanishing and the A-term becomes zero. In the harmonic oscillator approximation $<n_g|Q_k|v_e>$ and $<v_e|Q_k|m_g>$ are non-zero for $n = v\pm1$ and $v = m\pm1$. Thus, under the conditions for which the B-term is of importance (diagonal overlap integrals) the products $<n_g|Q_k|v_e><v_e|m_g>$ and $<n_g|v_e><v_e|Q_k|m_g>$ are restricted to the cases where $n = m\pm1$ and, in the low temperature limit ($m = 0$), only two such products would contribute, viz. $<1_g|Q_k|0_e><0_e|0_g>$ and $<1_g|1_e><1_e|Q_k|0_g>$. It is usually the case that coupling to only one other state $|s>$ needs to be considered and the summation may be dropped, giving the following expression for the B-term:

$$B = \frac{1}{h^2c^2}[\mu_\rho]^0_{gs}[\mu_\sigma]^0_{eg}\frac{h^k_{se}}{\Delta\tilde{\nu}_{se}}\left(\frac{<1_g|Q_k|0_e><0_e|0_g>}{\tilde{\nu}_{e0,g0}-\tilde{\nu}_0+i\Gamma_{e0}}\right) +$$

$$+\frac{1}{h^2c^2}[\mu_\rho]^0_{ge}[\mu_\sigma]^0_{sg}\frac{h^k_{es}}{\Delta\tilde{\nu}_{es}}\left(\frac{<1_g|1_e><1_e|Q_k|0_g>}{\tilde{\nu}_{e1,g0}-\tilde{\nu}_0+i\Gamma_{e1}}\right)$$

Clearly, B-term scattering occurs only for fundamentals ($n = 1$) and the symmetry selection rules are constrained by the symmetries of the states $|e>$ and $|s>$. The vibronic coupling integral h^k_{es} is non-zero if the irreducible representation of Q_k (i.e. Γ_k) is contained in the direct product $\Gamma_e \otimes \Gamma_s$. It follows that B-term activity may be expected for both totally symmetric and non-totally symmetric vibrations, but for the former to be realized it is necessary that $|e>$ and $|s>$ have the same symmetry.

2.5 C- and D-Term Resonance Raman Scattering

The C-term differs from the B-term in that the ground state, rather than the resonant excited state, is vibronically coupled to other excited states. It is normally assumed that the C-term is negligibly small, due to the large energy separation between the ground state and excited states.

The D-term is also very much smaller than the B-term. It may give rise to first overtones (k = k′) and binary combination tones (k ≠ k′) of both totally symmetric and non-totally symmetric modes. The observation of weak overtones and binary combination tones in the RR spectra of some metalloporphyrin species has been ascribed to D-term scattering.

2.6 Resonance Raman Scattering Involving Totally Symmetric Modes

Raman bands attributed to totally symmetric fundamentals may acquire intensity under resonance conditions from either the A- or B-terms of the transition polarizability. It is generally the case, however, that only the A-term contribution is important, for the reasons mentioned earlier. In this section the properties of A- and B-term RR scattering for totally symmetric fundamentals and their overtones are discussed. To begin with the simplest case of a species possessing only one totally symmetric mode is considered, in which case only diagonal components of the transition polarizability, $[\alpha_{\rho\rho}]_{gn,gm}$, are non-zero. If it is further assumed that (a) only a single resonant excited state $|e\rangle$ makes a significant contribution, and (b) all molecules are initially in the m = 0 vibrational level of the ground electronic state (low temperature limit) then

$$[\alpha_{\rho\rho}]_{gn,g0} = \frac{1}{hc}|[\mu_\rho]^0_{ge}|^2 \sum_v \frac{\langle n_g|v_e\rangle\langle v_e|0_g\rangle}{\tilde{\nu}_{ev,g0}-\tilde{\nu}_0+i\Gamma_{ev}} \qquad (15).$$

For a system with only one totally symmetric normal coordinate, Q_1, the vibrational overlap integrals are one-dimensional. Assuming harmonic potentials for the states $|g\rangle$ and $|e\rangle$, the overlap integrals may be calculated using the recurrence formulae of Manneback[5] in terms of the dimensionless shift parameter Δ, where

$$\Delta = \left(\frac{4\pi^2 c}{h}\right)^{\frac{1}{2}} \Delta Q_1 \left(\frac{\tilde{\nu}^e_1\tilde{\nu}^g_1}{\tilde{\nu}^e_1+\tilde{\nu}^g_1}\right)^{\frac{1}{2}} \qquad (16).$$

The displacement, ΔQ_1, of the excited-state potential minimum is related by the displacement along the symmetry coordinate ΔS_1 by the equation

$$\Delta Q_1 = \mu^{\frac{1}{2}} \Delta S_1 \qquad (17)$$

where μ denotes the reduced mass associated with the vibration.

The Franck-Condon overlap integrals may possess appreciable magnitudes for up to several quanta of n if ΔQ_1 is sufficiently large, in which case several overtones of ν_1 may appear in the RR spectrum. The intensity distribution within such an overtone progression depends on the magnitude of ΔQ (and hence on the bond length change Δr) and, in general, the larger the displacement the greater is the extent of the progression. Thus, excitation resonant with charge-transfer or π-π^* transitions of small molecules typically produces RR spectra that are characterized by an intense overtone progression $v_1\nu_1$.

Substitution of Equation (15) into Equation (1) and expansion of the square modulus of $[\alpha_{\rho\rho}]_{gn,g0}$ gives the following expression for Stokes A-term RR intensities:

$$I_{gn,g0}(\pi/2) = K(\tilde{\nu}_0-\tilde{\nu}_{gn,g0})^4|[\mu_\rho]^0_{ge}|^4 \times \sum_v \sum_{v'} \frac{\langle n_g|v_e\rangle\langle v_e|0_g\rangle\langle n_g|v'_e\rangle\langle v'_e|0_g\rangle[\varepsilon_v\varepsilon_{v'}+\Gamma_v\Gamma_{v'}]}{(\varepsilon_v^2+\Gamma_v^2)(\varepsilon_{v'}^2+\Gamma_{v'}^2)} \tag{18}$$

where $\varepsilon_v = \tilde{\nu}_{ev,g0}-\tilde{\nu}_0$ and K incorporates all of the constants appearing in Equations (1) and (15). From this expression it is possible to calculate the variation of the intensities of the fundamental and overtone bands throughout the region of the resonant absorption band. Such a plot, which is called an excitation profile (EP), differs markedly from the absorption spectrum if the vibrational fine structure of the latter is resolved (small Γ). Whereas absorption profiles are simply given by the superposition of the individual vibronic bands, the double summation in Equation (18) involves cross-terms ($v \neq v'$) which produce interference between the contributions from different vibronic states. These interference effects may result in an increase or decrease in scattered intensity compared with that arising solely from the sum of the individual resonances ($v = v'$). However, when Γ is large these interferences are not apparent and both the excitation profiles and the absorption spectrum are smooth and devoid of fine structure. The maxima of the excitation profiles are then blue-shifted for successive overtones, the difference in peak position being approximately equal to the vibrational wavenumber of the excited state.

By fitting calculated excitation profiles to experimental data the parameters Δ and Γ may be extracted. Thus, the technique provides a sensitive method for the determination of excited state geometries and lifetimes. It is, however, important to appreciate that lifetime, i.e. homogeneous, broadening (Γ) is not the only contribution to bandwidths and that the influence of inhomogeneous broadening (γ) must also be taken into account. Inhomogeneous broadening is due to the variation of molecular environment (i.e. molecules occupying different sites) and the "spread" of electronic energies resulting from it is described by a Gaussian or Lorentzian distribution about the mean value. Although a Gaussian distribution would normally be the correct description of inhomogeneous broadening, the influence of the band shape is not very great and a Lorentzian function is usually chosen for simplicity. Homogeneous broading and inhomogeneous broadening have markedly different effects on the excitation profiles[6,7], as discussed in detail in Ref.4. The principal differences are that (a) γ plays no part in the aforementioned interference phenomena, and (b) increasing the Γ contribution (for constant $\Gamma + \gamma$) decreases the intensity overtones relative to that of the fundamental. By measuring EPs of molecules in different environments (i.e. different solvents or host lattices) it is possible to determine Γ, and hence the excited-state lifetime, since this is largely independent of molecular environment. This information cannot be obtained from absorption spectra because these are insensitive to variation in the relative contributions of Γ and γ, the bandwidths being given by the sum $\Sigma = \Gamma + \gamma$.

The treatment given can be extended to molecules for which there is more than one excited state which contributes to the polarizability. The theory may also be extended to molecules with several totally symmetric modes if it is assumed that the normal coordinates of the excited state are identical to those of the ground state. The multi-dimensional overlap integrals $\langle \bar{n}_g | \bar{v}_e \rangle$ may then be expressed as products of one-dimensional ones:

$$\langle \bar{n}_g | \bar{v}_e \rangle = \langle (n_1 n_2 n_3 \ldots n_N)_g | (v_1 v_2 v_3 \ldots v_N)_e \rangle$$

$$= \prod_{i=1}^{N} \langle n_{ig} | v_{ie} \rangle \qquad (19).$$

Substitution of these products into Equation (18) or the equivalent multistate expression enables excitation profiles to be calculated. By fitting these to experimental data, values for ΔQ_1, ΔQ_2, etc., can be deduced, i.e. the displacement of the excited-state potential minimum along each of the totally symmetric normal coordinates. The relationship between these displacements and the geometric changes attendant upon electronic excitation (bond-length and bond-angle changes) is less straightforward than in the single-mode case since a knowledge of the relationship between symmetry and normal coordinates is required.

RR spectra of molecules possessing more than one totally symmetric mode may display overtone progressions for each of the active vibrations, i.e. those for which there are normal coordinate displacements. Additionally there may also be combination band progressions involving quanta of more than one mode. In general it is true that the most displaced coordinate will give rise to the greatest intensity enhancement and the longest overtone progression. On this basis it is still possible to make a qualitative assessment of the excited-state geometry from RR data when detailed calculations cannot be performed.

A special case of the A-term scattering regime exists when each of the displacements, ΔQ, is very small but non-zero. In this situation (e.g. for ruthenium red) it can be seen from Equation (16) that, assuming $\tilde{\nu}_i^e = \tilde{\nu}_i^g$, the overlap integrals have the values

$$\langle 1_g | 1_e \rangle = \langle 0_g | 0_e \rangle = 1$$

$$\langle 1_g | 0_e \rangle = -\langle 0_g | 1_e \rangle = -\Delta$$

Integrals in which either n or v has a value of 2 or more and $n \neq v$ may be approximated to zero because they involve higher powers of Δ ($\Delta \leqslant 1$ and so $\Delta^2 \leqslant \Delta$). The summation over the levels v is therefore restricted to the v = 0 and v = 1 contributions, and Franck-Condon factors will be non-zero only for n = 0 (Rayleigh scattering) and n = 1 (fundamental Raman scattering). Thus, in the small displacement approximation, overtones are forbidden.

Further details of these analyses are given elsewhere[4], particularly with respect to consideration of multidimensional Franck-Condon factors and A-term/B-term interference[8].

2.7 Resonance Raman Scattering Involving Non-Totally Symmetric Modes

RR bands attributed to non-totally symmetric fundamental vibrations may acquire intensity via one of three scattering mechanisms:

(a) A-term activity due to a change of molecular symmetry in the resonant excited state;

(b) A-term activity due to excited-state Jahn-Teller (JT) coupling;

(c) B-term scattering, involving the vibronic coupling of the resonant state to a second electronic excited state.

If a molecule undergoes a change of symmetry upon excitation then the A-term active modes are those that are totally symmetric in the subgroup formed by the symmetry operations common to the ground- and excited-state point groups (the common group). A good example is provided by the molecule ethylene which has a planar D_{2h} geometry in the ground state and a twisted D_{2d} geometry in the first $^1B_{1u}$ excited state. The common group is D_2 and, by correlation of D_2 to D_{2h}, it is seen that the A-term active modes are those of a_g and a_u symmetry. The a_u fundamental (ν_7, the torsion about the C=C bond) is Raman inactive but even quanta thereof, i.e. $2\nu_7$, $4\nu_7$, etc., have A_g symmetry and accordingly are Raman active. Excitation within the contour of the π-π* transition of ethylene using 193.4 nm irradiation[9] yields RR spectra displaying progressions in the ν(C=C) mode and even harmonics of $\nu_7(a_u)$. This recent study thus provided the first unambiguous evidence for the twisted geometry of the first excited state for ethylene.

In cases where the resonant excited state is degenerate, A-term activity may arise for some non-totally symmetric modes. The non-totally symmetric vibrations belonging to the irreducible representations contained in the symmetric direct product $\{\Gamma_e \otimes \Gamma_e\}$ may be responsible for JT coupling in the excited state. Some common examples of excited states so influenced are $T_2(T_d)$, $T_{1u}(O_h)$ and $E_u(D_{4h})$, for which the direct products are as follows:

$$T_2 \otimes T_2 = \{A_1 + E + T_2\} + [T_1]$$

$$T_{1u} \otimes T_{1u} = \{A_{1g} + E_g + T_{2g}\} + [T_{1g}]$$

$$E_u \otimes E_u = \{A_{1g} + B_{1g} + B_{2g}\} + [A_{2g}]$$

Vibrations transforming as the antisymmetric part of the direct product (given in square brackets) are not JT active because they do not lift the degeneracy of the excited state. This point is best illustrated by reference to an ion belonging to the D_{4h} point group, e.g. $[Pt(CN)_4]^{2-}$, for which the b_{1g} and b_{2g} vibrations remove the C_4-axis (which is responsible for the double degeneracy), but the a_{2g} vibration does not; thus the b_{1g} and b_{2g} modes, but not the a_{2g} mode, are JT active.

The consequence of excited-state JT effects on RR spectra may be summarized as follows:

(a) The appearance of combination band progressions involving quanta of totally symmetric modes plus one quantum of a JT-active vibration (small JT effect);

(b) the appearance of combination band progressions involving quanta of totally symmetric modes plus multiple quanta of JT-active vibrations, as well as overtone progressions of the latter (strong JT effect).

It is expected that the excited-state JT effect is a dynamic one and that no permanent distortion of the molecule occurs. Thus the major geometric changes accompanying excitation to a degenerate excited state are those involving totally symmetric modes and case (b) is something of a rarity.

For many polyatomic molecules B-term scattering is the major source of RR activity for non-totally symmetric modes. As pointed out earlier, its magnitude is dependent on the vibronic coupling integral h^k_{es}. It is therefore necessary that, for a non-totally symmetric coordinate Q_k to be B-term active, it must be effective in coupling the resonant state $|e\rangle$ to a second excited state $|s\rangle$. Expressed group theoretically, Γ_k must be contained in the product $\Gamma_e \otimes \Gamma_s$. From Equation (14) the transition polarizability is given by

$$[\alpha_{\rho\sigma}]_{g1,g0} = \frac{1}{h^2c^2}[\mu_\rho]^0_{gs}[\mu_\sigma]^0_{eg}\frac{h^k_{se}}{\Delta\tilde{\nu}_{es}}\langle 1|Q_k|0\rangle(\varepsilon_0+i\Gamma)^{-1} +$$

$$+ \frac{1}{h^2c^2}[\mu_\rho]^0_{ge}[\mu_\sigma]^0_{sg}\frac{h^k_{es}}{\Delta\tilde{\nu}_{es}}\langle 1|Q_k|0\rangle(\varepsilon_1+i\Gamma)^{-1} \qquad (20)$$

where $\langle 1|Q_k|0\rangle = \langle 1_e|Q_k|0_g\rangle = \langle 1_g|Q_k|0_e\rangle$

Close examination of this expression reveals that $\alpha_{\rho\sigma} \neq \alpha_{\sigma\rho}$, i.e. that the polarizability tensor is asymmetric. The latter may be expressed as the sum of symmetric and antisymmetric tensors, viz.

$$\alpha_{\rho\sigma} = \left(\frac{\alpha_{\rho\sigma}+\alpha_{\sigma\rho}}{2}\right) + \left(\frac{\alpha_{\rho\sigma}-\alpha_{\sigma\rho}}{2}\right) \qquad (21)$$

where the matrix elements are given by

$$[\alpha_{\rho\sigma}]_{g1,g0} \pm [\alpha_{\sigma\rho}]_{g1,g0} = \frac{1}{h^2c^2}\frac{h^{k}_{es}}{\Delta\tilde{\nu}_{es}}\langle 1|Q_k|0\rangle$$

$$\times\ \{[\mu_\rho]^0_{gs}[\mu_\sigma]^0_{eg}[\mu_\rho]^0_{ge}[\mu_\sigma]^0_{sg}\}\left(\frac{1}{\varepsilon_0+i\Gamma} \pm \frac{1}{\varepsilon_1+i\Gamma}\right) \tag{22}$$

Excitation profiles for Raman bands involving symmetric (+) and antisymmetric (-) tensor contributions are given by

$$I^{(\pm)}_{g1,g0} = K\,\frac{\left\{\frac{\Sigma}{\Gamma}[\varepsilon_0^2+\varepsilon_1^2+2\Sigma^2][(\varepsilon_1-\varepsilon_0)^2+4\Gamma^2] \pm 2[\varepsilon_0\varepsilon_1+\Sigma^2][(\varepsilon_0-\varepsilon_1)^2+4\Gamma\Sigma]\right\}}{(\varepsilon_0^2+\Sigma^2)(\varepsilon_1^2+\Sigma^2)[(\varepsilon_0-\varepsilon_1)^2+4\Gamma^2]} \tag{23}$$

The antisymmetric tensor EP resembles that arising from the small displacement A-term, i.e. there is constructive interference between the 0-0 and 0-1 resonances and destructive interference elsewhere. Thus, the antisymmetric tensor contribution becomes zero outside of the resonance region and normal, i.e. off-resonance, vibrational Raman scattering is controlled exclusively by a symmetric polarizability tensor. By contrast, electronic Raman transitions can involve antisymmetric scattering even in the off-resonance case. The most notable observations of antisymmetric scattering by non-totally symmetric modes have been for metalloporphyrin systems. Resonance Raman spectra obtained by excitation within the Q-band (E_u symmetry) are dominated by B-term scattering by vibrations which couple the Q-state to the higher energy Soret state (also of E_u symmetry). The B-term active vibrations are given by

$$E_u \otimes E_u = A_{1g} + A_{2g} + B_{1g} + B_{2g}$$

and the antisymmetric part of the product, A_{2g}, gives rise to antisymmetric scattering. Weak overtones of a_{1g}, a_{2g}, b_{1g}, and b_{2g} modes are sometimes observed in metalloporphyrin RR spectra and represent the only known examples of D-term scattering.

Two further complications may arise:

(a) Interference between contributions to the transition polarizability from JT-activity (intramanifold coupling) and B-term activity (intermanifold coupling);

(b) Non-adiabatic coupling, which may arise when the states $|e\rangle$ and $|s\rangle$ are close together, i.e. when $\Delta\tilde{\nu}_{es}$ is comparable with vibrational band wavenumbers.

The effect of either of these two processes is that the 0-0 and 0-1 peaks in the excitation profiles display unequal intensities. Modes of a_{2g} symmetry are JT inactive and show only the non-adiabatic effect, which favours the 0-1 resonance. This contrasts with the behaviour of other modes which tend to produce stronger 0-0 scattering, arising from interference between the inter- and intramanifold contributions.

2.8 Depolarization Ratios

The depolarization ratio for 90° Raman scattering excited by linearly polarized incident radiation is given by

$$\rho(\pi/2) = \frac{{}^{\perp}I_{\parallel}(\pi/2)}{{}^{\perp}I_{\perp}(\pi/2)} \tag{24}$$

where $\perp$ and $\parallel$ are defined by reference to the scatter plane (the plane containing the directions of propagation of the incident and scattered radiation) and the superscripts and subscripts refer to incident and scattered polarizations, respectively. For an assembly of randomly orientated molecules $\rho(\pi/2)$ may be expressed in terms of the tensor invariants $\bar{\alpha}$, γ_s, and γ_{as} thus:

$$\rho(\pi/2) = \frac{3\gamma_s^2+5\gamma_{as}^2}{45\bar{\alpha}^2+4\gamma_s^2} \tag{25}$$

where

$$\gamma_s^2 = \tfrac{1}{2}\{(\alpha_{xx}-\alpha_{yy})^2+(\alpha_{yy}-\alpha_{zz})^2+(\alpha_{zz}-\alpha_{xx})^2 + {}+\tfrac{3}{2}[(\alpha_{xy}+\alpha_{yx})^2+(\alpha_{yz}+\alpha_{zy})^2+(\alpha_{zx}+\alpha_{xz})^2]\}$$

$$\gamma_{as}^2 = \tfrac{3}{4}[(\alpha_{xy}-\alpha_{yx})^2+(\alpha_{yz}-\alpha_{zy})^2+(\alpha_{zx}-\alpha_{xz})^2]$$

For vibrational non-RR scattering γ_{as}^2 is zero, because, as stated previously, the antisymmetric polarizability tensor vanishes under these conditions. The following behaviour is then observed:

(a) For non-totally symmetric modes only off-diagonal polarizability tensor elements, $\alpha_{\rho\sigma}$, and the combination $(\alpha_{xx}-\alpha_{yy})$ are non-zero. Therefore $\bar{\alpha} = 0$ for all non-totally symmetric modes and $\rho(\pi/2) = \tfrac{3}{4}$;

(b) for cubic molecules $\alpha_{xx}=\alpha_{yy}=\alpha_{zz}$, hence $\gamma_s^2 = 0$ for totally symmetric modes and $\rho(\pi/2) = 0$. Both $\bar{\alpha}$ and γ_s^2 may be non-zero for the totally symmetric modes of non-cubic molecules and $0 \leqslant \rho(\pi/2) < \tfrac{3}{4}$.

Depolarization ratios for RR scattering are determined by the symmetry of the resonant transition and can provide a valuable tool for the determination of electronic band assignments. For totally symmetric modes of molecules not belonging to the point groups C_1, C_i, C_s, C_2 or C_{2h}, only the diagonal polarizability components, $\alpha_{\rho\rho}$, are non-zero. If the resonant transition is non-degenerate then only one of these will be non-zero (say α_{xx} if the transition is x-polarized) and, from Equation (25), $\rho(\pi/2) = \tfrac{1}{3}$. Similarly it can be shown that a doubly-degenerate resonant transition (x,y-polarization) gives rise to $\rho(\pi/2) = 1/8$, and that a cubic molecule for which only triply-degenerate transitions are electric-dipole-allowed (x,y,z-polarization), $\rho(\pi/2) = 0$. Two further cases may arise when either two non-degenerate transitions lie close together or when a non-degenerate transition is in close proximity to a doubly-degenerate one. In the former case the $\rho(\pi/2)$ value will depend on the relative magnitudes of, say α_{xx} and α_{yy}, and will vary as a function of the excitation wavenumber $\tilde{\nu}_0$ (polarisation dispersion). Likewise, in the second case the $\rho(\pi/2)$ values will depend on the relative magnitudes of say $\alpha_{xx} = \alpha_{yy}$ and α_{zz}.

If the molecular point group is either C_1, C_i, C_s, C_2 or C_{2h} then off-diagonal components, $\alpha_{\rho\sigma}$, may also be non-zero if B-term scattering occurs. Thus, the $\rho(\pi/2)$ is most likely to be within the range 0 to $3/4$.

Departure from the above-mentioned behaviour is expected if the electronic ground state is degenerate, for then the Raman transition is not a purely vibrational one. The Raman transition is $|g\rangle|n\rangle \leftarrow |g\rangle|0\rangle$, in the low temperature limit, and the irreducible representation for this process is given by $\Gamma_g \otimes \Gamma_n \otimes \Gamma_g \otimes \Gamma_1$ (the v = 0 level is necessarily totally symmetric). It is clear that, for a non-degenerate ground state, the product is simply Γ_n because $\Gamma_g \otimes \Gamma_g = \Gamma_1$. However, it is equally clear that if the ground state is degenerate then this will affect the symmetry of the transition. As an example, consider a doubly degenerate E_g'' ground state of an octahedral molecule (O_h point group). The product $E_g'' \otimes E_g'' = A_{1g} + T_{1g}$, and thus the Raman bands attributed to $\nu_1(a_{1g})$ and its overtones will transform as $A_{1g} + T_{1g}$. Both $\bar{\alpha}$ and γ_{as}^2 will be non-zero and $\rho(\pi/2)$ may take any value between zero and infinity, depending on the relative magnitudes of $\bar{\alpha}$ and γ_{as}^2, and will show dispersion throughout the resonance region. Such behaviour has been observed for $[IrBr_6]^{2-}$ and $[IrCl_6]^{2-}$,[10] where the ground state is the E_g'' spin-orbit component of the $^2T_{2g}$ term.

If the ground state is orbitally non-degenerate but possesses spin degeneracy then strong spin-orbit coupling may cause mixing of the ground state with degenerate excited states of the same spin degeneracy. Thus, the ground state acquires some degenerate character and the depolarization ratios of totally symmetric modes may become anomalously large. The only reported observation of this is for $[FeBr_4]^-$ where spin-orbit coupling causes the $^6A_1$ ground state to be mixed with sextet excited states, and the $\rho(\pi/2)$ value for the ν_1 band at resonance is observed to be 0.15[11,12].

For non-totally symmetric modes $\bar{\alpha}$ is zero and the $\rho(\pi/2)$ value is dependent on the relative sizes of γ_s^2 and γ_{as}^2. If, for a given symmetry class, there are only symmetric or antisymmetric tensor components, then $\rho(\pi/2) = 3/4$ or ∞, respectively, and does not display dispersion. Examples of such vibrational modes are the b_{1g} and b_{2g} modes (symmetric tensor) and a_{2g} modes (antisymmetric tensor) of metalloporphyrins. By contrast, for cases in which both symmetric and antisymmetric tensor components transform as the same irreducible representation (e.g. B_{1g}, B_{2g} or B_{3g} in the D_{2h} point group), $\rho(\pi/2)$ lies between $3/4$ and ∞ and displays dispersion throughout the resonance region. The depolarization ratio $\rho(\pi/2)$ maximizes at the mid-point between the 0-0 and 0-1 resonances and decreases asymptotically towards

¾ on either side. Its maximum value is a function of Γ and γ.

The overtones of non-degenerate non-totally symmetric modes all transform as the totally symmetric representation and their $\rho(\pi/2)$ values will be the same as for totally symmetric fundamentals. Overtones of degenerate fundamentals transform as the symmetric part of the direct product (e.g. in the O_h point group $\{E_g \otimes E_g\} = A_{1g} + E_g$ and $\{T_{2g} \otimes T_{2g}\} = A_{1g} + E_g + T_{2g}$).

2.9 Time-Dependent Theory of Resonance Raman Scattering

An alternative treatment of RR data with certain advantages over the Franck-Condon treatment is that based on the transformation of the Kramers-Heisenberg-Dirac formula to the time domain, viz. the Heller treatment.[13,14] The calculation of Raman band intensities in the time domain is considered to be very efficient since it avoids the evaluation of both Franck-Condon factors and the sum over states.

The transition polarisability is, in this treatment, given by

$$[\alpha]_{fi} = \frac{i}{\hbar}\int \langle\phi_f|\phi(t)\rangle \exp\{(i\omega' - \Gamma)t\}dt \tag{26}$$

where $|\phi_f\rangle = \mu|\chi_f\rangle$ is the final vibrational state of the ground electronic surface multiplied by the transition dipole moment, $|\phi_i(t)\rangle = \exp(-iH_{ex}t/\hbar)|\phi_i\rangle$ is a moving wavepacket propagated by the excited state Hamiltonian, $|\phi_i\rangle = \mu|\chi_i\rangle$ is the initial vibrational state of the ground electronic surface multiplied by the transition dipole moment, $\hbar\omega' = \hbar\omega_i + \hbar\omega_I$, $\hbar\omega_i$ is zero-point energy of the ground electronic surface, $\hbar\omega_I$ is the energy of the incident radiation and Γ is the damping factor. The Raman scattering amplitude is governed by the motion of a wavepacket on a multidimensional hypersurface representing the electronic state potential.

The initial wavepacket, ϕ, makes a vertical transition onto the potential surface of the excited state which, in general, is displaced relative to that of the ground state. The displaced wavepacket is not a stationary state and evolves according to the time-dependent Schrödinger equation. The quantity of interest is the overlap of the moving wavepacket $\phi(t)$ with the final state of interest ϕ_f. If it is assumed that (a) only one electronic excited state is involved, (b) the potential surfaces are harmonic, (c) the normal coordinates are not mixed in the excited state, and (d) the force constants do not change in the excited state, then the overlap has the form

$$\langle\phi_f|\phi(t)\rangle = \prod_k \left\{\exp\left(-\frac{\Delta_k^2}{2}(1-\exp(-i\omega_k t)) - \frac{i\omega_k t}{2}\right) \times (1-\exp(-i\omega_k t))^{n_k} \times \frac{(-1)^{n_k}\Delta_k^{n_k}}{(2^{n_k} n_k!)^{\frac{1}{2}}}\right\}\exp(-i\omega_0 t) \tag{27}$$

where ω_0 is the wavenumber difference between the quantum mechanical zero-point energy on the ground-state surface and the classical zero-point energy on the excited-state surface, ω_k and Δ_k are respectively the wavenumber and the displacement of the kth normal mode, and n_k is the vibrational quantum number of the kth normal mode in the ground electronic state (n = 0, 1, 2 etc.). Equation (27) is used to calculate the intensities of bands attributed to fundamentals, and to all the various overtone and combination tones. For example, in order to calculate the cross section of the combination band $(\nu_1+\nu_2)$ in a three mode case, $n_1 = 1$, $n_2 = 1$ and $n_3 = 0$.

The transition polarizability in the frequency domain is the half Fourier transform of the overlap in the time domain. The Raman intensity $I_{f\leftarrow i}$ into a particular mode f is given by the expression $I_{f\leftarrow i} \propto \omega_I \omega_s^3 |\alpha_{fi}|^2$, where ω_s is the wavenumber of scattered radiation.

There are now a number of applications of this treatment to the calculation of distortions of molecules in excited states, notably to molecules such as $W(CO)_5$pyridine, $W(CO)_5$piperidine, $Mo_2(O_2CCF_3)_4$, and $K_3Cr(CN)_5(NO)$.[15] There are also a set of molecules $Rh_2(O_2CCH_3)_4L_2$, where L = PPh_3, $AsPh_3$, or $SbPh_3$, for which an excellent fit is obtained between the calculated spectrum and that observed; the latter contains many progressions involving both the ν(RhRh) as well as ν(RhO) totally symmetric modes.[17]

3. EXPERIMENTAL TECHNIQUES

The apparatus required for the excitation, dispersion and detection of RR scattering is the same as that used for normal Raman spectroscopy, although sample illumination procedures are more critical. In order that RR spectroscopy may be applied to a wide range of molecules it is necessary that laser excitation throughout the visible and near UV spectral regions be available. Most RR studies to date have involved excitation provided by CW argon-ion and krypton-ion lasers and CW dye lasers, the latter providing continuously tunable radiation from 395 to 800 nm with seven different dyes (stilbene 1, stilbene 3, coumarin 30, sodium fluorescein, rhodamine 6G, rhodamine 101, and LD 700). Unfortunately, ion lasers provide only a small number of lines in the near-UV (330-370 nm), so it has been necessary to use different laser sources in order to obtain tunable radiation down to 250 nm. Three systems are in current use:

(a) Frequency-doubled output of a mode-locked ion or dye laser. With a repetition rate of ca. 40 MHz the radiation is quasi-CW, but only modest average power can be obtained (ca. 10-20 mW);

(b) Frequency-narrowed excimer lasers and excimer-pumped dye lasers. These have a repetition rate of ca. 10 Hz and average power of the order of 1-10 W; they are, however, very expensive;

(c) Frequency-tripled and -quadrupled pulsed Nd:YAG lasers which may be used to pump a dye laser. Such systems have found considerable application in time-resolved resonance Raman (TR^3) studies.

For most RR studies the scattered light is detected by photomultiplier tubes using photon-counting methods, in many cases employing dedicated computers, which also control the spectrometer. Many experiments have also been performed using multichannel detection involving either a TV camera (intensifier-vidicon) or diode array. Multichannel detection reduces the time required to obtain a spectrum by several orders of magnitude, thus permitting RR studies of photochemically labile species. In combination with picosecond pulsed lasers, multichannel detection facilitates TR^3 experiments in which transient species and molecules in optically (or radiolytically) pumped excited states may be investigated. The most recently introduced detection systems are based on two-dimensional charge-coupled devices (CCDs),[17] originally developed for astronomy.

Sample illumination procedures present certain problems associated with strong absorption of the laser beam. These are:[3]

(a) The need for optimization of the concentration of the scattering species in order to achieve a compromise between absorption and scattering processes.

(b) The avoidance of overheating the sample at the laser focus. Various rotating cells for solids, liquids, and gases have been employed as well as devices which enable the focused laser beam to be swept over the sample in a linear or circular fashion. The use of flowing samples for TR^3 studies of transients has been described.

(c) The elimination of photolysis and luminescence. These can now be discriminated against by the TR^3 techniques mentioned above. Luminescence rejection can be achieved by using single-channel spectrometers in conjunction with pulsed lasers by time-adjusted gating of the photon-counting electronics.

Hot band contributions to intensities are minimised by measuring RR spectra at as low temperatures as possible. This procedure not only results in a sharpening of the spectral features, thereby improving resolution and signal-to-noise ratio, but it also minimizes thermolysis. Several cryostats are available for Raman studies, the most popular being those in which liquid nitrogen (77 K) or liquid helium (4 K) are used as coolants, and also closed-cycle helium gas cryostats (10-12 K). The latter are widely used for matrix-isolation studies.

The most interesting of the recent technical development involves the use of high power Nd:YAG lasers, non-linear optical crystals, and stimulated Raman scattering of hydrogen gas in RR studies in the vacuum ultraviolet region to wavelengths as short as 141 nm.[18] This has permitted the study of a wide range of compounds for the first time, notably benzene, butadiene, carbon disulphide and oxygen.

4. APPLICATIONS OF RESONANCE RAMAN SPECTROSCOPY

The examples taken are illustrative of the wide range of applications of resonance Raman spectroscopy. Most obvious of these is that of detection and identification of chromophores at very low ($< 10^{-3}$ M) concentrations, a feature which is not only of great importance in the study of aqueous solutions of biological species such as haemoglobin, carotenoproteins, etc., but also of explosives (e.g. trinitrotoluene, nitroglycerine, PETN, RDX, etc.)[19] in forensic applications. Similar applications arise with geological samples, for which the identification of the chromophore is frequently a matter of considerable interest. This matter will be discussed with particular reference to lapis lazuli.

4.1 Lapis Lazuli and the Ultramarines

The intense royal blue colour of the mineral lapis lazuli and its synthetic equivalent, ultramarine blue (idealized formula $Na_8Al_6Si_6O_{24}.S_n$) had long been prized, perhaps for over 5000 years. Not only was the semi-precious gemstone admired as such, but the pigmentary properties of the material were highly valued even in the middle ages. The origin of the colour was not, however, understood until a combination of resonance Raman and electron spin resonance studies were carried out which led to the identification of the chromophores as sulphur radical anions trapped in the cubic holes of the sodalite structure but present only in low (ca. 1% by mass) proportions.[20-24] Careful studies indicate that the key chromophore present in ultramarine blue is S_3^- (λ_{max} = 610 nm, ω_1 = 550.3 cm^{-1}), although some S_2^- (λ_{max} = 380 nm, ω_e = 590.4 cm^{-1}) is also present. The key observations are the progressions detected in the ν_1 band of each species at the appropriate resonance, observations closely in keeping with the known spectroscopic features of these radical anions when inserted as substitutional impurities into alkali halide matrices. The bond length changes attendant upon excitation to the resonant excited state for S_2^- and Se_2^- (which may also be inserted into the sodalite lattice) are 0.30 and 0.32 Å, respectively[24]; these changes reflect the decrease in bond order from 1.5 in the $^2\Pi_u$ ground state to 0.5 in the $^2\Pi_g$ excited state. Ultramarine green (used in eye shadow) contains the same two chromophores as the blue, but in comparable proportions, while ultramarine pink (used in talcum powder) contains these same two species together with a third one, as yet unidentified.[24]

4.2 Ammonia and Water

Recent investigations of gaseous NH_3 have involved the use of the fifth harmonic of a pulsed Nd:YAG laser (212.8 nm) to irradiate within the contour of the singlet electronic transition of lowest energy (ca. 190 nm) assigned to the transfer of an electron from the nitrogen lone pair to a 3s Rydberg-type orbital.[25] This excitation brings about a change from C_{3v} to D_{3h} geometry, as made evident by the development of a long overtone progression in the deformation mode ν_2 in the absorption spectrum. In agreement with this conclusion, the RR spectrum displays a progression $v_2\nu_2$ to v_2 = 4. It also displays a combination band progression $\nu_1 + v_2\nu_2$ to v_2 = 4 which demonstrates that the NH bond length also changes slightly upon excitation to this state.

RR studies have also recently been carried out on water using excitation wavelengths in the far ultraviolet and vacuum ultraviolet regions;[26] anti-Stokes Raman shifted lines of the second, third and fourth harmonics of a Nd:YAG laser provided the necessary excitation wavelengths. Irradiation into the contour of the lowest singlet excited state (a Rydberg, directly dissociative state) with 160 nm excitation yields progressions exclusively in ν_1 (the symmetric stretch) to a maximum of $v_1 = 6$, implying that the dissociation of the A state proceeds initially along the symmetric stretching coordinate alone.

4.3 Tetrahedral Species

Many tetrahedral tetraoxo species have now been studied by RR spectroscopy, the most notable being $[MnO_4]^-$, $[MnO_4]^{2-}$, $[MoS_4]^{2-}$ and $[WS_4]^{2-}$. In particular, the excitation profile of the ν_1, $2\nu_1$ and $3\nu_1$ bands of $[MnO_4]^-$ are highly structured. Franck-Condon analysis of these excitation profiles indicates that the MnO bond length changes by 0.046 Å on ${}^1T_2 \leftarrow {}^1A_1$ excitation, a change which is obviously an increase owing to the bonding → antibonding nature of the resonant transition.[27] Many tetrahedral species have now been studied in this way;[28] in all cases, great intensification of the $\nu_1(a_1)$ band, coupled with the development of long overtone progressions in ν_1, indicate that the principal geometric change attendant upon excitation into the lowest allowed transition is along this coordinate.

4.4 Metal-Metal Bonded Species

A large number of metal-metal bonded species have now been studied by RR spectroscopy, of particular interest being the quadruply bonded ions $[Mo_2Cl_8]^{4-}$, $[Mo_2Br_8]^{4-}$, and $[Re_2X_8]^{2-}$, X = F, Cl, Br or I. Irradiation in each case within the contour of the ${}^1A_{2u} \leftarrow {}^1A_{1g}$, $\delta^*(b_{1u}) \leftarrow \delta(b_{2g})$, transition yields a RR spectrum dominated by a progression in the metal-metal stretch, whereas irradiation within the contour of the next allowed electronic transition, the $b_{1u}(\delta^*) \leftarrow X, e_g(\pi)$ transition yields a RR spectrum dominated by progressions in the metal-halogen stretch. These observations demonstrate conclusively the distinctly different nature of each resonant transition and illustrate the value of RR spectroscopy as a means for making or confirming electronic band assignments. It is also important to note that the transition moment of the first transition would be A_{2u} and of the second E_u in D_{4h} symmetry. These requirements can be demonstrated conclusively at resonance since the depolarisation ratio of ν_1 should be ⅓ and of the latter ⅛ at resonance in each case (see Section 2.8); this is indeed what is found.[29]

4.5 Linear Chain Complexes

Extensive RR studies of linear-chain complexes have been carried out since ca. 1974. Of particular interest have been those on Wolffram's red, $[Pt^{II}L_4][Pt^{IV}L_4Cl_2][ClO_4]_4$, where L = $C_2H_5NH_2$. Such complexes are mixed valent and form as stacked linear chains of the general sort

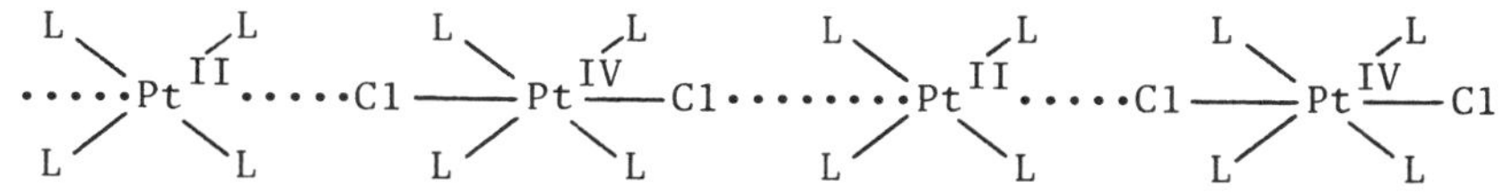

A great many variants on this complex are known, ones in which Cl is changed to Br or I, Pt to Pd, and with many different equatorial ligands L and counterions.[30,31]

The intense colours of such complexes are caused by the $Pt^{II} \rightarrow Pt^{IV}$ intervalence transition; these are axially (i.e. chain) polarized and occur in the regions 25,000-18,200 cm^{-1} (X = Cl), 23,600-14,300 (X = Br) and 20,600-7,500 (X = I). The shorter the $Pt^{II} \cdots Pt^{IV}$ distance the lower in wavenumber is the intervalence transition.

The Raman spectra of such halogen-bridged chain complexes demonstrate that the ν_1 mode (ν_1 = the symmetric Cl-Pt^{IV}-Cl stretch) is strongly vibronically coupled to the resonant (i.e. intervalence) transition. Thus both ν_1 and its overtones are tremendously enhanced at resonance, the number of detectable harmonics reaching as far as 17 in some cases. This implies a very substantial (0.05-0.10 Å) change in Pt^{IV}-Cl bond length on changing from the ground to the intervalence state.

There are many matters of great current interest in the RR spectra of these complexes; in particular studies are aimed at (a) establishing the nature of the dependence of ν_1 and of the absorption edge of the intervalence band on pressure, (b) understanding the origin of the luminescence emitted by these complexes, (c) understanding the origin of the fine structure to ν_1 and its overtone bands and (d) probing the basis for the apparent dispersion of ν_1 with ν_0, the wavenumber of the exciting line used to gather the RR spectra. In the last matter, a most interesting analogy has been drawn between Wolffram's red and trans-polyacetylene; in each case it is thought that the chains are broken into segments of differing correlation length, in the one case by trace (1 in 10^4) platinum(III), in the other by solitons in the chain. It has been argued that the presence of these entities in the chain may account for the apparent dispersion in ν_1 with change of ν_0 for Wolffram's red[32] and in ν(C=C) for trans-polyacetylene.[33]

RR studies have also played a major role in unravelling the nature of complexes based on stacking of $[Pt_2(H_2P_2O_5)_4]^{4-}$ units into chains with halide ions as bridging agents. Such studies are important for understanding the relationships between structure, spectroscopy, electrical (chain) conductivity, and bonding in linear-chain semiconductors.[34]

4.6 Electronic Raman Spectroscopy

The Raman effect is most usually associated with studies of spectroscopic transitions between rotational and vibrational levels of the ground electronic state of molecules and ions. It is, however, also possible to detect Raman transitions between the ground and low lying excited electronic states. The first experimental observation of the electronic Raman (ER) effect was by Rasetti in 1930, only two years after the discovery of the vibrational Raman effect. A band at 122 cm^{-1} was detected in the Raman spectrum of NO and correctly assigned to the $^2\Pi_{3/2} \leftarrow {}^2\Pi_{1/2}$ electronic transition.[36] However, over 30 years passed before another report appeared in the literature, of Pr^{3+} in a $PrCl_3$ crystal at 77 K. Subsequently Koningstein observed ER scattering from the Eu^{3+} ion doped into single crystals of yttrium gallium garnet $Y_3Ga_5O_{12}$. Since this time, the technique has grown in importance, and a major review has appeared.[37]

The importance of ER spectroscopy lies in the nature of the selection rules, which differ from those operative in absorption or emission electronic spectroscopy. Thus transitions between states of the same parity, though forbidden by electric dipole selection rules, may be Raman active. By ER spectroscopy it is thus possible to measure transition frequencies which might otherwise be inaccessible spectroscopically. Many ER studies have now been carried out on small molecules, transition metal and lanthanide ions in solids, semiconductors, transition-metal halide complexes, and metallocenes.[37] It is possible here merely to give one recent illustration.

An ER transition ($^3E_g \leftarrow {}^3A_g$) has been detected at 1940 cm^{-1} between the trigonally split components of the $^3T_{1g}(F)$ ground (in O_h symmetry) term of the hexa-aquavanadium(III) ion in $CsV(SO_4)_2.12H_2O$ and $[NH_4]V(SO_4)_2.12H_2O$.[34] This is an important observation, since it confirms the implications of the temperature variation of the magnetic moment of this salt, first reported in 1960.[39] This direct spectroscopic determination of the trigonal splitting component to the ligand field is valuable since the knowledge of its magnitude reduces the number of adjustable parameters required for fitting the magnetic data obtained on d^2 complexes. Similar ER spectra have been obtained for $KV(SO_4)_2$, the $^3E_g \leftarrow {}^3A_g$ transition being detected at 1560 cm^{-1} in this case.[40]

5. CONCLUSION

Raman spectroscopy is a very rapidly expanding subject, with many technological developments fuelling this expansion. Many further studies, particularly of very short-lived species and of species with transitions only in the vacuum ultraviolet can be anticipated. Moreover, the use of absorption and emission spectroscopy of photodissociating molecules as a means of probing details of dissociation dynamics in extremely short-lived transients is of great interest; recent studies on methyl iodide and ozone illustrate how emission spectra yield information relevant to the study of reaction dynamics.[41]

Studies of such sophistication should not be taken to indicate that everything worthwhile is known about diatomics. The recent series of papers[42-44] on the RR spectra of iodine in various solvents has provided much fundamental and interesting new information on this superb Raman scatterer.

6. REFERENCES

1. *'Proceedings of the Eleventh International Conference on Raman Spectroscopy'*, R.J.H. Clark and D.A. Long (eds.), Wiley, Chichester, 1988, pp. liv + 1034.
2. Long, D.A. (1977) Raman Spectroscopy, McGraw-Hill, London.

3. *'Advances in Infrared and Raman Spectroscopy'*, R.J.H. Clark and R.E. Hester (eds.), Wiley, Chichester, Vols. 1-17, 1975-1989.
4. Clark, R.J.H. and Dines, T.J. (1986) *Angew. Chem.*, 25, 131-158.
5. Manneback, C. (1951) *Physica*, 17, 1001.
6. Penner A.P. and Siebrand, W. (1976) *Chem. Phys. Lett.*, 39, 11.
7. Siebrand, W. and Zgierski, M.Z. (1982) *J. Phys. Chem.*, 86, 4718.
8. Clark, R.J.H. and Dines, T.J. (1981) *Chem. Phys. Lett.*, 79. 321.
9. Ziegler, L.D. and Hudson, B. (1983) *J. Chem. Phys.*, 79, 1197.
10. Hamaguchi, H. (1977) *J. Chem. Phys.*, 66, 5757; (1978) *ibid*, 69, 569.
11. Clark, R.J.H. and Turtle, P.C. (1976) *J. Chem. Soc. Faraday 2*, 72, 1885.
12. Clark, R.J.H. and Dines, T.J. (1982) *Chem. Phys.*, 70, 269.
13. Heller, E.J. (1981) *Acc. Chem. Res.*, 14, 368.
14. Tannor, D. and Heller, E.J. (1982) *J. Phys. Chem.*, 27, 2225.
15. Zink, J.I. (1985) *Coord. Chem. Rev.*, 64, 93.
16. Clark, R.J.H. and Hempleman, A.J. (1988) *Inorg. Chem.*, 27, 2225; Shin, K.-S., Clark, R.J.H. and Zink, J.I., submitted for publication.
17. Batchelder, D.N. (1988) *European Spectroscopy News*, 80, 28.
18. Hudson, B., Sension, R.J., Brudzynski, R.J. and Li, S., in Ref.1, p.51.
19. Clark, R.J.H. and Dines, T.J. (1986) *Analyst*, 111, 411.
20. Holzer, W., Murphy, W.F. and Bernstein, H.J. (1969) *J. Mol. Spectrosc.*, 32, 13.
21. Clark, R.J.H. and Franks, M.L. (1975) *Chem. Phys. Lett.*, 34, 69.
22. Clark, R.J.H. and Cobbold, D.G. (1978) *Inorg. Chem.*, 17, 3169.
23. Holzer, W., Murphy, W.F. and Bernstein, H.J. (1970) *J. Chem. Phys.*, 52, 399.
24. Clark, R.J.H., Dines, T.J. and Kurmoo, M. (1983) *Inorg. Chem.*, 22, 2766.
25. Ziegler, L.D. and Hudson, B. (1984) *J. Phys. Chem.*, 88, 1110.
26. Sension, R.J., Brudzynski, R.J. and Hudson, B., in Ref.1, p.567; (1988) *Phys. Rev. Lett.*, 61, 694.
27. Clark, R.J.H. and Stewart, B. (1981) *J. Amer. Chem. Soc.*, 103, 6593.
28. Clark, R.J.H., Dines, T.J. and Doherty, J.M. (1985) *Inorg. Chem.*, 24, 2088.
29. Clark, R.J.H. and Stead, M.J. (1983) *Inorg. Chem.*, 22, 1214.
30. Clark, R.J.H. (1984) *'Advances in Infrared and Raman Spectroscopy'*, R.J.H. Clark and R.E. Hester (eds.), Wiley-Heyden, Chichester, Vol.11, p.95.
31. Clark, R.J.H. (1984) *Chem. Soc. Rev.*, 13, 219.
32. Clark, R.J.H. in *'Vibrational Spectra and Structure'*, J.R. Durig (ed.) Elsevier, Amsterdam, Vol.18, in press.
33. Tiziani, R., Brivio, G.P. and Mulazzi, E. (1985) *Phys. Rev. B.*, 31, 1985.
34. Kurmoo, M. and Clark, R.J.H. (1985) *Inorg. Chem.*, 24, 4420.
35. Butler, L.G., Zietlow, M.H., Che, C.-M., Schaefer, W.P., Sridhar, S., Grunthaner, P.J., Swanson, B.I., Clark, R.J.H. and Gray, H.B. (1988) *J. Amer. Chem. Soc.*, 110, 1155.

36. Rasetti, F. (1930) *Z. Phys.*, 66, 646.
37. Clark, R.J.H. and Dines, T.J. (1982) in *'Advances in Infrared and Raman Spectroscopy'*, R.J.H. Clark and R.E. Hester (eds.), Wiley-Heyden, Chichester, Vol.9, p.282.
38. Best, S.P. and Clark, R.J.H. (1985) *Chem. Phys. Lett.*, 122, 401.
39. Figgis, B.N., Lewis, J. and Mabbs, F.E. (1960) *J. Chem. Soc.*, 2480.
40. Fehrmann, R., Krebs, B., Papatheodorou, G.N., Berg, R.W. and Bjerrum, N.J. (1986) *Inorg. Chem.*, 25, 1571.
41. Imre, D., Kinsey, J.L., Sinha, A. and Krenos, J. (1984) *J. Phys. Chem.*, 88, 3956.
42. Sension, R.J. and Strauss, H.L. (1986) *J. Chem. Phys.*, 85, 3791.
43. Sension, R.J., Kobayashi, T. and Strauss, H.L. (1987) *J. Chem. Phys.*, 87, 6221, 6233.
44. Sension, R.J. and Strauss, H.L. (1988) *J. Chem. Phys.*, 88, 2289.

THE LIGAND POLARISATION MODEL FOR d-d AND f-f INTENSITIES

Brian Stewart
Department of Chemistry
Paisley College of Technology
PAISLEY, Renfrewshire PA1 2BE
UNITED KINGDOM

ABSTRACT. The ligand polarisation (LP) model for Laporte-forbidden transition probabilities is introduced and developed in some detail for the f-f case. Expectations are compared with those of the crystal field model. The LP model is extended to cover vibronic transitions and the symmetry implications are investigated.

1. INTRODUCTION

1.1 Laporte - forbidden transitions

The set of states arising from a d^n or from an f^n electron configuration are labelled by definite values of the parity. Radiative transitions amongst them are therefore forbidden by the Laporte selection rule [1] since they involve no change of parity. The Laporte rule arises from the ungerade nature of the leading, electric dipole, term in the radiation field. The so-called d-d transitions are however allowed via the even multipole parts of the radiation field: quadrupole (2^2-pole) and hexadecapole (2^4-pole). f-f transitions have in addition a 2^6-pole moment. The pure quadrupole mechanism is expected to give transition probabilities which are about 10^{-6} of those in the electric dipole mechanism. Experimentally however, d-d and f-f transitions are not observed with such low intensities but typically with an attenuation in the range 10^{-2} - 10^{-6} relative to a fully electric dipole allowed transition.

In principle, the answer to this problem has been known for some time [2]: The Laporte rule is relaxed because the d or f orbitals do not have a definite parity when the transition metal ion finds itself at a site lacking a centre of inversion. Alternatively, the parity restriction is removed at the turning points of appropriate non-totally symmetric vibrations if the equilibrium nuclear configuration is centrosymmetric.

C. D. Flint (ed.), Vibronic Processes in Inorganic Chemistry, 327–345.
© 1989 by Kluwer Academic Publishers.

With the removal of the parity restriction there still remains the question of the ultimate source of the electric dipole intensity. There are three possibilities: metal-based transitions, ligand-based transitions and metal$\leftrightarrow$ligand charge-transfer transitions.

The traditional 'crystal field' (CF) model, for d-d [3-5] and f-f [6,7] transitions, is based on the neglect of metal-ligand overlap. The ground state charge distribution of the ligands provides the requisite odd-parity potential to mix the Laporte-forbidden transition of interest with electric dipole allowed transitions based on the metal ion. Within this level of approximation, and considering only the Coulombic coupling of non-overlapping metal and ligand charge distributions, there is however another contribution in which the ligands are allowed to become excited by the radiation field. This is the ligand polarisation (LP) model [8-11] which is the subject of this article. In the LP model the source of electric dipole transitions is located entirely on the ligands, in the CF model it is on the metal ion. The LP and CF contributions arise naturally together in the context of the so-called Independent Systems Approach to be described in section 1.2.

The charge-transfer contributions involve recognition of metal-ligand overlap and so fall outside the independent systems approach. There has been no reliable assessment of charge-transfer contributions to d-d intensities. In the case of f-f transitions, charge-transfer contributions are not expected to be important since f orbital-ligand overlap is so small. The available evidence suggests that this is indeed the case [12].

1.2. The Independent Systems Approach

In complexes of the lanthanides the interaction of the f electrons with the ligand environment is small. The metal-ligand interaction can be treated to a good approximation by first-order perturbation theory as a Coulombic interaction, V_{ML}, between two non-overlapping charge distributions. In the absence of metal-ligand interaction, the states of the composite system consist of the set of products $|M_k L_l\rangle$ of the eigenstates $|M_k\rangle$ of the isolated metal ion (Hamiltonian H_M) and $|L_l\rangle$ for the isolated ligand (Hamiltonian H_L). These products are the zero order states to which the perturbation V_{ML} is applied.

This approach is outlined on the next page where, beginning at the top, a typical arrangement of energy levels is shown for a metal complex.

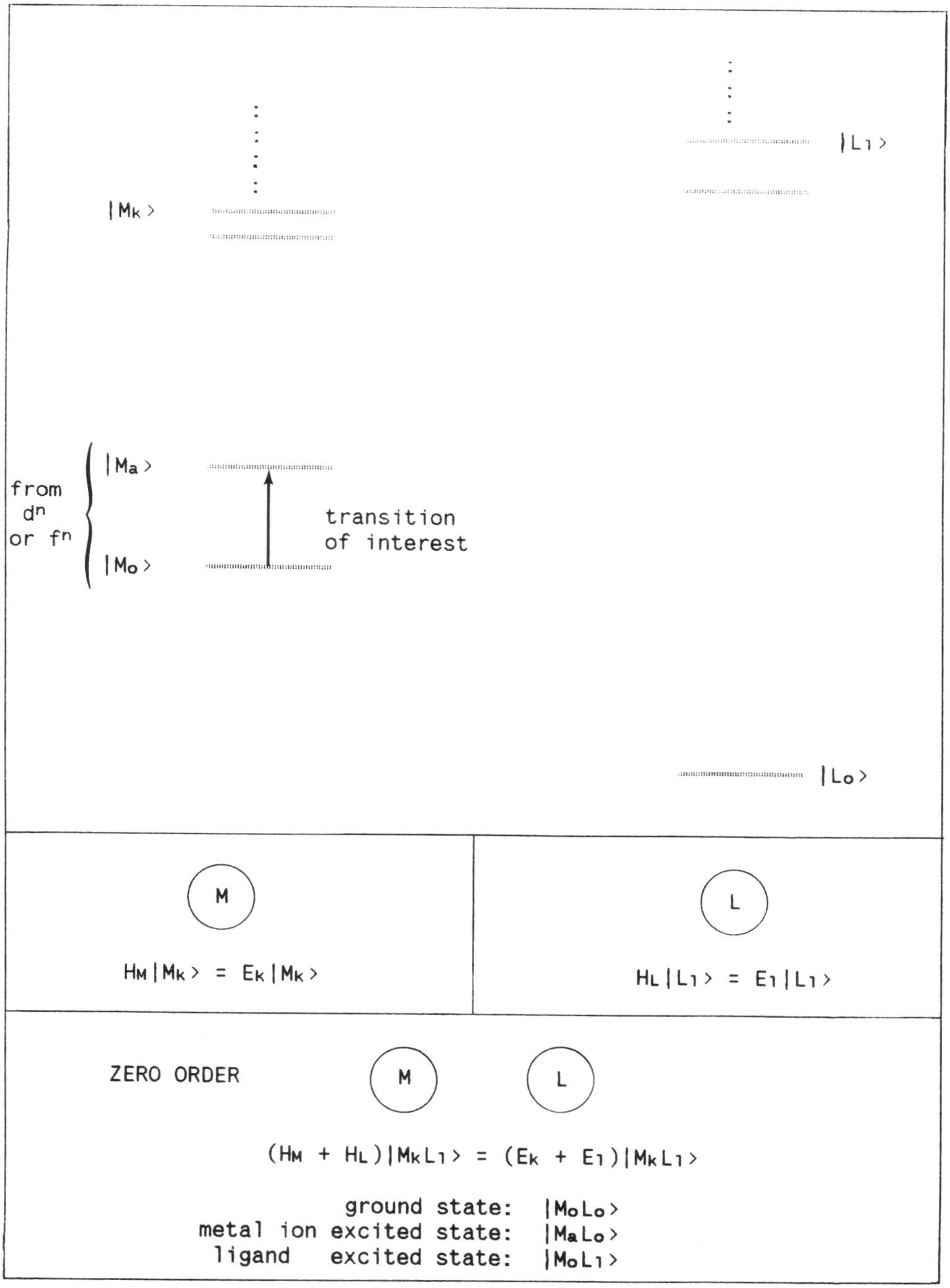
|Mk>
|L1>
|Ma>
from
dn
or fn
|Mo>
transition
of interest
|Lo>
M
L
HM|Mk> = Ek|Mk>
HL|L1> = E1|L1>
ZERO ORDER
M
L
(HM + HL)|MkL1> = (Ek + E1)|MkL1>
ground state: |MoLo>
metal ion excited state: |MaLo>
ligand excited state: |MoL1>

Now, 'switching on' the metal-ligand interaction potential:

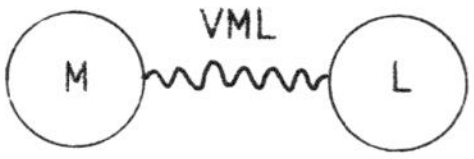

the total Hamiltonian becomes $H = (H_M + H_L) + V_{ML}$, where $(H_M + H_L)$ is the zero-order Hamiltonian and

$$V_{ML} = \sum_{i(M)} \sum_{j(L)} e^2 / r_{ij} \tag{1}$$

for a single ligand. A sum over L is introduced in the multi-ligand case.

The ground and excited states of the transition of interest are, corrected to the first order in V_{ML}:

$$\langle o| = \langle M_o L_o| + \sum_{k,l} (-E_k - E_l)^{-1} \langle M_o L_o | V_{ML} | M_k L_l \rangle \langle M_k L_l | \tag{2}$$

$$|a\rangle = |M_a L_o\rangle + \sum_{k,l} (E_a - E_k - E_l)^{-1} | M_k L_l \rangle \langle M_k L_l | V_{ML} | M_a L_o \rangle \tag{3}$$

(The zero of energy is taken as that of $|M_o L_o\rangle$ with both metal and ligand in their zero-order ground states.)

The potential of equation (1) is a sum of two-electron operators and therefore the matrix elements in equations (2) may connect states differing by up to two one-electron wavefunctions. However, the electric dipole moment operator (or any other transition moment operator) is a one-electron operator:

$$\mu_{TOTAL} = \mu_M + \mu_L = \sum_{i(M)} e r_i + \sum_{j(L)} e r_j \tag{4}$$

This leads to restrictions when evaluating $\langle o|\mu|a\rangle$ using equations (2) and (3).

Four terms arise here: The zero-order term in $\langle o|\mu|a\rangle$ is $\langle M_o L_o |\mu| M_a L_o\rangle = \langle M_o |\mu_M| M_a\rangle$, which vanishes for a Laporte-forbidden transition. The second order term, which connects the first-order corrections in (2) and (3), is neglected. This leaves the two first-order cross terms.

Consider the transition dipole connecting $\langle M_o L_o|$ with $|M_k L_l\rangle$:

$$\langle M_o L_o | \mu_M + \mu_L | M_k L_l \rangle = \langle M_o |\mu_M| M_k\rangle \langle L_o | L_l \rangle + \langle L_o |\mu_L| L_l \rangle \langle M_o | M_k \rangle$$

$$= \langle M_o |\mu_M| M_k\rangle \, \delta_{lo} + \langle L_o |\mu_L| L_l \rangle \, \delta_{ko} \tag{5}$$

where the delta functions arise from the orthonormality of the basis states. The result is that in equation (4) either the μ_M term contributes or the μ_L term, but not both. Similarly, for the other first-order cross term in $\langle o|\mu|a\rangle$:

$$\langle M_k L_l|\mu|M_a L_o\rangle = \langle M_k|\mu_M|M_a\rangle\langle L_l|L_o\rangle + \langle L_l|\mu_L|L_o\rangle\langle M_k|M_a\rangle$$

$$= \langle M_k|\mu_M|M_a\rangle\,\delta_{lo} + \langle L_l|\mu_L|L_o\rangle\,\delta_{ka} \qquad (6)$$

The choice l=o in (5) and (6) corresponds to the CF model in which the ligands remain unexcited and the μ_M terms contribute the electric dipole intensity.

$$\langle o|\mu|a\rangle_{CF} = \sum_{k\neq a} (E_a-E_k)^{-1}\langle M_o|\mu|M_k\rangle\langle M_k L_o|V_{ML}|M_a L_o\rangle$$

$$+ \sum_{k\neq o} (-E_k)^{-1}\ \langle M_o L_o|V_{ML}|M_k L_o\rangle\langle M_k|\mu|M_a\rangle \qquad (7)$$

The choice k=o and k=a in (5) and (6) respectively is that of the LP model in which a set of ligand based transition moments constitute the source of electric dipole intensity.

$$\langle o|\mu|a\rangle_{LP} = \sum_{l\neq o} (E_a-E_l)^{-1}\langle L_o|\mu|L_l\rangle\langle M_o L_l|V_{ML}|M_a L_o\rangle$$

$$+ \sum_{l\neq o} (-E_a-E_l)^{-1}\langle M_o L_o|V_{ML}|M_a L_l\rangle\langle L_l|\mu|L_o\rangle \qquad (8)$$

Equation (7) forms the basis of the CF model of intensities. For example, in the approach of Ballhausen and Liehr [3,5] for d-d transitions the configuration $3d^n$ is mixed with $3d^{n-1}4p$ by the potential arising from the ground state charge distribution of the ligands. The matrix elements of V_{ML} may be formally rearranged to display this: eg $\langle M_k|\langle L_o|V_{ML}|L_o\rangle|M_a\rangle$. The analogous approach for f-f intensities, due to Judd [6] and Ofelt [7], involves the mixing of $4f^n$ with excited configurations of opposite parity such as $4f^{n-1}5d$.

The CF model will not be developed in detail here but the expectations for intensities will be compared with those of the LP model in section 3. What has been said so far applies equally to the treatment of d-d or f-f transition intensities. From now on the LP model will be developed specifically for the f-f transitions.

1.3. f-f transitions

There are several advantages in treating the f-f transitions:

(a) f electron states are associated to a good approximation with well-defined values of total angular momentum. This allows maximum use of the parent spherical symmetry in treating the effects of ligand-f electron interactions. The methods are introduced in section 1.4.

(b) Quantitative intensity data are readily available.

(c) There is a striking qualitative dependence of intensities on symmetry and ligand. This is the phenomenon of hypersensitivity [13,14] which manifests itself as a unique sensitivity to environment in transitions which are formally quadrupole allowed in the free ion.

1.4. Spherical Irreducible Tensor Operators.

In evaluating matrix elements of dipole or other multipole operators and perturbation operators between f electron states it is a considerable advantage to be able to treat the operators and the states on the same footing. The f-electron state basis consists of eigenfunctions of the total angular momentum and these form bases for the irreducible representations of the group O_3^+ of an isolated atom or ion. For these reasons spherical irreducible tensor operators are particularly useful [15,16], and they are introduced here.

In quantum mechanics it is always possible to express operators in a way which makes explicit use of the symmetry of the quantities they represent. A tensor is a mathematical object which is constructed so as to have a well-defined symmetry with respect to transformations of the basic coordinates of which it is constituted. An important property of tensors is that they combine together to form new tensors and the rules for doing this are contained in the tensor algebra. The spherical irreducible tensors to be used here are those of Racah. They are defined in terms of the standard spherical harmonic functions by

$$C^{(k)}_q = (4\pi/2k+1)\ Y^{(k)}_q \qquad q = k,k-1,\ldots.-k \tag{9}$$

The compact notation $C^{(k)}$ will often be used, implying the set of 2k+1 operators defined in (9). k is the rank of the tensor.

$C^{(0)}_0 = 1$ and forms the basis of all scalar quantities or numbers. For example, the metal-ligand potential energy is a scalar which can formally be written $V^{(0)}$ whenever it is necessary to draw attention to its scalar nature.

$C^{(1)}_1 = -(2)^{-½}\sin\theta e^{i\phi}$, $C^{(1)}_0 = \cos\theta$, $C^{(1)}_{-1} = -(2)^{-½}\sin\theta e^{-i\phi}$ and forms the basis for all vectors. The radius vector is given by $r(1) = |r|C^{(1)}$ and therefore the dipole moment vector operator is $\boldsymbol{\mu}(1) = -e|r|C^{(1)}$ (a sum over electrons being implied in a many-electron system) Generally there is a set of multipole moment operators for an electronic charge distribution with an origin at M:

$$D^{(\lambda)} = -e\sum_i (r_{iM})^{\lambda} C^{(\lambda)}(\theta,\phi)_{iM}$$

The sum is taken over the electrons i associated with centre M.

The fundamental relation of the tensor algebra is the coupling of two irreducible tensor operators to give a resultant:

$$C^{(c)} = [A^{(a)}B^{(b)}]^{(c)} \tag{10}$$

The 2c+1 components of the irreducible tensor $C^{(c)}$ are given by

$$C^{(c)}_{\gamma} = [A^{(a)}B^{(b)}]^{(c)}_{\gamma} = (-1)^{c-\gamma}\sum_{\alpha,\beta}(2c+1)\begin{pmatrix} a & b & c \\ \alpha & \beta & -\gamma \end{pmatrix} A^{(a)}_{\alpha} B^{(b)}_{\beta} \quad (11)$$

where $\begin{pmatrix} a & b & c \\ \alpha & \beta & -\gamma \end{pmatrix}$ is a Wigner 3-j symbol [16]. The process is entirely analogous to the coupling of two states with angular momentum quantum numbers a and b to give a resultant c. Restrictions on the allowed values of the tensor ranks a, b and c and symmetry properties of the coupling are embodied in the 3-j symbol which vanishes unless the triangle rule, $a+b \geq c \geq |a-b|$, is obeyed and again unless $\alpha+\beta+\gamma = 0$.

An important point in the analysis which follows concerns the parity of the tensor operators. The parity of a Racah tensor (or any so-called true tensor) of rank k is $(-1)^k$. However, the parity of a coupled tensor such as $C^{(c)}$ in (10) is given by the product of the parities of the tensors being coupled, ie $(-1)^{a+b}$ in this example.

Consider the coupling of two vectors, $A^{(1)}$ and $B^{(1)}$. The triangle rule gives 0, 1 and 2 for the possible ranks of the resultant. Rank zero corresponds to the scalar product of the two vectors, rank one is the vector product and rank two is a tensor product. In this case the vector product has even parity and as such it is not a true vector but is referred to as a pseudovector. The angular momentum vector $l^{(1)} = [r^{(1)}p^{(1)}]^{(1)}$ is an example. If (a+b+c) is even(odd) in (11), then the coupling is symmetric(antisymmetric) with respect to interchange of $A^{(a)}$ and $B^{(b)}$. The vector product just referred to is antisymmetric, as is the rank one part of the polarisability tensor (see section 2.2).

Another important tool in dealing with matrix elements of irreducible tensor operators is the Wigner-Eckart theorem [17]. The useful result here is a factorisation of the matrix element:

$$\langle \alpha JM | A^{(k)}_{q} | \alpha' J' M' \rangle = (-1)^{J-M}\begin{pmatrix} J & k & J' \\ -M & q & M' \end{pmatrix}\langle \alpha J \| A^{(k)} \| \alpha' J' \rangle \quad (12)$$

in which all geometric information refering to the components of operators and states is contained in the 3-j symbol. The reduced matrix elements on the right hand side of (12) are a set of numbers characterising the effect of the operator in the chosen basis.

2. LIGAND POLARISATION MODEL FOR f-f TRANSITIONS.

2.1. Treatment of the Metal-Ligand Interaction

In order to evaluate equation (8) it is necessary to deal with the matrix elements: $\langle M_o L_1 | V_{ML} | M_a L_o \rangle = \langle M_o L_o | V_{ML} | M_a L_1 \rangle$. In the product basis it is a considerable advantage to make a separation of V_{ML} using a bipolar expansion in products of multipole moments of the two non-overlapping charge distributions. (See eg Judd [18]). It is sufficient to restrict this expansion to the lowest order non-vanishing multipole in each subsystem. In the ligand system this is a dipole. (The ligand monopole or charge term is dealt with in the CF model, equation (7)). For the metal system, the allowed multipole orders are determined by the restriction to transitions within the f^n configuration. At the one-electron level, $\langle f|C^{(\lambda)}|f\rangle$ is non-vanishing for $\lambda = 2,4$ or 6, corresponding to quadrupole, hexadecapole and 2^6-pole transition moments respectively.

The restricted bipolar expansion of V_{ML} is given by

$$V^{(0)}{}_{ML} = -\sum_{\lambda}[(2\lambda+3)(2\lambda+1)(\lambda+1)]^{1/2}\ [D^{(1)}D^{(\lambda)}G^{(\lambda+1)}]^{(0)} \tag{13}$$

$$D^{(1)}{}_L = -e\sum_j r_{jL}C^{(1)}(\theta,\phi)_{jL} \tag{14}$$

$$D^{(\lambda)}{}_M = -e\sum_i (r_{iM})^{\lambda}C^{(\lambda)}(\theta,\phi)_{iM} \tag{15}$$

$$G^{(\lambda+1)}{}_{LM} = (R_{LM})^{-\lambda-2}C^{(\lambda+1)}(\Theta,\Phi)_{LM} \tag{16}$$

The key part of (13) is the geometric tensor $G^{(\lambda+1)}{}_{LM}$ which depends on the disposition of the ligands relative to the metal centre. The coordinate system used in this expansion is shown in Figure 1.

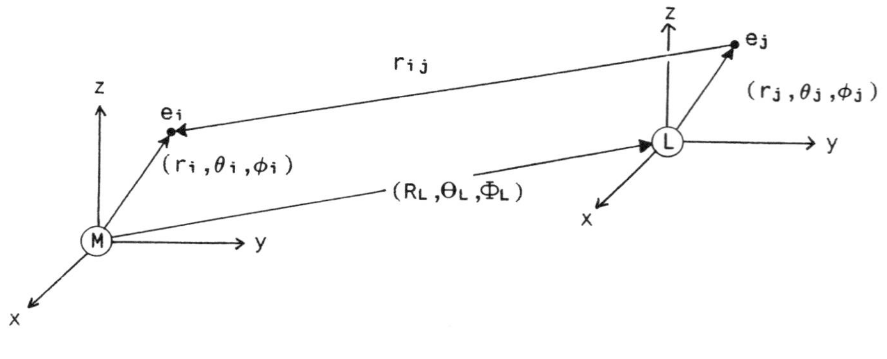

Figure 1. Two-centre coordinate frame for the expansion of the metal-ligand interaction potential.

The expression (13) is identical to that governing the electrostatic potential between permanent moments. In the present context it has an electrodynamic interpretation since it describes the correlation of transition moments in the metal and ligand subsystems. This correlation process may be described diagrammatically so as to allow a qualitative assessment of the resultant electric dipole intensities.

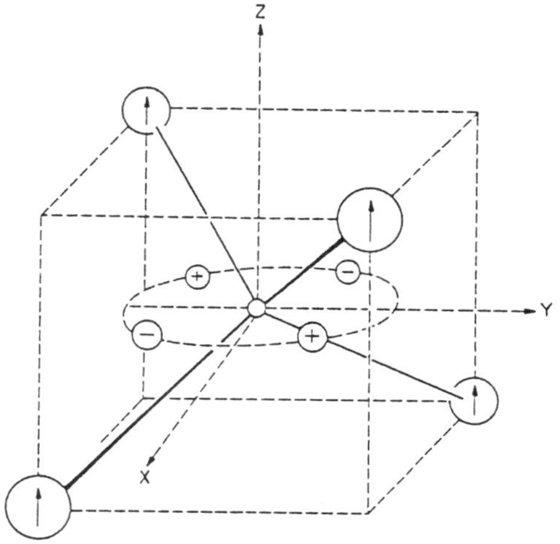

Figure 2. The Coulombic correlation between the z component of the electric dipole transition moment in each ligand and the xy component of a d-d quadrupole transition moment in a tetrahedral metal complex.[19,20]

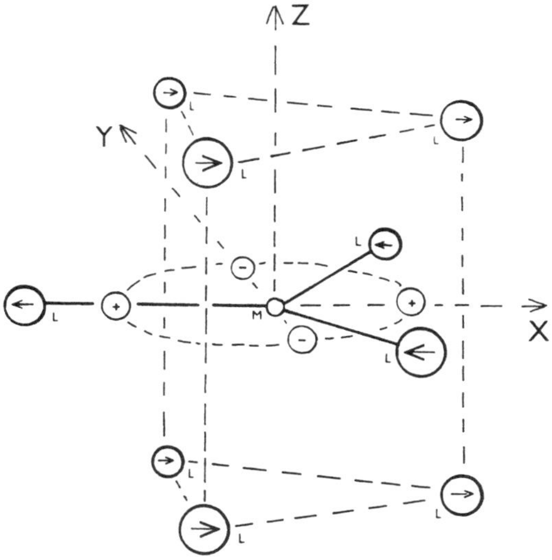

Figure 3. The electric dipole moments correlated in the ligands of a 9-coordinate trigonal Ln(III) complex by the (x^2-y^2) component of the electric quadrupole f-f transition moment of the metal ion, showing the negative interference between the resultant dipole of the equatorial $[ML_3]$ ligand set and that of the $[ML_6]$ trigonal prismatic ligand set.

Since the correlation results from a first-order perturbation correction it must lead to a lowering of the energy of the composite metal-ligand system. The diagrams in Figures 2 and 3 therefore show the instantaneous phases of transition moments in their lowest energy arrangement. The head of a ligand dipole arrow is attracted towards negative lobes (and repelled by positive lobes) of the transient multipolar charge distribution of the metal. It is difficult to resist imposing a time sequence of events when thinking about this process, although this is not appropriate in the context of time-independent perturbation theory. Perhaps the best description is that, in the presence of the radiation field (at the frequency of a metal ion transition), a resonant transition moment is present on the metal ion, and non-resonant dipoles are present in the ligands (induced via their polarisabilities); LP theory concerns the key aspect of the correlation between these moments leading to a net electric dipole transition moment for the complex. It is this explicit treatment of correlation which was missing from the earlier inhomogeneous dielectric model of f-f hypersensitivity [21].

The qualitative predictions of Figure 3 are in agreement with the observed dipole strengths of the $^7F_0 \rightarrow {}^5D_2$ transition in a number of trigonal 6- and 9-coordinate Eu(III) complexes [22,23]. Some data are presented in Table 1 (I, II and III) where the environmentally insensitive magnetic dipole transition $^7F_0 \rightarrow {}^5D_1$ can be used as a basis of comparison. There is a marked increase in dipole strength for the 6-coordinate site in the borate crystal (III) in comparison with the 9-coordinate cases (I and II), from which the destructive interference effect of the equatorial ligand set is apparent. (The D_4 case (IV) will be discussed in section 3.2.)

TABLE 1. Wavenumber $\bar{\nu}$/cm^{-1} and dipole strength D/10^{-66} C^{2}m^2 of the zero-order magnetic dipole and electric quadrupole transitions of Eu(III) observed in (I) $[Eu(H_2O)_9](EtSO_4)_3$, (II) $Na_3[Eu\{O(CH_2COO)_2\}_3].NaClO_4.6H_2O$, (III) $GdEu_xAl_{3-x}(BO_3)_4$ and (IV) $[Eu(bipyO_2)_4][ClO_4]$.

		I ($C3h$)		II (D_3)		III (D_3)		IV (D_4)	
Transition		$\bar{\nu}$	D	$\bar{\nu}$	D	$\bar{\nu}$	D	$\bar{\nu}$	D
$^7F_0 \rightarrow {}^5D_1$	$\Delta M \pm 1$	19020	30	18981	31	18966	38	18900	23
	ΔM 0	19024	26	18965	24	18995	23		
$^7F_0 \rightarrow {}^5D_2$	$\Delta M \pm 2$	21499	10	21433	14	21419	230	–	–
	$\Delta M \pm 1$	–	–	21484	62	21491	220	21470	680

2.2. The effective dipole transition moment operator within f^n.

By employing the methods of tensor recoupling algebra, full advantage may be taken of the separation of metal and ligand coordinates introduced in (13). The result is that all ligand dependent quantities can be gathered together into an environmental tensor $F^{(t)}$ [24], leaving only the metal multipole operator $D^{(\lambda)}$ to act within the states of f^n. In this process the electric dipole polarisability tensor, $\boldsymbol{\alpha}^{(k)}$, of the ligand emerges from the coupling of the ligand dipole operators and the perturbation sum over ligand excited states:

$$\boldsymbol{\alpha}^{(k)} = -\sum_{l}\{(-1)^k[\langle L_0|D^{(1)}|L_l\rangle\langle L_l|D^{(1)}|L_0\rangle]^{(k)}(E_a-E_l)^{-1}$$

$$-[\langle L_0|D^{(1)}|L_l\rangle\langle L_l|D^{(1)}|L_0\rangle]^{(k)}(E_a+E_l)^{-1}\} \qquad (17)$$

The electric dipole polarisability is perhaps more familiar as a second rank cartesian tensor α_{ij}. In spherical symmetry the cartesian form is reducible, giving rise to three irreducible tensors expressed by (17):

The isotropic polarisability with k = 0,
the antisymmetric anisotropy with k = 1 and
the symmetric anisotropy with k = 2.

In many cases $\boldsymbol{\alpha}^{(1)}$ may be neglected but it makes significant contributions in resonance Raman scattering [25] involving degenerate vibrational modes and degenerate electronic states as well as systems with ground state spin degeneracy.

The emergence of the polarisability is a very convenient aspect of the LP model in that the problematic sum over ligand excited states has been incorporated into an experimentally accessible quantity.

As a result of the recoupling transformations an effective electric dipole transition moment operator can be defined:

$$\boldsymbol{\mu}^{(1)}{}_{eff} = \sum_{L}\sum_{t,\lambda} c(\lambda)[F^{(t)}D^{(\lambda)}]^{(1)} \qquad (18)$$

where $c(\lambda) = [(2\lambda+3)(2\lambda+1)(\lambda+1)]^{1/2}$ and

$$F^{(t)} = (-1)^k\,[(2k+1)(2t+1)/3]^{1/2} \begin{Bmatrix} \lambda+1 & k & t \\ 1 & \lambda & 1 \end{Bmatrix} [\boldsymbol{\alpha}^{(k)}G^{(\lambda+1)}]^{(t)} \qquad (19)$$

$$= c(\lambda,t,k)\,[\boldsymbol{\alpha}^{(k)}G^{(\lambda+1)}]^{(t)} \text{ for convenience} \qquad (20)$$

For the isotropic polarisability contribution,

$$F^{(t)}(k=0) = (1/3)\boldsymbol{\alpha}^{(0)}G^{(\lambda+1)} = -(3)^{-1/2}\,\bar{\boldsymbol{\alpha}}\,G^{(\lambda+1)} \qquad (21)$$

with $\bar{\boldsymbol{\alpha}} = (1/3)(\alpha_{xx} + \alpha_{yy} + \alpha_{zz})$ being the mean polarisabilty of a ligand.

So for an electronic transition, $i \rightarrow f$, between initial and final states, both from f^n,

$$\langle i|\mu^{(1)}_{eff}|f\rangle = \sum_{L}\sum_{t,\lambda} c(\lambda)[F^{(t)}\langle i|D^{(\lambda)}|f\rangle]^{(1)} \tag{22}$$

The form of the effective dipole operator is common to all mechanisms of the forced electric dipole type, including the CF model. The nature of $F^{(t)}$ and the allowed values of t are mechanism dependent. The consequences of this are discussed in section 3.

3. COMPARISON OF THE CF AND LP INTENSITY MODELS.

3.1. Hypersensitivity.

The connection between the multipolarity, λ, of a transition and the allowed values of the rank t of the environmental tensor $F^{(t)}$ is shown below:

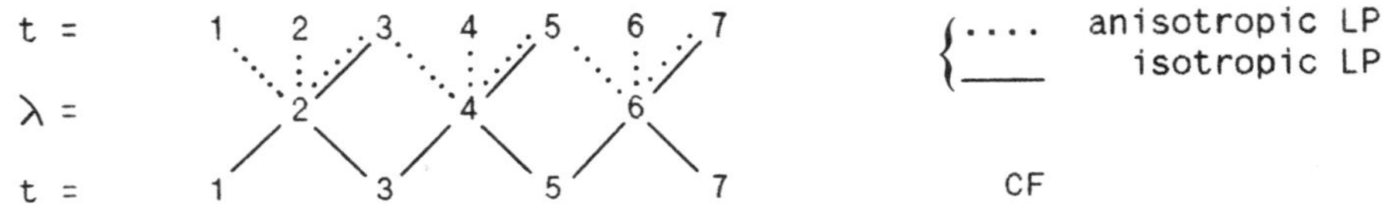

This comes from the triangle rule restrictions on the coupling of the effective dipole operator in (17): $\Delta(t,\lambda,1) \rightarrow t = \lambda-1, \lambda, \lambda+1$. Further restrictions arise from a consideration of the parities of the coupled tensors in equation (17). The dipole operator has odd parity and therefore the coupled form of it on the right side also has odd parity. The multipole operator is of even parity (λ is even) and thus $F^{(t)}$ must have odd parity.

	CF	isotropic LP	general LP
$F^{(t)}_p$	$\sim \Xi(t,\lambda)A^t_p$	$\sim \bar{\alpha} G^{(\lambda+1)}$	$\sim [\alpha^{(k)} G^{(\lambda+1)}]^{(t)}$
	$t = \lambda \pm 1$	$t = \lambda + 1$	$t = \lambda, \lambda \pm 1$

In the CF model, $F^{(t)} \sim \Xi(t,\lambda)A^t_p$ where A^t_p are the crystal field parameters with parity $(-1)^t$. Therefore t must be odd and given by $t = \lambda \pm 1$. In the isotropic LP model $t = \lambda + 1$ arises from the geometric tensor alone and this has the requisite odd parity $(-1)^{\lambda+1}$.
The key point here is that a rank 3 parameter is shared between $\lambda = 2$ and $\lambda = 4$ transitions in the CF model whereas it uniquely affects $\lambda = 2$ transitions in the isotropic LP model. The CF model can only account for the unique sensitivity of the 'quadrupolar' transitions through the polar (rank one) CF parameters [26] since the rank 3 parameters are shared

between $\lambda = 2$ and $\lambda = 4$. However, hypersensitivity is not restricted to sites possessing polar CF parameters (C_1, C_s, C_n, C_{nv}) [14]. Notable cases from the dihedral symmetries are referred to in Table 1 in section 2.1. (The polar tensors may of course contribute to a vibronic mechanism.)

3.2. Anisotropic ligand polarisabilities.

In the general LP model, the coupling of the even parity polarisability leads to the full range of t values in the environmental tensor $F^{(t)}$; $t = \lambda, \lambda \pm 1$ where notably $t = \lambda$ is specifically associated with polarisability anisotropy. Actually, these $t = \lambda$ terms vanish if the anisotropy is radially aligned (with the principal axis of the polarisability ellipsoid along a metal-ligand bond) and so it is more accurate to say that they arise from non-cylindrically symmetric metal-ligand interactions. These interactions (not present in the CF model) are necessary in order to generate the number of intensity parameters required group theoretically for a given multipole transition [24,27].

Some specific consequences of the inclusion of polarisability anisotropy are observable in dihedral symmetries. To see these, the matrix elements of $\boldsymbol{\mu}^{(1)}{}_{eff}$ in equation (22) are written in the $4f^n\alpha JM$ basis:

$$\langle \alpha JM | \boldsymbol{\mu}^{(1)}{}_{eff} | \alpha' J' M' \rangle = \sum_{L} \sum_{t,\lambda} c(\lambda) [F^{(t)} \langle \alpha JM | D^{(\lambda)} | \alpha' J' M' \rangle]^{(1)} \qquad (23)$$

Selection rules for these matrix elements can be derived from the product of two 3-j symbols:

$$\begin{pmatrix} t & \lambda & 1 \\ p & (M-M') & -q \end{pmatrix} \begin{pmatrix} J & \lambda & J' \\ -M & (M-M') & M' \end{pmatrix}$$

The first comes from the irreducible product $[F^{(t)}D^{(\lambda)}]^{(1)}$ and the second from the application of the Wigner-Eckart theorem (see section 1.4) to the matrix elements of $D^{(\lambda)}$. Those sums over ligands of the environmental tensor which do not vanish are the ones in which some component of $F^{(t)}$ transforms under the totally-symmetric irreducible representation in the point group of a given metal complex. For the series of europium(III) complexes in Table 1. the relevant $F^{(t)}{}_p$ for the transition $^7F_0 \leftrightarrow {}^5D_2$ ($\lambda = 2$) are given below:

C_{3h}	D_3	D_4	
	$F^{(2)}{}_0$	$F^{(2)}{}_0$	anisotropic LP only
$F^{(3)}{}_{+3}$	$F^{(3)}{}_3 + F^{(3)}{}_{-3}$		CF and LP

The component sum rule, $p + (M-M') - q = 0$, leads to $\Delta M = \pm 2$ only for C_{3h} symmetry since the dipole polarisation q is restricted to $0, \pm 1$. In D_3, $\Delta M = \pm 1$ becomes allowed solely through the presence of $F^{(2)}{}_0$ which is specifically a result of the non-cylindrically symmetric interactions of anisotropic ligand polarisation. This case is not a clear-cut one since in principle $\Delta M = \pm 1$ and ± 2 transitions may be mixed via the D_3 crystal field, as they both possess E symmetry. The D_4 case (IV in Table 1.) is a crucial one in that it has a unique contribution from $F^{(2)}{}_0$. Only the $\Delta M = \pm 1$ transition is observed, in agreement with the component selection rule and the dipole strength, which is one of the largest observed for Eu(III), is well accounted for by the anisotropic LP model [28].

4. VIBRONIC LIGAND POLARISATION MODEL

4.1. Introduction

Most vibronic intensity theories of Laporte-forbidden electronic transitions [3] treat only the centrosymmetric symmetries (notably O_h) where the activity of non-totally symmetric vibrational modes destroying the equilibrium geometry is the sole source of electric dipole transition moment. However, many non-centrosymmetric systems exhibit extensive vibronic activity and generally in Laporte-forbidden transitions totally symmetric and non-totally symmetric modes are of comparable importance [29]. A general vibronic treatment should consider all point symmetry groups and all available vibrational modes.

As in the previous treatments, vibronic terms arise from a Taylor expansion of the metal-ligand interaction potential V_{ML} in the vibrational displacements of the ligands from the equilibrium geometry of the complex. The alternative approach will be adopted here in which the Taylor expansion is applied to the effective electric dipole transition moment operator. The expansion in the vibrational space is linked to the multipolarity of the transitions in the electronic space so that selection rules may be sought which determine specific mode activity for a given multipole transition. This does not turn out to be a very fruitful approach for the simpler complexes because the limited number of available odd-parity modes are all active. There are connections between vibrational mode activity and the detailed intensity mechanism, for example the deviation from the isotropic polarisability approximation.

4.2. Symmetry aspects of vibronic activity.

As was seen in section 2.2, the effective dipole operator has the form:

$$\boldsymbol{\mu}^{(1)}{}_{eff} = \sum_t c(\lambda)[F^{(t)}D^{(\lambda)}]^{(1)} \qquad (24)$$

for a transition within the configuration f^n allowed via an even multipole of order λ . In converting (24) into an effective vibronic operator the Condon approximation is assumed, in which $D^{(\lambda)}$ is independent of vibrational coordinates. Thus all the dependence on ligand positions is contained in $F^{(t)}$ and it is to this tensor that the Taylor expansion in vibrational coordinates is applied. In order to indicate the general approach in a compact way, a Taylor expansion will be represented by the application of a scalar operator T. (A Taylor expansion does not alter the rank of a tensor.) Thus,

$$T\mu^{(1)}_{eff} = \sum_t c(\lambda)[(TF^{(t)})D^{(\lambda)}]^{(1)} \qquad (25)$$

Now when this vibronic operator is evaluated within an adiabatic Born-Oppenheimer basis the result is a separation between vibrational and electronic spaces:

$$\langle M_0 X_0 | T\mu^{(1)}_{eff} | M_a X_v\rangle = \sum_t c(\lambda)[\langle X_0 | TF^{(t)} | X_v\rangle\langle M_0 | D^{(\lambda)} | M_a\rangle]^{(1)} \qquad (26)$$

X_v are the vibrational eigenfunctions of the metal complex with vibrational quantum number v.

$TF^{(t)}$ is an effective operator <u>in the vibrational space</u> and the symmetries of active modes derive from it. At this stage, without going into any details of how to evaluate (25) (see section 4.3), the symmetry criterion for mode activity can be stated:

> Vibronically active modes will belong to those irreducible representations of the point group of the complex contained in the intersection, $\{\Gamma_{3N-6}\} \cap \{\Gamma_{t-}\}$, between the set of 3N-6 internal degrees of freedom and the set arising from the odd parity spherical tensors of rank t.

So vibronic activity is ultimately determined by the allowed values of t, which are dependent on the type of transition and on the intensity mechanism. Two points must be remembered here. First, <u>all</u> possible values of t must be included for a given mechanism, even those which are excluded by equilibrium symmetry considerations in the non-vibronic model. (Only those parts of $F^{(t)}$ which transform under the totally-symmetric irreducible representation make a contribution to the static mechanism.) Second, in a case where the polarisability anisotropy vanishes at the equilibrium structure, its derivatives with respect to vibrational displacements will generally not vanish. Anisotropy terms must therefore be included in a general vibronic treatment.

Consider the specific case of a $\lambda = 2$ transition in a selection of centrosymmetric coordination geometries.

	t = 1	t = 2	t = 3	
$\{\Gamma_{t-}\}$	T_{1u}	$E_u + T_{2u}$	$A_{2u} + T_{1u} + T_{2u}$	(O_h)
	T_u	$E_u + T_u$	$A_u + 2T_u$	(T_h)
	T_{1u}	H_u	$T_{2u} + G_u$	(I_h)
$\{\Gamma_{3N-6}\}_u$	MX_6	$2T_{1u} + T_{2u}$		(O_h)
	$M(XY_2)_6$	$A_u + E_u + 8T_u$		(T_h)
	MC_{60}	$A_u + 4T_{1u} + 5T_{2u} + 6G_u + 7H_u$		(I_h)

In an MX_6 complex, only T_{1u} and T_{2u} modes are available and both are active in the CF and in the LP models. The CF model, through the more favourable radial dependence of the $TF^{(1)}$ term, might be expected to give dominant T_{1u} mode activity.

It is necessary to go to $M(XY_2)_6$ type systems (eg the lanthanide hexanitrites) of T_h symmetry in order to find E_u modes which may be active via $t = 2$ terms. These modes are obviously associated with the presence of non-cylindrically symmetric interactions in these complexes. Within a crystal lattice however, the latter interactions can arise from remote counter ions. For instance, the cube of eight Cs^+ ions which lie adjacent to the trigonal faces of the $[UBr_6]^{2-}$ ion in $Cs_2[UBr_6]$ [30]. The motions of these Cs^+ ions contain a mode of E_u symmetry and this mode may induce modulation in the orientation of the polarisability anisotropy of the coordinated chloride ions.

In icosahedral symmetry the five-fold degenerate H_u modes should be specific to vibrational modulation of the polarisability anisotropy. This class is represented by the recently prepared LaC_{60} [31]. A laser-induced fluorescence spectrum of EuC_{60} (not yet available!) would be interesting to see.

4.3. Vibronic expansion of the effective dipole operator

4.3.1. Taylor expansions

The Taylor expansion of the effective dipole operator in a normal coordinate Q of a complex is

$$\mu^{(1)}_{eff}(Q) = \sum_{n=0}^{\infty}(1/n!)Q^n(\partial/\partial Q)^n\mu^{(1)}_{eff}(Q) \tag{27}$$

Since $\mu^{(1)}_{eff}$ is, through $F^{(t)}$, a known function of ligand position it is more convenient to use the alternative expansion in terms of the vibrational displacement vector, r_L, of an individual ligand L from its equilibrium position R^o_L:

$$\sum_L \mu^{(1)}_{eff}(R_L) = \sum_L \sum_{n=0}^{\infty} \frac{(-1)^n}{n!} (r_L \cdot \nabla)^n \mu^{(1)}_{eff}(R_L) \tag{28}$$

where the factor $(-1)^n$ arises from the definition of the tensor scalar product. The effect of the differential operators is to be evaluated at the equilibrium position, $R_L = R^o_L$, of each ligand. The term n=0 corresponds to the equilibrium contribution already considered. Due to the relatively weak coupling of f electrons with the environment it is sufficient to take the n=1 term which creates a single vibrational quantum. The Taylor operator introduced in section 4.1 is therefore given by $T = -(r^{(1)} \cdot \nabla^{(1)})$, to this approximation, in which $r^{(1)}$ and $\nabla^{(1)}$ are spherical vector operators. The evaluation of $TF^{(t)}$ in (25) and (26) proceeds as follows:

$$TF^{(t)} = c(\lambda,t,k)T[\alpha^{(k)}G^{(\lambda+1)}]^{(t)}$$

$$= c(\lambda,t,k)\{[[T\alpha^{(k)}]^{(k)}G^{(\lambda+1)}]^{(t)} + [\alpha^{(k)}[TG^{(\lambda+1)}]^{(\lambda+1)}]^{(t)}\} \quad (29)$$

Only the second term in (29), involving the derivative of the geometric tensor, has so far been used in vibronic treatments based on the LP model [32,33]. The reasons for believing that the polarisability derivative terms, $T\alpha^{(k)}$, are at least as important will be discussed in section 4.4.

Four terms arise from (29) when the antisymmetric $\alpha^{(1)}$ is neglected. When these are substituted into (25) they give four types of contribution to the effective vibronic dipole operator. When substituted into the vibrational matrix element, $\langle X_0|TF^{(t)}|X_v\rangle$ in (26), the contributions to the vibronically active modes are obtained:

$$TF^{(t)} = c(\lambda,\lambda+1,0)\ \{\ [(T\alpha^{(0)})G^{(\lambda+1)}]^{(\lambda+1)} \qquad \ldots A$$

$$+\ [\alpha^{(0)}(TG^{(\lambda+1)})]^{(\lambda+1)}\ \} \qquad \ldots B$$

$$+\ c(\lambda,t,2)\ \{\ [(T\alpha^{(2)})G^{(\lambda+1)}]^{(t)} \qquad \ldots C$$

$$+\ [\alpha^{(2)}(TG^{(\lambda+1)}]^{(t)}\ \} \qquad \ldots D \qquad (30)$$

4.3.2. Gradient operators.

In differentiating harmonic functions the gradient formula can be used in which the effect of ∇ is to lower or to raise by one the rank of a regular or an irregular harmonic respectively. The gradient formula is normally given in uncoupled form (eg in [34]). The coupled forms are sometimes more convenient and are given here since they do not seem to be generally available.

For regular harmonics:

$$[\ \nabla^{(1)}r^kC^{(k)}]^{(j)} = -\,\delta_{j,k-1}[k(2k+1)]^{1/2}\,r^{k-1}C^{(k-1)} \quad (31)$$

and for irregular harmonics:

$$[\ \nabla^{(1)}r^{-k-1}C^{(k)}]^{(j)} = -\,\delta_{j,k+1}[(k+1)(2k+1)]^{1/2}\,r^{-k-2}C^{(k+1)} \quad (32)$$

The geometric tensor $G^{(\lambda+1)}$ defined in equation (16) is an example of an irregular harmonic and so the gradient can immediately be evaluated from (32) with $k = \lambda + 1$, ie

$$[\ \nabla^{(1)}G^{(\lambda+1)}]^{(j)} = -\,\delta_{j,\lambda+2}[(\lambda+2)(2\lambda+5)]^{1/2}\,G^{(\lambda+2)} \quad (33)$$

First of all the gradient operator is recoupled from T onto $G^{(\lambda+1)}$:

$$TG^{(\lambda+1)} = -(r^{(1)}\cdot\nabla^{(1)})G^{(\lambda+1)} = 3^{1/2}[[r^{(1)}\nabla^{(1)}]^{(0)}G^{(\lambda+1)}]^{(\lambda+1)}$$

$$= \sum_j (-1)^j[(2j+1)/(2\lambda+3)]^{1/2}[r^{(1)}[\ \nabla^{(1)}G^{(\lambda+1)}]^{(j)}]^{(\lambda+1)} \quad (34)$$

Then substituting for the gradient from (33),

$$TG^{(\lambda+1)} = -[(2\lambda+5)(\lambda+2)]^{1/2} [r^{(1)}G^{(\lambda+2)}]^{(\lambda+1)} \quad (35)$$

4.4 Vibrational modes from $TF^{(t)}$.

It should be recalled (see section 4.2) that $TF^{(t)}$ is the generator of vibrational modes in the vibronic picture.
It turns out that all terms arising from $TF^{(t)}$ can be written in a general coupled form:

$$TF^{(t)} \longrightarrow \sum_j c(\lambda,j)\ [r^{(1)}{}_L X^{(j)}]^{(t)} \quad (36)$$

of which (35) is a specific example. When summed over ligands, the $X^{(j)}$ generate the coefficients of the vibrational displacements $r^{(1)}{}_L$ at each ligand position. This closely follows the approach of Judd [35].

4.5. Polarisability derivatives

Polarisability derivatives make contributions, through $T\boldsymbol{\alpha}^{(k)}$, to terms A and C in equation (30) whereas terms B and D involve the gradient of the geometric tensor, $TG^{(\lambda+1)}$. The relative orders of magnitude of (say) A and B may be estimated by the following argument.

Using (35) to evaluate the Taylor expansion of $G^{(\lambda+1)}$, (30)B becomes

$$-[(2\lambda+5)(\lambda+2)]^{1/2}\ [r^{(1)}{}_L \boldsymbol{\alpha}^{(0)} G^{(\lambda+2)}]^{(\lambda+1)} \quad (37)$$

Recoupling of (30)A leads to terms of the type

$$\sum_j (-1)^j\ [(2j+1)/(2\lambda+1)]^{1/2}\ [r^{(1)}{}_L[(\nabla^{(1)}\boldsymbol{\alpha}^{(0)})G^{(\lambda+1)}]^{(j)}]^{(\lambda+1)} \quad (38)$$

Comparing (37) and (38), the coefficients on the left have similar magnitudes and the vibrational amplitude, $r^{(1)}{}_L$, is a common factor. The mean polarisability of a ligand, $\bar{\alpha} = -(1/3)^{1/2}\boldsymbol{\alpha}^{(0)}$, is of the same order of magnitude as its derivative $\nabla^{(1)}\bar{\alpha}$ determined from Raman intensity data in the bond polarisability approximation [36]. The angular parts of the geometric tensors are of order unity for non-vanishing terms and the radial dependences are given by (16):

$$G^{(\lambda+2)} \sim (R_{LM})^{-\lambda-3} \text{ and } G^{(\lambda+1)} \sim (R_{LM})^{-\lambda-2}\ ,$$

being more favourable in the case of (38).

5. REFERENCES

[1] Laporte O: Z Physik **51**, 512 (1924)
[2] Van Vleck J H: J Phys Chem **41**, 67 (1937)
[3] Liehr A D and Ballhausen C J: Phys Rev **106**, 1161, (1957)
[4] Koide S and Pryce M H L: Phil Mag **3**, 607 (1958)
[5] Ballhausen C J and Liehr A D: J Mol Spectrosc **2**, 342 (1958) and J Mol Spectrosc **4**, 190 (1960)
[6] Judd B R: Phys Rev **127**, 750 (1962)
[7] Ofelt G S: J Chem Phys **37**, 511 (1962)
[8] Mason S F, Peacock R D and Stewart B: Chem Phys Lett **29**, 149 (1974)
[9] Mason S F, Peacock R D and Stewart B: Molec Phys **30**, 1829 (1975)
[10] Mason S F: Structure and Bonding **39**, 43 (1980)
[11] Mason S F: Inorg Chim Acta **94**, 313 (1984)
[12] Peacock R D: Molec Phys **33**, 1239 (1977)
[13] Jorgensen C K and Judd B R: Molec Phys **8**, 281 (1964)
[14] Peacock R D: Structure and Bonding, **22**, 83 (1975)
[15] Condon E U and Odabasi H: 'Atomic Structure', p169ff, Cambridge University Press, 1980
[16] Rotenberg M, Bivins R, Metropolis N and Wooten J K: 'The 3-j and 6-j Symbols', MIT Press, 1959
[17] Condon E U and Odabasi H: 'Atomic Structure', p173, Cambridge University Press, 1980
[18] Judd B R: 'Angular Momentum Theory for Diatomic Molecules', p98; Academic Press, 1975
[19] Gale R, Godfrey R E, Mason S F, Peacock R D and Stewart B: J Chem Soc Chem Commun, 329 (1975)
[20] Mason S F: Structure and Bonding **39**, p59 (1980)
[21] Jorgensen C K and Judd B R: Molec Phys **8**, 281 (1964)
[22] Kuroda R, Mason S F and Rosini C: Chem Phys Lett **70**, 11 (1980)
[23] Kuroda R, Mason S F and Rosini C: J Chem Soc Faraday Trans II, **77**, 2125 (1981)
[24] Stewart B: Molec Phys **50**, 161 (1983)
[25] Clark R J H and Stewart B: Structure and Bonding **36**, 2 (1979)
[26] Judd B R: J Chem Phys **44**, 839 (1966)
[27] Newman D J and Balasubramaniam G: J Phys C, **8**, 37 (1975)
[28] Mason S F and Stewart B: Molec Phys **55**, 611 (1985)
[29] Bird B D, Cooke E A, Day P and Orchard A F: Phil Trans Roy Soc London **A276**, 277 (1974)
[30] Chodos S L and Satten R A: J Chem Phys **62**, 2411 (1975)
[31] Zhang Q L, O'Brien S C, Heath J R, Liu Y, Kroto H W and Smalley R E: J Phys Chem **90**, 525 (1986)
[32] Faulkner T R and Richardson F S: Molec Phys **35**, 1141 (1978)
[33] Acevedo R and Flint C D: Molec Phys **58**, 1033 (1986)
[34] Brink D M and Satchler G R: 'Angular Momentum', Clarendon Press, Oxford, 1968
[35] Judd B R: Physica Scripta **21**, 543 (1980)
[36] Clark R J H: in 'Advances in Infrared and Raman Spectroscopy' **1**, p152, Heyden 1975

NON-RESONANT ENERGY TRANSFER BETWEEN INORGANIC IONS IN SOLIDS

Marco Bettinelli
Department of Inorganic Chemistry
University of Padua
Via Loredan 4
35131 Padova
Italy

ABSTRACT. The non-resonant transfer of optical excitation between inorganic ions in solids is discussed and reviewed. When there is no pairwise resonance between electronic transitions of the donor and the acceptor, higher-order phonon-assisted and many-body processes become important. The theories are briefly described and the relative importance of the two mechanisms is discussed. Attention is given to the dependence of the non-resonant transfer probability on the temperature, the energy gap and the concentration of the species involved in the process, particularly in the case they are trivalent lanthanide ions. Experimental examples concerning the Tb^{3+}-->Eu^{3+} non-resonant transfer in $Tb_{1-x}Eu_xP_5O_{14}$ and $Cs_2NaTb_{1-x}Eu_xCl_6$ are reported, in order to show how the theoretical concepts can be applied to real cases.

1. INTRODUCTION

The non-radiative energy transfer processes between ions of transition metals, lanthanides and actinides in solids are very important from both a theoretical and a technological point of view. The experimental observations and their theoretical interpretation have been reviewed [1-7]. Particular attention has normally been given to transfer processes involving exact resonance between electronic emission transitions of the donor centre and electronic absorption transitions of the acceptor centre. In this contribution an account is given of the more limited area of the non-resonant energy transfer phenomena, which are less well understood than the resonant processes. The reason lies in the fact that the theory does not allow sufficiently precise calculations of the energy transfer probability, and moreover the non-

C. D. Flint (ed.), Vibronic Processes in Inorganic Chemistry, 347–369.
© 1989 by Kluwer Academic Publishers.

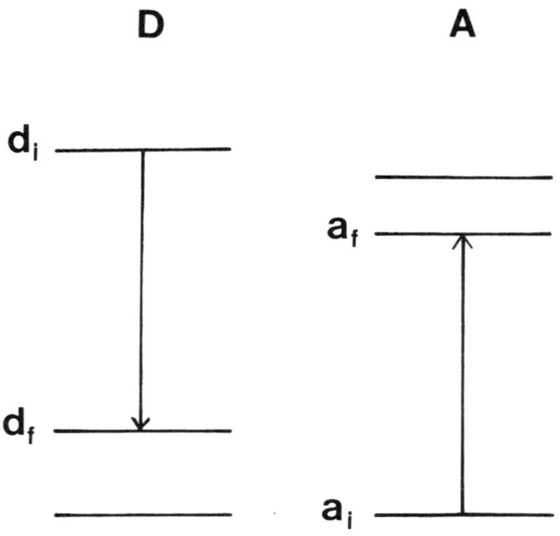

Figure 1. Resonant energy transfer from the donor ion D to the acceptor ion A.

resonant transfer is often more difficult to study from an experimental point of view, because of its intrinsically poor efficiency.

This contribution is divided in two parts. In the first part, a brief review of the theory underlying the phonon-assisted energy transfer and the many-body processes is reported. Particular attention is given to the transfer of excitation between trivalent lanthanide ions in solids. In the second part, experimental examples are reported, in which the theoretical concepts of the first part are applied to "real" situations. The examples in the second part deal entirely with the typically non-resonant energy transfer processes between the lanthanide ions Tb^{3+} and Eu^{3+}, when they are present in different crystalline hosts. This fact gives rise to important differences in the experimental behaviour of the energy transfer, which are explained and discussed.

2. THEORETICAL BACKGROUND

The theory of the non-radiative resonant transfer of excitation from a donor centre (D) to an acceptor centre (A) (fig. 1) was first formulated by Förster [8] for the transfer between organic molecules and by Dexter [9] for the transfer between ions in solids. According to this theory, the probability per unit time $W_{D\text{-->}A}$ of the process can be expressed as:

$$W_{D\text{-->}A} = 2\pi/\hbar^2 |\langle d_i\ a_i | H_T | d_f\ a_f \rangle|^2 \int f_D(\omega) f_A(\omega) d\omega \qquad (1)$$

where the transfer hamiltonian H_T is given by one of the following interactions:

a) electrostatic multipolar interaction (electric dipole-electric dipole, electric dipole-electric quadrupole, electric quadrupole-electric quadrupole);
b) magnetic dipole-magnetic dipole interaction;
c) exchange interaction;

d_i, a_i, d_f and a_f are the initial and final electronic states of D and A, and $f_D(\omega)$ and $f_A(\omega)$ are the line shape functions of the frequency for the relevant electronic transitions in the donor and the acceptor, respectively, and are normalized in the sense $\int f(\omega)d\omega = 1$. The magnetic dipole-magnetic dipole interaction is very weak and usually not important for the transfer of excitation in the optical region, so just the electrostatic and the exchange interactions are needed to describe the energy transfer between luminescent levels of inorganic ions in solids. When the process is not resonant, i.e. when the difference in energy of the two electronic transitions is much larger than the sum of their homogeneous widths (the effect of inhomogeneous broadening will not be considered) (fig. 2), it is clear that the overlap integral in eq.(1) becomes negligible. In this case the processes responsible for the transfer of the excitation must be of a different nature. This section deals with the two different processes which can allow the conservation of energy when the resonant pairwise transfer is not possible. The section is divided in two parts. In the first the phonon-assisted transfers are described, in the second an account of the processes involving more than two ions (the "many-body" processes) is given.

2.1. Phonon-assisted energy transfer

In the case that the energy transfer is not resonant, the energy difference ΔE between the electronic emission transition of the donor and the electronic absorption transition of the acceptor can be bridged by the emission or the absorption of lattice vibrations. An excellent exposition of the phonon-assisted energy transfer has been given by Orbach [10]. In this brief account, the processes involving one phonon and many phonons are treated separately.

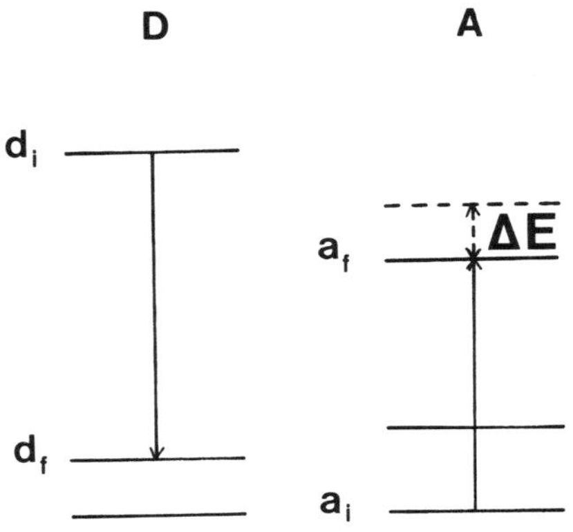

Figure 2. Non-resonant energy transfer from the donor ion D to the acceptor ion A. The energy ΔE is dissipated through the emission of phonons or the excitation of other ions.

2.1.1. One-phonon processes. When the energy transfer between electronic states requires the assistance of one phonon to occur, vibronic wavefunctions for the initial and final states of the donor and acceptor ions are needed in order to calculate the energy transfer probability. Vibronic wavefunctions in the crude adiabatic approximation scheme [11] are used:

$$\Psi_i^k(q,Q) = \psi_k^0(q)\ \chi_{ki}(Q) \tag{2}$$

where q stands for the electronic and Q for the nuclear coordinates. The electronic wavefunction is given by the static Schrodinger equation at the reference nuclear configuration Q_0:

$$H_{el}(Q_0)\ \psi_k^0\ (q) = \varepsilon_k^0\ \psi_k^0\ (q) \tag{3}$$

and the vibrational wavefunction is given by the Schrodinger equation:

$$[-1/2 \sum_i \partial^2/\partial Q_i^{\ 2} + \varepsilon_k^0 + V'_{kk}(Q)]\ \chi_{ki}(Q) = E_i^k\ \chi_{ki}(Q)\ . \tag{4}$$

$V'_{kk}(Q)$ is the diagonal element of the vibronic interaction:

$$V'_{kk}(Q) = \langle\psi_k^0(q)|\Delta H_{el}(Q)|\psi_k^0(q)\rangle \tag{5}$$

and $\Delta H_{el}(Q)$ expresses the linear vibronic coupling:

$$\Delta H_{el}(Q) = \sum_i (\partial V_c/\partial Q_i)_0 \, Q_i \tag{6}$$

where V_c is the crystal field around the ion in the solid.

The matrix element of H_T is therefore calculated using vibronic wavefunctions like (2) for the initial and the final states of the donor and acceptor ions. When one phonon of energy $\hbar\omega_k$ is absorbed or emitted in the final states of the donor or the acceptor ion, in order to conserve the energy in the transfer process, the matrix element M_T of the interaction hamiltonian H_T is shown to be [5]:

$$M_T = \sum_{m=D,A} \langle d_f \; a_f \; n_k{\pm}1 | H_T | d_i \; a_i \; n_k{\pm}1\rangle\langle d_i \; a_i \; n_k{\pm}1 | \; \Delta H_{el}(m) | d_i \; a_i \; n_k\rangle \; x$$

$$x \; (\mp\hbar\omega_k)^{-1} \; + \tag{7}$$

$$+ \sum_{m=D,A} \langle d_f \; a_f \; n_k{\pm}1 | \; \Delta H_{el}(m) | d_f \; a_f \; n_k\rangle\langle d_f \; a_f \; n_k | H_T | d_i \; a_i \; n_k\rangle \; (\Delta E)^{-1}$$

where the factor n_k is the occupation number for phonons having wave vector k and the hamiltonians $\Delta H_{el}(D)$ and $\Delta H_{el}(A)$ refer to the donor and the acceptor respectively. The sign + applies to the emission of a phonon of energy $\hbar\omega_k$ and the sign - to its absorption. In this case M_T is in general different from zero.

The probability W_{PAET} of the energy transfer assisted by one phonon is then [10]:

$$W_{PAET} = 2\pi/\hbar^2 |M_T|^2 \int f_D(\omega) \; f_A(\omega \mp \omega_k) \; d\omega \tag{8}$$

where the overlap integral of the line shape factors includes now the phonon sidebands (compare with (1)).

The actual calculation of the one-phonon transfer probability is possible, with some difficulties, only when the frequency of the phonon is much less than the cut-off frequency in the vibrational spectrum, otherwise the problem becomes intractable [10]. It is more interesting to take into account some important features related to the one-phonon processes, such as the dependence of the transfer probability on the energy gap ΔE and on the temperature.

Orbach [10] has shown that the probability for the transfer of excitation W_{PAET} is proportional to the energy gap ΔE for the one-phonon processes, and this leads to the rather peculiar result that the larger is the energy mismatch, the more probable is the transfer,

In the case of the emission of a phonon, the transfer probability is proportional to $n(\omega)+1$, where $n(\omega)$ is the occupation number of the phonon having energy $\hbar\omega$. Therefore, using the Bose-Einstein average for $n(\omega)$, it is possible to write for the phonon-assisted energy transfer probability W_{PAET}:

$$W_{PAET}(T)=W_{PAET}(0)\ [1-\exp(-\hbar\omega/kT)]^{-1} \tag{9}$$

where $W_{PAET}(0)$ is a constant, independent of temperature.

When one phonon of energy $\hbar\omega$ is absorbed in the transfer process, the probability of the event is proportional to $n(\omega)$, so:

$$W_{PAET}(T)=W_{PAET}(0)\ [\exp(\hbar\omega/kT)-1]^{-1} \tag{10}$$

The dependence on temperature of (9) is fairly weak: $W_{PAET}(T)$ is independent of T at low T, and increases as T at high T. Following (10), $W_{PAET}(T)$ increases as T at high T, but is proportional to $\exp(-\hbar\omega/kT)$ at low T. The analysis of the experimental transfer rates using equations (9) and (10) can allow the determination of the energy of the phonon involved in the process, when it is not known.

When the energy transfer occurs between two Ln^{3+} ions, the energy levels of the donor and the acceptor, between which the excitation is non-radiatively transferred, are J levels belonging to the $4f^n$ configurations of two Ln^{3+} ions (see fig. 3). These J levels are split by the crystal field, with a spread in energy of typically some hundred cm^{-1}. Thermal equilibrium within the sets of crystal field states is reached much more rapidly than the energy transfer process occurs [12]. The energy transfer probabilities between the individual crystal field states can be different, because phonons of different energies can be needed to match the various energy gaps between acceptor and donor, and therefore the observed energy transfer rate between J levels (W_{obs}) must be expressed as the weighted average of the rates between the individual crystal field states:

$$W_{obs}=\sum_{\substack{all\\comb.}} g_{d_i}\, g_{d_f}\, g_{a_i}\, g_{a_f}\, W_{PAET}(d_i a_i,\ d_f a_f)\, \exp(-E_{d_i}/kT)\ \times \tag{11}$$
$$\times\ \exp(-E_{a_i}/kT)\ [\ \sum_{d_i} g_{d_i}\, \exp(-E_{d_i}/kT) \sum_{a_i} g_{a_i}\, \exp(-E_{a_i}/kT)]^{-1}$$

where the sum is performed over all the possible combinations, $W_{PAET}(d_i a_i,\ d_f a_f)$ is the individual transfer rate from the initial

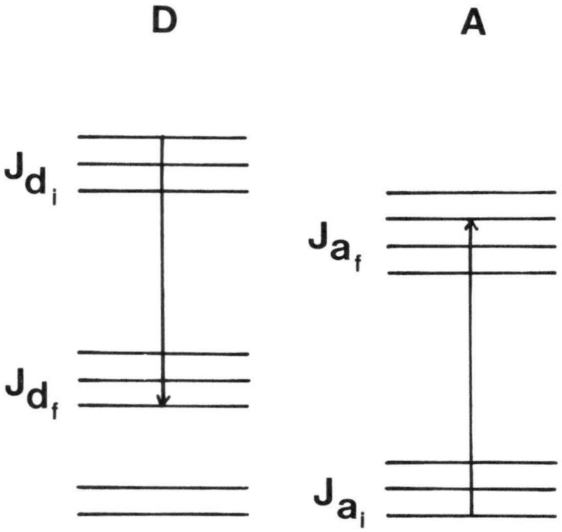

Figure 3. Energy transfer between crystal field-split J levels of two trivalent lanthanide ions.

$|d_i\ a_i\rangle$ states to the final $|d_f\ a_f\rangle$ states, g_r is the degeneracy of the single ion state r and E_r is its energy separation from the bottom state of the J level. It is clear that equation (11) introduces another source of temperature dependence for the observed energy transfer probability.

2.1.2. Many-phonon processes. The quantitative evaluation of the energy transfer probability for a process involving the emission or the absorption of N phonons requires, in principle, calculations using the linear vibronic coupling hamiltonian (6) in the N-th order of perturbation. As shown by Orbach [10], the algebraic difficulties in this case become prohibitive. The alternative procedure of using a many-phonon interaction hamiltonian in a low order of perturbation is not suitable because in this case the coupling coefficients are impossible to evaluate. For these reasons, in this section only the dependence of the many-phonon assisted energy transfer rate on the energy gap and on the temperature will be considered.

Miyakawa and Dexter [13] were able to show that the probability of many-phonon relaxation of the excited state of an ion depends exponentially on the energy gap between the excited state and the lower-lying state. So, if the energy gap is ΔE, and the process is dominated by the emission of N phonons of energy $\hbar\omega$ ($N=\Delta E/\hbar\omega$), the many-phonon relaxation probability W_{MPR} at the temperature T is:

$$W_{MPR}(\Delta E)=W_{MPR}(0)\ \exp(-\alpha\Delta E) \qquad (12)$$

where α is given by:

$$\alpha = \{\ln[N\ G^{-1}\ (n(\omega)+1)^{-1}]-1\}/\hbar\omega \qquad (13)$$

and G is an electron-lattice coupling constant, characteristic of the host in which the ion is present. Riseberg and Moos [14] have reported a phenomenological derivation of equation (12).

Following Miyakawa and Dexter, the probability for the transfer of excitation assisted by the emission of N phonons of energy $\hbar\omega$ ($\Delta E=N\hbar\omega$) between two different ions can be written in a very similar way:

$$W_{PAET}(\Delta E)=W_{PAET}(0)\ \exp(-\beta\Delta E) \qquad (14)$$

where β is again a function of the temperature and of the strength of the electron-lattice coupling. β and α are related by:

$$\beta = \alpha - \gamma \qquad (15)$$

where:

$$\gamma = \ln(1-G_A/G_D)/\hbar\omega \qquad (16)$$

In (16), G_A and G_D are electron-lattice coupling constants for the acceptor and the donor.

It is peculiar that equations (12) and (14) show that the many-phonon relaxation probability and the energy transfer probability for the many-phonon processes are dependent only on the energy gap ΔE and on the strength of the electron-lattice coupling in the particular host under consideration, and do not depend on the individual nature of the transitions involved in the relaxation and energy transfer processes. Due to the high order process, the characteristic features of the individual phonons and energy levels involved will be very effectively averaged out, and therefore the resulting equations (12) and (14) will not contain information about them. Experimental confirmations of the "energy gap laws" (12) and (14) have been reported [14,15].

The temperature dependence of the energy transfer probability, when N phonons are active in conserving the energy, is analogous to to the dependence on the temperature of the one-phonon processes. For the emission of N phonons of energy $\hbar\omega$:

$$W_{PAET}(T)=W_{PAET}(0)\ [1-\exp(-\hbar\omega/kT)]^{-N} \qquad (17)$$

and for the absorption of N phonons:

$$W_{PAET}(T)=W_{PAET}(0)\ [\exp(\hbar\omega/kT)-1]^{-N} \quad (18)$$

The assumption that the phonons needed to bridge the gap have the same energy is reasonable, because even if their energies were spread around to some central value, the behaviour would not be very different. The critical factor, as far as the temperature dependence is concerned, is the order of the process, and not the energy of the individual phonons. For an experimental confirmation of this behaviour, see [14].

If the energy transfer occurs between the J levels, split by the crystal field, of two Ln^{3+} ions, the many-phonon energy transfer probability will be again the weighted average of the probabilities of the individual transfers between the crystal field states. The overall energy transfer probability will be therefore described by an equation like (11).

2.2. Many-body processes

When no simple resonant energy transfer process between a donor ion and an acceptor ion is possible, i.e. when the difference between the energies of the two relevant electronic transitions is large, it is possible to devise another mechanism responsible for the transfer of the excitation, without invoking the involvement of vibrations in order to conserve the energy. This latter mechanism requires the participation in the process of one or more other ions in the lattice, and their function is to bridge the energy gap between the donor and the acceptor by being promoted to a higher-lying electronic state or by relaxing to a lower-lying electronic state. The energy is obviously conserved in this kind of processes, but the transfer does not involve just a pair of ions. These phenomena are therefore classified as "many-body processes". The one or more ions whose intervention is crucial in order to conserve the energy can be donor (D) or acceptor (A) ions, in their ground state or in a thermally populated state. Two examples of this behaviour for a three-particle system are reported in figs. 4 and 5. These many-body processes require a fairly high concentration of the donor and/or the acceptor ions, if they are to be reasonably probable, because the three or more ions involved in the process must be close enough in the crystal, to interact simultaneously [16].

In order to gain more insight in the nature of the general class of the many-body processes, and to understand if they can be sufficiently probable to be observed, it is instructive to examine the process described in fig. 4. In this case, the excitation is non-radiatively transferred from the initial excited electronic level of

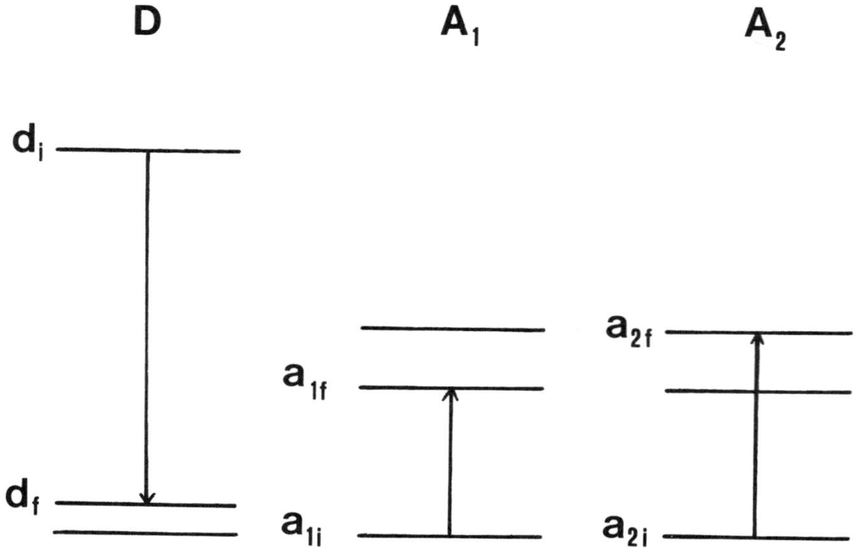

Figure 4. Three-body process: the energy is transferred from the donor ion D to two acceptor ions A_1 and A_2.

the donor ion D, to two different final excited electronic levels of the acceptor ion A. The probability $W_{i\to f}$ of the energy transfer process between the initial state and a quasicontinuous range of final states can in general be written following the Fermi golden rule:

$$W_{i\to f} = 2\pi/\hbar \sum_{i,f} |\langle i|\tau|f\rangle|^2 \; p_i(E_i) \;\; \delta(E_i - E_f) \qquad (19)$$

where $|i\rangle$ and $|f\rangle$ are the wavefunctions for the initial and the final state of the system, constructed as the product of the wavefunctions of the levels of the ions involved; E_i and E_f are the energies of the initial and final state of the system, given as the sum of the energies of the single ion levels, and $p_i(E_i)$ is the probability that the system is in the initial state $|i\rangle$.

The hamiltonian τ can be expressed taking the interaction hamiltonian H' at the second order of the perturbation [16]:

$$\tau = H' + \sum_m H'|m\rangle\langle m|H'(E_i - E_m)^{-1} \qquad (20)$$

where E_m is the energy of an intermediate state $|m\rangle$.

In the case described in fig. 4, the initial state is defined as:

$$|i\rangle = |d_i \; a_{1i} \; a_{2i}\rangle \qquad (21)$$

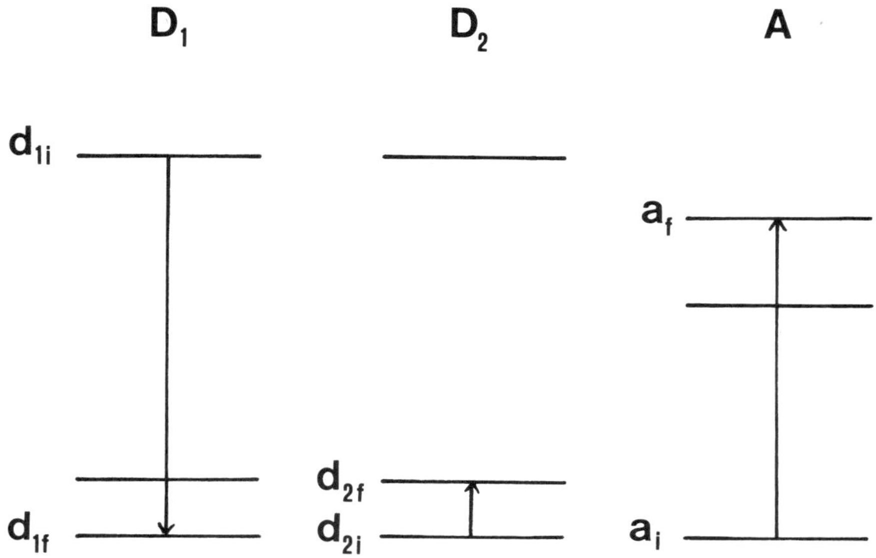

Figure 5. Three-body process: the process is similar to the one reported in fig. 4, but one of the two acceptor ions (D_2) is identical to the donor ion (D_1).

and the final state is defined as:

$$|f\rangle = |d_f\ a_{1f}\ a_{2f}\rangle \tag{22}$$

and the interaction hamiltonian is simply given by the sum of the pairwise interaction hamiltonians involving the three ions:

$$H' = H'_{DA_1} + H'_{DA_2} + H'_{A_1A_2} \tag{23}$$

The initial and final energies are perfectly matched:

$$E_i = E_{d_i} + E_{a_{1i}} + E_{a_{2i}} = E_f = E_{d_f} + E_{a_{1f}} + E_{a_{2f}} \tag{24}$$

In general, if there is no pairwise resonance between D and A, the first order contribution to the transfer probability will vanish, so that only the second order contribution will be of interest:

$$\langle i|\tau|f\rangle = \sum_m \langle i|H'|m\rangle\langle m|H'|f\rangle\ (E_i - E_m)^{-1} \tag{25}$$

A typical term of (25) is explicitly:

$$\sum_{d_m} \sum_{a_{1m}} \sum_{a_{2m}} \langle d_i\, a_{1i}\, a_{2i} | H'_{DA_1} | d_m\, a_{1m}\, a_{2m} \rangle \; x$$

$$x \quad \langle d_m\, a_{1m}\, a_{2m} | H'_{DA_2} | d_f\, a_{1f}\, a_{2f} \rangle \; (E_i - E_m)^{-1} \tag{26}$$

where d_m, a_{1m} and a_{2m} are intermediate states of the donor and the acceptor ions. (26) can be written as:

$$\sum_{d_m} \sum_{a_{1m}} \sum_{a_{2m}} \langle d_i\, a_{1i} | H'_{DA_1} | d_m\, a_{1m} \rangle \langle a_{2i} | a_{2m} \rangle \; x$$

$$x \quad \langle d_m\, a_{2m} | H'_{DA_2} | d_f\, a_{2f} \rangle \langle a_{1m} | a_{1f} \rangle \; (E_i - E_m)^{-1} \tag{27}$$

which by orthogonality becomes:

$$\sum_{d_m} \langle d_i\, a_{1i} | H'_{DA_1} | d_m\, a_{1f} \rangle \langle d_m\, a_{2i} | H'_{DA_2} | d_f\, a_{2f} \rangle \; (E_i - E_m)^{-1} \tag{28}$$

Terms of the form (28) in general will be nonzero, and therefore it is possible to have a second order contribution to the energy transfer process involving three ions, one donor and two acceptors. These procedure can be extended to a higher order of perturbation, accounting then for processes involving more than three ions.

The many-body processes can be regarded as important, in the case of the energy transfer between lanthanide ions in solids, for several reasons. In general, when the possibility of a pairwise resonant transfer is not present, it is often possible to construct a large number of combinations for the many-body transfer. Moreover, the intermediate electronic states $|m\rangle$ can have opposite parity with respect to the initial and final states, and this fact will greatly increase the importance of the strong and long-range electric dipole interactions in the second order terms, in the case of the energy transfer involving transitions between $4f^n$ levels of lanthanide ions. The many-body processes require a fairly high concentration of the donor and/or the acceptor ion, in order to become important. Grant [17] has shown that, when the resonant migration of energy among the donor or the acceptor ions is much faster than the D-->A transfer, i.e. when the concentration of D or A are sufficiently high, the many-body processes can be treated on the basis of rate equations. In this case, the dependence of the energy transfer probability on the concentration of the ions reflects the number of particles involved in the microscopic process. Schematically, a process involving S ions of the species B will occur with a probability proportional to n_B^S, where n is the concentration of the species under examination.

In the three-body example described in fig. 4, for which the interaction hamiltonian has been considered in some detail, if the energy migration among the donor ions D is very fast, as in the case of a crystal composed mainly of D ions and containing A ions as dilute impurities, it is possible to write the following rate equation:

$$-dn_{d_i}(t)/dt = \tau_D^{-1}\, n_{d_i}(t) + Kn_{a_{1i}}(t)n_{a_{2i}}(t)n_{d_i}(t) \tag{29}$$

where τ_D is the lifetime of the level d_i when isolated, and $n_r(t)$ defines the population of the level r in the crystal at the time t. The first term in (29) expresses the rate of the internal decay of the donor ion, and the second term expresses the rate of the energy transfer from the donor to the acceptors.

In the case that a_{1i} and a_{2i} are the ground states of the acceptor ions A, and if the exciting flash of light is weak, the solution of the rate equation is:

$$n_{d_i}(t)=C\ \exp[(-\tau_D^{-1}-KN_A{}^2)t] \tag{30}$$

where C is a constant and N_A is the total concentration of the acceptor ions, and therefore the decay curve of the luminescence from the level D after pulsed excitation is given by:

$$f(t) = a\ \exp(-t/\tau) \tag{31}$$

where a is another constant. So, the decay curve of the level D is still exponential, and can be characterized by an effective lifetime τ. The energy transfer probability is then given by:

$$W_{ET} = KN_A{}^2 \tag{32}$$

and is proportional to the square of the concentration of the A ions in the crystal. This kind of behaviour has been observed in many experimental cases [18] and will be further illustrated in the next section.

2.3. Conclusions

It is possible to conclude that the phonon-assisted and the many-body energy transfer are processes which can become important when there is no pairwise resonance between the electronic levels of the donor and the acceptor. Both processes require an interaction whose order of perturbation is higher than for the resonant energy transfer, and it is not possible to decide "a priori" which mechanism will be dominant when there is no resonance, especially when the concentrations of the donor and/or the acceptor ions are high. Only a careful examination of the energies of the electronic levels of D and A and of the vibrational spectrum of the crystal, and the analysis of the dependence of the transfer probability on the temperature and on the concentration of the two ions allow to disentangle this rather complicated problem.

3. EXPERIMENTAL EXAMPLES

A typical example of energy transfer process between lanthanide ions in solids, occurring in absence of pairwise resonance of the energies of the electronic levels, is given by the pair Tb^{3+}-Eu^{3+}. The transfer of excitation from the $^5D_4$ level of Tb^{3+} to the excited levels of Eu^{3+} was for the first time directly observed in $(Tb_{1-x}Eu_x)_2(WO_4)_3$ crystals [19], in which no overlap between the emission transitions of Tb^{3+} and the absorption transitions of Eu^{3+} was measured. Since then, after selective excitation of the Tb^{3+} ion in the $^5D_4$ level, luminescence has been observed from the level $^5D_1$ and/or the level $^5D_0$ of the Eu^{3+} ion (fig. 6) in several crystals. When the temperature is sufficiently low, so that the ground $^7F_0$ level of Eu^{3+} is heavily populated, the transfer can in general occur through two different phonon-assisted transfer mechanisms [20]:

$$^5D_4(Tb)+^7F_0(Eu)+\Delta E_{ph}(1) \dashrightarrow {}^7F_4(Tb)+^5D_0(Eu) \tag{33}$$

and:

$$^5D_4(Tb)+^7F_0(Eu) \dashrightarrow {}^7F_6(Tb)+^5D_1(Eu)+\Delta E_{ph}(2) \tag{34}$$

where $\Delta E_{ph}(1)$=20-300 cm^{-1} and $\Delta E_{ph}(2)$=1000 cm^{-1}.

Besides these two phonon-assisted processes, the electronic structures of the ions in general allow, in crystals containing sufficiently high concentrations of the two ions, several many-body transfer processes.

At temperatures sufficiently high, such that the $^7F_1$ first excited level of Eu^{3+} is significantly populated, phonon-assisted energy transfer mechanisms involving the latter level become possible, such as:

$$^5D_4(Tb)+^7F_1(Eu) \dashrightarrow {}^7F_4(Tb)+^5D_0(Eu)+\Delta E_{ph}(3) \tag{35}$$

and:

$$^5D_4(Tb)+^7F_1(Eu) \dashrightarrow {}^7F_5(Tb)+^5D_1(Eu)+\Delta E_{ph}(4) \tag{36}$$

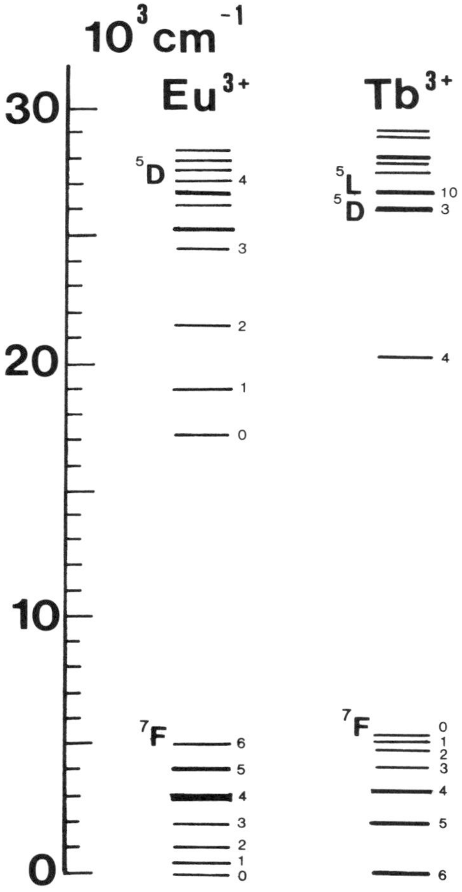

Figure 6. A schematic diagram of the energy levels of Eu^{3+} and Tb^{3+} up to 30000 cm^{-1}.

The magnitude of the energy gap bridged by the phonons will be characteristic of the crystal under investigation.

It is clear that the details of the transfer mechanisms will vary from crystal to crystal, and will be in general dependent on the energy of the individual crystal fields states of the levels involved, and on the phonon spectrum of the crystal. In all cases, the phonon-assisted and the many-body processes will compete. The dominant mechanism will be decided by the nature of the crystal in which the Ln^{3+} ions are present, by the temperature and by the concentration of the two ions involved, as the following examples will show.

3.1. Energy transfer from Tb^{3+} to Eu^{3+} in $Tb_{1-x}Eu_xP_5O_{14}$

The lanthanide pentaphosphates LnP_5O_{14} have been widely studied because the interesting mini-laser crystal NdP_5O_{14} belongs to this class of compounds [21]. The time evolution of the luminescence has been investigated for TbP_5O_{14} [22] and for EuP_5O_{14} [23], and several studies concerning the energy transfer from Tb^{3+} to Eu^{3+} have been published [24-26]. Particular attention has been given to the composition $Tb_{0.66}Eu_{0.33}P_5O_{14}$, whose luminescent behaviour has been studied in detail as a function of temperature [20]. At low temperatures, monochromatic excitation of the $^5D_4$ level of Tb^{3+}, whose center of gravity lies at 20550 cm^{-1} [22], gives rise to an emission spectrum in the range 586-595 nm, which clearly shows the $^5D_4 \rightarrow {}^7F_4$ transition of Tb^{3+} and the $^5D_0 \rightarrow {}^7F_1$ transition of Eu^{3+}. It is evident that the excitation is transferred from the $^5D_4$ level of Tb^{3+} to the $^5D_0$ level of Eu^{3+}. The luminescence spectra in the region 532-539 nm, excited into the $^5D_4$ level of Tb^{3+} and into the $^5D_2$ level of Eu^{3+}, are substantially identical and both show clearly the $^5D_1 \rightarrow {}^7F_1$ transition of Eu^{3+}. Therefore, the transfer of the excitation appears to occur from the level $^5D_4$ of Tb^{3+} to the level $^5D_1$ of Eu^{3+}, which then relaxes to the lower-lying $^5D_0$.

The decay curves of the luminescence from the $^5D_4$ level are exactly exponential in the 40-200 K range and this is indicative of a fast migration among the donor ions [2], which is reasonable because of the high concentration of the Tb^{3+} ions. The lifetime of the $^5D_1$ level of Eu^{3+} in $Tb_{0.66}Eu_{0.33}P_5O_{14}$ at 40 K, after excitation of $^5D_4$(Tb), is virtually identical to the lifetime of $^5D_4$(Tb) at the same temperature (2.2 ms), in agreement with the behaviour predicted by a rate equation which takes into account the transfer from a long-lived level ($^5D_4$) to a short-lived level ($^5D_1$).

The presence of strong features around 1000 cm^{-1} in the Raman spectrum of the related compound NdP_5O_{14} [27] has led to the assignment of the transfer at low temperatures to the process described by (34), assisted by the emission of one phonon having $\tilde{\nu} = 1000\ cm^{-1}$. At temperatures higher than 150 K, the level $^7F_1$ of Eu^{3+}, whose center of gravity lies 360 cm^{-1} above the ground level, becomes thermally populated and a phonon-assisted process via the $^7F_1$ level like (35) becomes possible. This second process is assisted by the emission of one phonon having $\tilde{\nu} = 25\ cm^{-1}$. By solving the relevant rate equation, the probability W of the energy transfer process is given by:

$$W_{ET} = 1/\tau_D - 1/\tau_D^0 \quad (37)$$

where τ_D is the lifetime of $^5D_4$ in $Tb_{1-x}Eu_xP_5O_{14}$ and τ_D^0 is its lifetime in TbP_5O_{14}. Due to the fact that the decay curves of the luminescence from the $^5D_4$ level are not exactly exponential for the mixed crystal at T>200 K, it is necessary to define a suitable effective lifetime for $^5D_4$ at high temperatures [20]. The observed energy transfer probability drops sharply when the temperature is decreased down to 150 K, and it is constant for T<100 K. It should then be possible to fit this temperature dependence with the theoretical dependence predicted for a combination of the phonon-assisted processes (34) and (35).

Laulicht and Meirman [20] have fitted the experimental curve with an equation of the form:

$$W(T) = an_1(T)T + bn_0(T) \quad (38)$$

where a and b are constants to be adjusted in the fitting and $n_1(T)$ and $n_0(T)$ are the thermal population of the $^7F_1$ and the $^7F_0$ levels. (38) is the application of (11) to the case under investigation. The first term describes the process via $^7F_1$, accompanied by the emission of one phonon having $\tilde{\nu} = 25\ cm^{-1}$. When $\hbar\omega < kT$, which is true in the present case, it is possible to approximate (9) with:

$$W_{PAET}(T) = W_{PAET}(0)kT/\hbar\omega \quad (39)$$

and this accounts for the proportionality to T. The second term describes the process via $^7F_0$, which occurs through the emission of one phonon, with $\tilde{\nu}=1000\ cm^{-1}$. When $\hbar\omega > kT$, W_{PAET} becomes independent of temperature, and this is valid for this second mechanism.

The calculated W(T) is in fair agreement with the experimental data, but in the fitting process the $^7F_1$ level has been considered as a non-degenerate state, lying at 360 cm^{-1}, which is not correct. Anyway, an equation of the form (38) should, in general, be the best way to describe the temperature dependence of the energy transfer probability in this case.

Laulicht and Meirman have not considered the possibility that the $Tb^{3+} \rightarrow Eu^{3+}$ energy transfer occurs through a many-body process. The concentration of the two ions in the crystal is high enough to allow such mechanisms, and the energy level structure of Tb^{3+} and Eu^{3+} is favourable. For instance, the energy difference $\Delta E = 1000\ cm^{-1}$ which must be bridged for the process described in (34) to occur, can be available with the promotion of a second Eu^{3+} ion to the $^7F_2$ level, which has crystal field components are at 937, 960, 1070, 1097 and 1180 cm^{-1} in EuP_5O_{14} [28]. It is clear that the temperature dependence of the energy transfer probability in this case would be completely different from the one observed (W(T) would decrease when the temperature increases), and so this process should not be operative in $Tb_{0.66}Eu_{0.33}P_5O_{14}$. Anyway, it would be interesting to study the dependence of the energy transfer probability on the concentration of Tb^{3+} and Eu^{3+} in order to check if a many-body process like the one mentioned (whose probability is proportional to $[n_0(T)]^2$, following (32)) could occur. A more detailed knowledge of the energy level structure of TbP_5O_{14} is needed in order to obtain a complete characterization of the energy transfer process.

3.2. Energy transfer from Tb^{3+} to Eu^{3+} in $Cs_2NaTb_{1-x}Eu_xCl_6$

The lanthanide hexachloroelpasolites $Cs_2NaLnCl_6$ have been the subject of considerable spectroscopic interest, because the lanthanide ions occupy an exactly octahedral site and because the crystals are much easier to prepare and to handle, compared to the isostructural hexabromoelpasolites $Cs_2NaLnBr_6$. Despite the fact that the energy levels of the Ln^{3+} ions in the hexachloroelpasolite lattice are fairly well known [29], the studies concerning the energy transfer processes in this lattice have been limited to some individual cases [30-34].

Laser-excited luminescence spectra and decay curves of the luminescence have been measured at 293 and 80 K for the crystals of stoichiometry $Cs_2NaTb_{1-x}Eu_xCl_6$, with x = 0.01, 0.02, 0.03, 0.04, 0.05, 0.07, 0.10 and 0.15, and the energy transfer processes have been investigated [35]. The luminescence spectra at 293 and 80 K excited at 488.0 nm, i.e. in the $^5D_4$ level of Tb^{3+}, show bands which are readily assigned to transitions from the $^5D_4$(Tb) and $^5D_0$(Eu) levels to their respective ^7F_J manifolds, whereas no transitions from $^5D_1$(Eu) are detected. This shows that energy transfer from Tb^{3+} to Eu^{3+} occurs either directly to the $^5D_0$ level, or via the $^5D_1$ level, followed by very fast relaxation from $^5D_1$, which is therefore non-luminescent, to $^5D_0$.

After excitation at 527.3 nm, i.e. directly in the $^5D_1$ level of Eu^{3+}, the transitions from $^5D_1$ to the ^7F_J manifold are intense and clearly measurable at both 293 and 80 K. The decay curves of the luminescence from $^5D_1$ (λ_{exc} = 527.3 nm, λ_{em} = 559.0 nm) are exponential for all the compositions. The lifetime of $^5D_1$ is almost independent of the composition and is about 23 μs at 293 K and 1.2 ms at 80 K (compared to

65 μs at 293 K and 0.54 ms at 80 K for $Cs_2NaEuCl_6$ and 5.0 ms at 293 and 7.0 ms at 80 K for $Cs_2NaY_{0.99}Eu_{0.01}Cl_6$). The $^5D_1$ level is therefore luminescent and this suggests that the energy transfer occurs directly via the $^5D_0$ level.

The decay curves of the luminescence from the $^5D_4$ level of Tb^{3+} (λ_{exc}= 488.0 nm, λ_{em}= 548.0 nm) are exponential at 293 and 80 K for all the compositions. This agrees with an energy transfer process preceded by a fast migration of the excitation among the Tb^{3+} donors [2]. The decay curves of the luminescence from the $^5D_0$ level of Eu^{3+}, when excited via the $^5D_4$ level of Tb^{3+} (λ_{exc}= 488.0 nm, λ_{em}= 593.0 nm) are very well fitted by an equation of the type:

$$f(t) = A\,[\exp(-k_1t)-\exp(-k_2t)] \tag{40}$$

where the risetime k_2^{-1} is experimentally identical to the corresponding lifetime of the $^5D_4$ level of Tb^{3+}, for all the compositions and for both 293 and 80 K. This agrees with the solution of the rate equation describing the direct transfer of excitation from the "short-lived" $^5D_4$ level to the "long-lived" $^5D_0$ level, which is what is expected in the case of a rapid migration of the excitation among the donors [17].

A measure of the efficiency η_T of the energy transfer can be obtained from the equation [36]:

$$\eta_T = 1-\tau_D/\tau_D^0 \tag{41}$$

where τ_D^0 is the radiative lifetime of the donor alone and τ_D is its lifetime in the presence of the acceptor. The efficiencies of the energy transfer are reported in Table I. The energy transfer process is

Table I. Overall efficiencies η_T of the energy transfer process from the $^5D_4$ level of Tb^{3+} in $Cs_2NaTb_{1-x}Eu_xCl_6$ at 293 and 80 K.

x	η_T(293 K)	η_T(80 K)
0.01	0.09	0.06
0.02	0.19	0.14
0.03	0.38	0.30
0.04	0.56	0.51
0.05	0.55	0.45
0.07	0.68	0.59
0.10	0.76	0.68
0.15	0.87	0.77

clearly not very efficient, compared with the efficiencies measured for resonant energy transfer between lanthanides in hexachloroelpasolites in the same conditions (for instance $\eta_T = 0.99$ at 80 K for Ho^{3+}-->Yb^{3+} in $Cs_2NaHo_{0.99}Yb_{0.01}Cl_6$ [30]) and this indicates that a higher order process is operative.

When the energy transfer probability W_{ET} calculated following (37) is plotted against the % fractional concentration of Eu^{3+} for the $Cs_2NaTb_{1-x}Eu_xCl_6$ crystals at 293 and 80 K, W_{ET} appears to increase linearly with x over the range of concentrations considered at both 293 and 80 K, apart from a deviation towards a higher power dependence of W_{ET} for x>0.10 at 293 K. This behaviour is in fair agreement with a pairwise Tb^{3+}-->Eu^{3+} transfer process for low values of x being dominant. The deviation towards a higher power dependence on x for x>0.10 indicates that, at 293 K and for relatively high concentrations of Eu^{3+}, a many-body process (probably involving one Tb^{3+} and two Eu^{3+} ions) becomes important. The ratio of the energy transfer probability at 293 K and 80 K ($W_{ET}(293)/W_{ET}(80)$) is roughly constant over the considered concentration range and on average is equal to 2.2. An energy transfer mechanism involving the thermally populated $^7F_1$ level (which is not split by the crystal field and lies at 352 cm^{-1} in $Cs_2NaEuCl_6$ [37]) would imply a much stronger temperature dependence of the energy transfer probability and therefore must be regarded as not important in the crystals under investigation. For these reasons, the dominant transfer mechanism can be assigned as (32) for low concentrations of Eu^{3+}.

The transition $^7F_0$-->$^5D_0$ of Eu^{3+} is strictly forbidden at first order in octahedral symmetry and is usually not observed [38]. It has never been detected in the hexachloroelpasolite lattice by means of absorption and luminescence spectroscopy. The excitation spectrum of $Cs_2NaEuCl_6$ (λ_{em}=705.6 nm) has been measured in the region 574-590 nm at 10 and 80 K [39]. The relatively strong features at 17312, 17333 and 17391 cm^{-1} are present in both the 80 and the 10 K spectrum and are assigned to vibronic origins involving the ν_4(TO) and ν_4(LO) internal modes and the Na motion mode, based on a very weak 0-0 transition at 17206 cm^{-1}. In the 80 K spectrum, hot bands appear to the low energy side of the origin at 17098, 17075 and 17019 cm^{-1}. They are assigned to hot vibronic origins promoted by the ν_4(TO), ν_4(LO) and Na motion modes respectively in the electronic ground state. The position of the 0-0 origin is in good agreement with the energy of the $^5D_0$ level as deduced from the luminescence spectrum (17210 cm^{-1}) [37] and calculated (17209 cm^{-1}) [29].

At 80 K several of the hot bands in the excitation spectrum overlap strongly with bands belonging to the $^5D_4$-->$^7F_4$ emission transition of $Cs_2NaTbCl_6$ [40]. At 293 K, the hot high energy tail of the emission from the $^5D_4$ level in $Cs_2NaTbCl_6$ overlaps almost completely the structured $^7F_0 \rightarrow {}^5D_0$ transition in $Cs_2NaEuCl_6$. Due to the uncertainties about the assignment of the individual bands for the $^5D_4 \rightarrow {}^7F_4$ transition in $Cs_2NaTbCl_6$ [41, 42], no detailed assignments can be made for the non-resonant transfer of the excitation between the crystal field states of the $^5D_4$ level of Tb^{3+} and the $^5D_0$ level of Eu^{3+}. For this reason, and also because of the complexity of the process, involving many different phonons, it is not possible to fit the observed temperature dependence of the energy transfer probability to an equation like (11). The experimental data are anyway in complete agreement with a mechanism similar to (32) for the pairwise energy transfer from Tb^{3+} to Eu^{3+}, for low concentration of the latter ion.

At higher fractional concentrations x of Eu^{3+}, the deviation from the linear dependence of W_{ET} against x can be explained by taking into account three-body mechanisms involving crystal field states of the $^5D_4$ level of Tb^{3+} and of two ^7F_J levels of Eu^{3+}, with a complete matching of the electronic energies of the donor and of the acceptors. Because of the many possibilities and of the complexity of the problem, it is neither possible to give a detailed assignment for the many-body transfer processes in $Cs_2NaTb_{1-x}Eu_xCl_6$, nor to account satisfactorily for the temperature dependence of the energy transfer probability when x is relatively high ($x>0.10$).

The analysis of the dependence of W_{ET} upon x, for $x<0.15$, cannot evidentiate the presence of many-body transfer processes involving two or more Tb^{3+} ions (one in the $^5D_4$ level and one or more in the $^7F_6$ level) and one Eu^{3+} ion in the $^7F_0$ level. These processes are indeed possible, but only a study of the energy transfer processes in crystals having a fractional concentration x of Eu^{3+} higher than 0.15 and in diluted crystals containing fairly high concentrations of Tb^{3+} and Eu^{3+}, possibly over a wider range of temperatures, can show their importance.

ACKNOWLEDGMENTS

The author thanks Professor Colin D. Flint (University of London) and Professor Gianluigi Ingletto (University of Parma) for valuable comments. This work has been carried out during the tenure of a CNR (Consiglio Nazionale delle Ricerche) Fellowship at the Department of Chemistry, Birkbeck College, University of London, United Kingdom.

REFERENCES

[1] F.E. Auzel, Proceedings of the IEEE 61, 758 (1973).

[2] R.K. Watts in Optical Properties of Ions in Solids, ed. by B. Di Bartolo (Plenum, N.Y., 1975) p. 307.

[3] L.A. Riseberg and M.J. Weber, 'Relaxation Phenomena in Rare Earth Luminescence' in Progress in Optics, ed. by E. Wolf, Vol. 14 (1976), p. 91.

[4] R. Reisfeld and C.K. Jørgensen, Lasers and Excited States of Rare Earths (Springer-Verlag, Berlin, 1977), chapter 4.

[5] R.C. Powell and G. Blasse, Structure and Bonding 42, 43 (1980).

[6] Energy Transfer Processes in Condensed Matter, ed. by B. Di Bartolo (Plenum, N.Y., 1984).

[7] W.M. Yen and P.M. Selzer, in Laser Spectroscopy of Solids, ed. by W.M. Yen and P.M. Selzer, (Springer-Verlag, Berlin, 1986), p. 141.

[8] Th. Förster, Ann. Physik 2, 55 (1948).

[9] D.L. Dexter, J. Chem. Phys. 21, 836 (1952).

[10] R. Orbach in Optical Properties of Ions in Solids, ed. by B. Di Bartolo (Plenum, N.Y., 1975) p. 355.

[11] C.J. Ballhausen and Aa.E. Hansen, Annu. Rev. Phys. Chem. 23, 15 (1972).

[12] W.B. Gandrud and H.W. Moos, J. Chem. Phys. 49, 2170 (1968).

[13] T. Miyakawa and D.L. Dexter, Phys. Rev. B 1, 2961 (1970).

[14] L.A. Riseberg and H.W. Moos, Phys. Rev. 174, 429 (1968).

[15] N. Yamada, S. Shionoya and T. Kushida, J. Phys. Soc. Japan 32, 1577 (1972).

[16] F.K. Fong and D.J. Diestler, J. Chem. Phys. 56, 2875 (1972).

[17] W.J.C. Grant, Phys. Rev. B 2, 648 (1971).

[18] See the literature quoted in [16] and [17].

[19] W.W. Holloway, Jr., M. Kestigian and R. Newman, Phys. Rev. Lett. 11, 458 (1963).

[20] I. Laulicht and S. Meirman, J. Lumin. 34, 287 (1986).

[21] T. Kobayashi, T. Sawada, H. Ikeo, K. Muto and J. Kai, J. Phys. Soc. Japan 40, 595 (1976).

[22] S. Colak and W.K. Zwicker, J. Appl. Phys. 54, 2156 (1983).

[23] I. Laulicht and I. Pe'er, J. Lumin. 24/25, 87 (1981).

[24] B. Blanzat, J.P. Denis and J. Loriers, Proc. Tenth Rare Earth Research Conf. 2, 1170 (1973).

[25] T.K. Anh, N.D. Hung, N.H. Chi and N.M. Son, Phys. Status Solidi (a) 84, K159 (1984).

[26] I. Laulicht, S. Meirman and B. Ehrenberg, J. Lumin. 31/32, 814 (1984).

[27] W.K. Unger, Solid State Comm. 29, 601 (1979).

[28] C. Brecher, J. Chem. Phys. 61, 2297 (1974).

[29] C.A. Morrison, R.P. Leavitt and D.E. Wortman, J. Chem. Phys. 73, 2580 (1980).

[30] A.K. Banerjee, F.S. Stewart-Darling, C.D. Flint and R.W. Schwartz, J. Phys. Chem. 85, 146 (1981).

[31] P.A. Tanner, Mol. Phys. 53, 813 (1984).

[32] P.A. Tanner, Mol. Phys. 54, 883 (1985).

[33] P.A. Tanner, Mol. Phys. 58, 317 (1986).

[34] P.A. Tanner, J. Chem. Soc., Faraday Trans. 2 83, 553 (1987).

[35] M. Bettinelli and C.D. Flint, to be published.

[36] R. Reisfeld and N. Lieblich-Sofer, J. Solid State Chem. 28, 391 (1979).

[37] C.D. Flint and F.L. Stewart-Darling, Mol. Phys. 44, 61 (1981).

[38] R.D. Peacock, Structure and Bonding 22, 83 (1975).

[39] C.D. Flint and M. Bettinelli, to be published.

[40] L.C. Thompson, O.A. Serra, J.P. Riehl, F.S. Richardson and R.W. Schwartz, Chem. Phys. 26, 393 (1977).

[41] T.R. Faulkner and F.S. Richardson, Mol. Phys. 36, 193 (1978).

[42] F.L. Stewart-Darling, Thesis, Birkbeck College, University of London (1982).

VIBRONIC INTERACTIONS IN THE ELECTRONIC GROUND STATE: VIBRATIONAL CIRCULAR DICHROISM SPECTROSCOPY

Philip J. Stephens
Department of Chemistry
University of Southern California
Los Angeles, California 90089-0482 USA

ABSTRACT: A theory of Vibrational Circular Dichroism (VCD) has been developed by the author and implemented at the ab initio self-consistent field (SCF) level of approximation. VCD spectra calculated from this theory for several small chiral organic molecules are in excellent overall agreement with experiment, when basis sets of adequate sophistication are used and when the ab initio SCF force fields are scaled following the procedure of Pulay and coworkers. VCD spectroscopy, together with a priori theoretical calculations, now affords a new technique for the study of the stereochemistry of chiral molecules.

Vibrational Circular Dichroism (VCD) is the circular dichroism of vibrational transitions. Circular dichroism is a phenomenon specific to chiral molecules. In the case of fundamental vibrational transitions of molecules, VCD occurs in the infrared spectral region. VCD was first observed in chiral organic molecules in 1974 [1]. Since that time the technology for the measurement of infrared circular dichroism has advanced very considerably [2]. It is now possible to measure VCD spectra from frequencies below 700 cm^{-1} [3] throughout the infrared spectral region.

The theoretical prediction of VCD spectra has advanced somewhat more slowly. A number of theoretical approaches have been proposed, leading to relatively simple equations, but have not proven to be satisfactory [4]. Very recently, however, a theory has been developed, placing the theory of VCD on the same level of rigor as the theory of vibrational absorption intensities [5]. Further, the calculational machinery required for the implementation of this theory at the ab initio self-consistent field (SCF) level of approximation has been developed [6] and applied to the prediction of the VCD spectrum of a number of small chiral organic molecules [7]. Where compared, predicted and experimental VCD spectra have been in remarkably good agreement.

C. D. Flint (ed.), Vibronic Processes in Inorganic Chemistry, 371–384.
© 1989 by Kluwer Academic Publishers.

At the present time, therefore, it is possible both to measure and to predict the VCD spectra of simple chiral molecules with substantial accuracy. Clearly, the potential of the VCD phenomenon to enable the stereochemistry of chiral molecules to be elucidated is now capable of realisation.

In this article, we will summarise the fundamental equations used in the prediction of vibrational absorption and circular dichroism spectra. The technology currently available for their implementation will be described. Spectra predicted for a small organic molecule will be presented and compared to experimental spectra. The limitations of current techniques and foreseeable developments in the prediction of VCD spectra will be commented upon.

Theory

The general form of the vibrational absorption spectrum of a molecule can be written [8]:

$$\epsilon(\nu) = \gamma \sum_i \nu \; D_i f_i(\nu_i,\nu) \tag{1}$$

where ϵ is the extinction coefficient, γ is a constant, D_i is the dipole strength of the **ith** transition, and f_i is a normalized line shape function:

$$\int f_i(\nu)d\nu = 1. \tag{2}$$

Equation 1 is appropriate to a dilute, isotropic solution of randomly oriented molecules, obeying Beer's Law. It ignores the effect of the solvent on the molecular absorption. The molecular property D_i is given by

$$D_i = |\langle g|\vec{\mu}_{el}|e\rangle|^2 \tag{3}$$

where g and e are the ground and excited states involved in the **ith** transition and $\vec{\mu}_{el}$ is the electric dipole moment operator. As is extremely wellknown, within the harmonic approximation for molecular vibrational motion, for the fundamental (0→1) transition in the **ith** mode D_i is given by

$$D_i^{0\to1} = \left(\frac{\hbar}{2\omega_i}\right) \left| \left(\frac{\partial \vec{\mu}_{el}^G}{\partial Q_i}\right)_0 \right|^2 \tag{4}$$

where $\hbar\omega_i$ is the energy of the **ith** mode and Q_i is its normal coordinate. $(\partial\vec{\mu}_{el}^G/\partial Q_i)_0$ is the derivative with respect to the normal coordinate Q_i of the electric dipole moment of the ground electronic state G of the molecule, evaluated at the equilibrium molecular geometry.

Using the relationship between the normal coordinates Q_i and the set of Cartesian nuclear displacement coordinates, $X_{\lambda\alpha}$(λ = nucleus; α = x,y,z):

$$X_{\lambda\alpha} = \sum_i S_{\lambda\alpha,i} Q_i \qquad (5)$$

where $S_{\lambda\alpha,i}$ is determined by the force field of the molecule, equation 4 can be written alternatively as

$$D_i^{0\to 1} = \left(\frac{\hbar}{2\omega_i}\right) \sum_\beta \left\{ \sum_{\lambda\alpha} P^\lambda_{\alpha\beta} S_{\lambda\alpha,i} \right\} \left\{ \sum_{\lambda'\alpha'} P^{\lambda'}_{\alpha'\beta} S_{\lambda'\alpha',i} \right\} \qquad (6)$$

where

$$P^\lambda_{\alpha\beta} = \left(\frac{\partial(\mu^G_{el})_\beta}{\partial X_{\lambda\alpha}}\right)_0 \qquad (\alpha,\beta = x,y,z) \qquad (7)$$

is referred to as the atomic polar tensor of nucleus λ [9].

The prediction of the fundamental absorption spectrum, $\epsilon(\nu)$, thus involves the prediction of the vibrational absorption frequencies, ω_i, the normal coordinates, Q_i, or (equivalently) the $S_{\lambda\alpha,i}$ matrix, and the atomic polar tensors $P^\lambda_{\alpha\beta}$. Both ω_i and $S_{\lambda\alpha,i}$ follow from the vibrational force field. If this is written in terms of the Cartesian displacement coordinates, $X_{\lambda\alpha}$:

$$W_G = W_G^0 + \frac{1}{2} \sum_{\lambda\alpha,\lambda'\alpha'} k_{\lambda\alpha,\lambda'\alpha'} X_{\lambda\alpha} X_{\lambda'\alpha'} + \ldots \qquad (8)$$

where W_G^0 is the energy at the equilibrium molecular geometry, ω_i and $S_{\lambda\alpha,i}$ are obtained via diagonalisation of the mass-weighted force constant matrix $k_{\lambda\alpha,\lambda'\alpha'}$. In the end, therefore, prediction of $\epsilon(\nu)$ requires calculation of the force constant matrix $k_{\lambda\alpha,\lambda'\alpha'}$ and the atomic polar tensors, $P^\lambda_{\alpha\beta}$. $k_{\lambda\alpha,\lambda'\alpha'}$ is the matrix of second-derivatives of the molecular energy:

$$k_{\lambda\alpha,\lambda'\alpha'} \equiv \left(\frac{\partial^2 W_G}{\partial X_{\lambda\alpha} \partial X_{\lambda'\alpha'}}\right) \qquad (9)$$

$P^\lambda_{\alpha\beta}$ is the matrix of first derivatives of the molecular electric dipole moment

$$\vec{\mu}^G_{el} = \langle \psi_G | \vec{\mu}_{el} | \psi_G \rangle \qquad (10)$$

where ψ_G is the electronic wavefunction of the ground state of the molecule.

The general form for the vibrational circular dichroism spectrum of a molecule parallels equation 1 and can be written [8]

$$\Delta\epsilon(\nu) = 4\gamma \sum_i \nu \; R_i f_i(\nu_i, \nu) \tag{11}$$

where R_i is the rotational strength of the **ith** transition, defined by

$$R_i = \mathrm{Im}[\langle g|\vec{\mu}_{el}|e\rangle . \langle e|\vec{\mu}_{mag}|g\rangle] \tag{12}$$

where $\vec{\mu}_{mag}$ is the magnetic dipole operator. As shown by Stephens [5], the expression for the rotational strength of the **ith** fundamental transition within the harmonic approximation, paralleling that of equation 6 for the dipole strength, is

$$R_i^{0\to1} = \hbar^2 \mathrm{Im} \sum_\beta \{\sum_{\lambda\alpha} P^\lambda_{\alpha\beta} S_{\lambda\alpha,i}\}\{\sum_{\lambda'\alpha'} M^{\lambda'}_{\alpha'\beta} S_{\lambda'\alpha',i}\} \tag{13}$$

where $M^\lambda_{\alpha\beta}$ is termed the atomic axial tensor [5b] and is a molecular property paralleling the atomic polar tensor. The expression for $M^\lambda_{\alpha\beta}$ is somewhat more complex than that for $P^\lambda_{\alpha\beta}$. It is

$$M^\lambda_{\alpha\beta} = I^\lambda_{\alpha\beta} + J^\lambda_{\alpha\beta}$$

$$I^\lambda_{\alpha\beta} = \langle (\frac{\partial\psi_G}{\partial X_{\lambda\alpha}})_0 | (\frac{\partial\psi_G(H_\beta)}{\partial H_\beta})_0 \rangle \tag{14}$$

$$J^\lambda_{\alpha\beta} = \frac{i}{4\hbar c} \sum_\gamma \epsilon_{\alpha\beta\gamma} (Z_\lambda e) R^0_{\lambda\gamma}$$

Here, $I^\lambda_{\alpha\beta}$ and $J^\lambda_{\alpha\beta}$ are electronic and nuclear contributions to $M^\lambda_{\alpha\beta}$. $I^\lambda_{\alpha\beta}$ involves the overlap of two derivative wavefunctions : $(\partial\psi_G/\partial X_{\lambda\alpha})_0$ and $(\partial\psi_G(H_\beta)/\partial H_\beta)_0$. The former represents the linear variation of the electronic wavefunction of the molecular ground state G with respect to nuclear displacement at the molecular equilibrium geometry. The latter represents the linear variation of the electronic wavefunction of the molecular ground state G at the equilibrium geometry with respect to magnetic field, H , when a magnetic field perturbation of the form

$$\mathcal{H}'(H_\beta) = -(\mu^e_{mag})_\beta H_\beta \tag{15}$$

is applied to the molecule. In equation 15, $\vec{\mu}^e_{mag}$ is the electronic contribution to the magnetic moment operator. $J^{\lambda}_{\alpha\beta}$ simply involves nuclear charges, $Z_{\lambda}e$, and equilibrium positions, $\mathbf{R}^{0}_{\lambda}$.

Atomic axial tensors differ from atomic polar tensors in being intrinsically origin dependent. When the origin dependence of $M^{\lambda}_{\alpha\beta}$ is examined, it transpires that it can be written [5b]

$$(M^{\lambda}_{\alpha\beta})^{0} = (M^{\lambda}_{\alpha\beta})^{0'} + \frac{i}{4\hbar c} \sum_{\gamma\delta} \epsilon_{\beta\gamma\delta} \Delta_{\gamma} P^{\lambda}_{\alpha\delta} \tag{16}$$

where 0 and 0′ are two origins and $\vec{\Delta}$ is the vector from 0 to 0 . Equation 16 is an example of the fascinating interrelationship between the electric and magnetic properties of a molecule. Stephens has used equation 16 to define a more sophisticated expression for the atomic axial tensors, with respect to an origin 0, in the form [5b]:

$$(M^{\lambda}_{\alpha\beta}) = (M^{\lambda}_{\alpha\beta})\lambda + \frac{i}{4hc} \sum_{\gamma\delta} \epsilon_{\beta\gamma\delta} R^{0}_{\lambda\gamma} P^{\lambda}_{\alpha\delta} \tag{17}$$

where $(M^{\lambda}_{\alpha\beta})^{\lambda}$ is given by equation 14 and the superscript indicates that each $(M^{\lambda}_{\alpha\beta})^{\lambda}$ tensor is evaluated with the origin at the equilibrium position of nucleus λ - the position $\mathbf{R}^{0}_{\lambda}$ relative to 0. Equation 17 is termed the equation for the atomic axial tensor in the Distributed Origin with origins at nuclei gauge. There are two reasons for the use of equation 17. First, when substituted into equation 13 it leads to rotational strengths that are independent of origin - a necessary property for any physical observable! Second, using the methods to be described below, it provides the most accurate algorithm for the calculation of $M^{\lambda}_{\alpha\beta}$ tensors yet known.

In sum, to predict the fundamental VCD spectrum, $\Delta\epsilon(\nu)$, requires all of the quantities involved in the prediction of the absorption spectrum, $\epsilon(\nu)$, together with the atomic axial tensors, $(M^{\lambda}_{\alpha\beta})^{\lambda}$. The absorption spectrum is automatically available whenever all quantities required for calculating the VCD spectrum are at hand.

Implementation

One of the most dramatic developments in quantum chemistry in the last decade has been the improvement in the techniques and computational programs for ab initio molecular orbital calculation of molecular properties [10]. It has become routine to calculate the first and second derivatives of the ground state energy at the self-consistent field (SCF) level of approximation, and thence to calculate equilibrium molecular geometries and vibrational force fields. In addition, not only dipole moments but also atomic polar tensors can be routinely calculated at the SCF level. Thus, it is possible to calculate the frequencies and intensities of a vibrational absorption spectrum entirely ab initio at the SCF level of approximation.

For very simple molecules, predicted vibrational frequencies and intensities recognisably resemble experiment [10]. However, frequencies calculated at the SCF level are typically 10-15% higher than experimental frequencies, implying that force constants are in error by as much as 20-30%. As molecular size and the number of vibrational frequencies increase, the relationship between theory and experiment becomes less perfect, reflecting increasing error in the normal coordinates due to the errors in the force field.

There are two ways in which the residual errors in vibrational force fields, frequencies and normal coordinates calculated at the SCF level can be removed. One is to improve the level of ab initio approximation. A variety of methods exist by which correlation corrections can be included, such as Møller-Plesset (MP) perturbation theory, configuration interaction (CI) and so on. Methods for the efficient calculation of vibrational frequencies at post-SCF levels of approximation are being developed and implemented at this time [11] and are starting to appear in the most up-to-date ab initio molecular orbital programs. Results reported so far are very impressive, in the extent to which the errors found at the SCF level are reduced and to which predicted frequencies are in agreement with experiment. In addition, post-SCF methods for the calculation of atomic polar tensors are being developed simultaneously [12] leading to improved agreement between predicted and experimental absorption intensities. It is very likely that post-SCF methods will soon become as routine as SCF methods are now. However, at this time, post-SCF calculations are still restricted to a small number of programs and are practical only for quite small molecules.

An alternative method by which SCF force fields, frequencies and normal coordinates can be improved has been popularised by Pulay and coworkers [13]. In this method the SCF force field is empirically refined to fit experimental frequencies, somewhat in the manner of traditional methods for deriving vibrational force fields. However, the procedure of Pulay is constrained in two ways which eliminate most of the ambiguity of the traditional empirical force fields. First, the fact that the initial guess is a SCF calculated force field, and therefore already quite close to the true force field, enormously constrains the outcome. Second, the refinement is carried out using a very simple algorithm, in which it is required that true force constant matrix elements, k_{ij}, expressed with respect to a basis of internal (not Cartesian) coordinates, are related to SCF ones, K_{ij}, by

$$k_{ii} = \alpha_i K_{ii}; \qquad k_{ij} = \sqrt{\alpha_i \alpha_j}\, K_{ij} \tag{18}$$

where α_i are so-called scaling factors. This procedure has by now been used to analyse the vibrational spectra of a substantial number of molecules. It has the advantage that post-SCF calculations are not required. Its disadvantages are that it is empirical and non-unique. Not only does the force field depend on an assignment of the experimental spectrum, but also on the specific choice of coordinates in terms of which the force field is expressed and on the number of scaling factors adopted. In addition, the Pulay procedure is not applicable to the atomic polar tensors and does not address the errors in intensities arising from errors in the SCF atomic polar tensors.

Predictions of VCD spectra require in addition calculation of atomic axial tensors. Following the introduction of these tensors by Stephens, their calculation has been implemented at the SCF level of approximation [6]. At this time, the most efficient implementation is contained in the CADPAC program, an ab initio molecular orbital program developed at Cambridge University by Drs. R.D. Amos, N.C. Handy and their coworkers [14]. CADPAC is a pnemonic for Cambridge Analytical Derivatives Package. In addition to atomic axial tensors, CADPAC also enables state-of-the-art calculations of energy first and second derivatives with respect to nuclear coordinates and of atomic polar tensors at the SCF level. All molecular properties required for the prediction of vibrational absorption and circular dichroism spectra based on SCF calculations are thus available from the CADPAC program.

Application

We describe here a single application of the theory described above. The molecule of interest is propylene oxide (epoxypropane)

a small, simple, chiral organic molecule. The VCD of this molecule has probably been more thoroughly studied than any other at this time [15].

VCD spectra have been obtained [15] over the frequency ranges 650-1600 cm^{-1} and 2800-3200 cm^{-1}, which span all but three of the fundamental transitions of propylene oxide (those lying < 500 cm^{-1}). The spectra have been studied for the neat liquid and for dilute solutions in CS_2 and CCl_4. The spectra are qualitatively identical and quantitatively very similar. The absorption and VCD spectra of the neat liquid over the range 700-1600 cm^{-1} are shown in Figures 1 and 2.

VCD and absorption spectra have been calculated [15] using SCF atomic polar and axial tensors and force fields obtained from SCF force fields using the methodology of Pulay. In doing so, the following choices exist. First the equilibrium molecular geometry can be either taken from experiment, calculated or obtained by empirical correction of calculated geometries. In the case of propylene oxide, an experimental geometry does not exist and we have used the latter two approaches. The correction of theoretically calculated geometries using corrections obtained from comparison of calculated and experimental geometries for other molecules has been advocated by Pulay and coworkers in the derivation of vibrational force fields [13]. Second, in deriving a scaled force field from an ab initio calculation, a set of internal coordinates and, subsequently, the number and distribution of independent scaling factors must be chosen. In our work, we have carried out scaling in stages, each one incrementing the number of scaling factors over the previous number. We start with one scaling factor, which scales all frequencies by the same amount and does not change the normal coordinates. Then a small number of independent scaling factors is introduced, such that chemically similar internal coordinates (e.g. all C-H stretching coordinates) receive identical scaling factors. Following this the number of scaling factors is gradually incremented until no further improvement in the fit between calculated and experimental frequencies is obtained. In the case of propylene oxide we have used a maximum of 10 scaling factors. Thirdly, an assignment of the vibrational spectrum must be chosen before the scaled force field can be derived. This involves the differentiation of fundamental and overtone/combination transitions and the evaluation of the relationship between the ab initio calculated and observed frequencies (as it does in the development of a vibrational force field by any empirical method). In our work on propylene oxide we were assisted by the availability of matrix-isolation spectra in N_2 and Ar matrices, in which line widths were much less than in liquid phase spectra and spectral resolution consequently superior [15]. Fourthly, in calculating geometry, force field and atomic polar and axial tensors a basis set must be chosen. This need not be the same for different properties. In our work on propylene oxide we have used predominantly the basis sets developed by Pople and coworkers: STO-3G, 3-21G, 4-31G, 6-31G, 6-31G*, 6-31G** and 6-311G** [10]. STO-3G is a minimal basis set. 3-21G, 4-31G and 6-31G are split-valence basis sets and those bearing asterisks are split-valence basis sets to which polarisation functions are added. One asterisk indicates polarisation functions are added only to "heavy" atoms (i.e. the first row atoms C, N, O etc.). Two asterisks implies polarisation functions are also added to H atoms. Lastly, there may be a practical choice of computer program and machine. In our studies of propylene oxide, for pragmatic, historical reasons, geometry optimisation and vibrational force field calculations were carried out using the GAUSSIAN 82 program [10], while atomic polar and axial tensors were calculated using version 3.0 of CADPAC. In both cases, calculations were executed using the San Diego Supercomputer Center CRAY-XMP machine.

Selected results from our studies are shown in Figures 1 and 2. In all cases, vibrational force fields were obtained by fitting absorption frequencies. Absorption and VCD intensities were then calculated with no further use of experimental data. Predicted spectra assume Lorentzian line shapes and use line widths obtained by fitting experimental absorption spectra. (Note that absorption spectra are measured at 1 cm^{-1} resolution, while VCD are measured at lower resolution in order to achieve adequate signal-to-noise ratios).

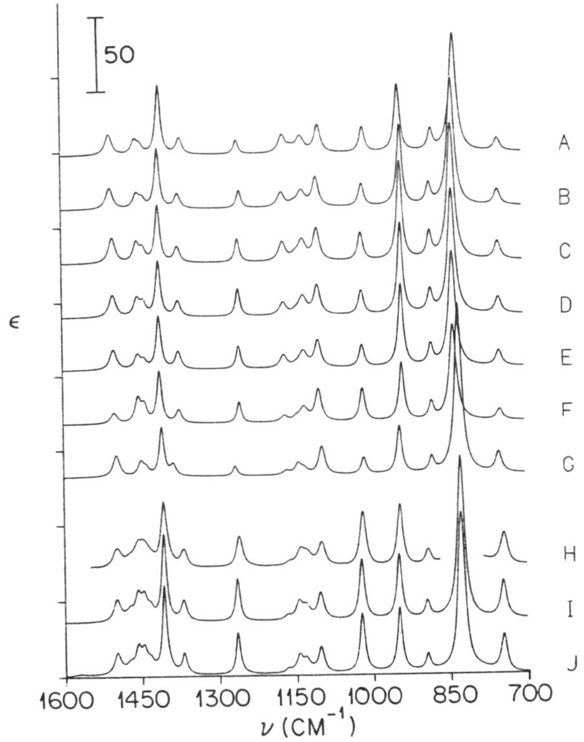

Figure 1: Calculated (A-G) and experimental (H-J) absorption spectra of propylene oxide. **A.** STO-3G optimised geometry, 6-31G** vibrational force field and 6-31G** atomic polar and axial tensors: STO-3G/6-31G**/6-31G**. **B.** 6-31G**/6-31G**/6-31G**. **C.** CORR/6-31G**/DZ/IP. **D.** CORR/6-31G**/6-311G**. **E.** CORR/6-31G**/6-31G**. **F.** CORR/6-31G**/4-31G. **G.** CORR/4-31G/6-31G**. (CORR is a corrected 4-31G optimised geometry.) **H** and **J**. Experimental spectra at resolution used for VCD measurement (Figure 2) and 1 cm^{-1} respectively. **I.** Lorentzian fit to spectrum **J**. Spectra A-G are calculated using Lorentzian line shapes obtained from fit I.

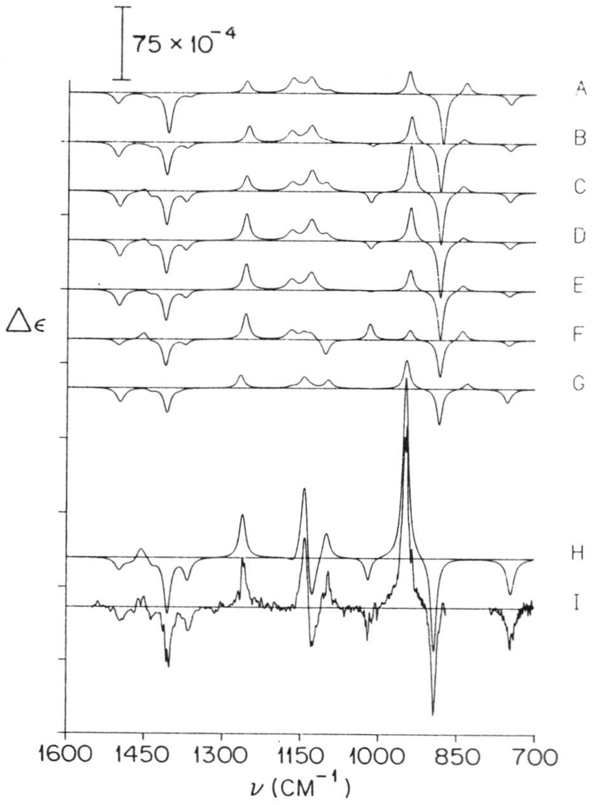

Figure 2: Calculated (A-G) and experimental (H,I) VCD spectra of S-(-)-propylene oxide. A-G as in Figure 1. I. Experimental VCD spectrum (resolution varying from 6-10 cm^{-1}). H. Lorentzian fit to spectrum I. Spectra A-G are calculated using Lorentzian line shapes obtained from fit H.

We find that

1) all force fields give similar results, with the exception of STO-3G force fields which yield a worse fit to experimental frequencies and absorption and VCD spectra in worse agreement with experiment;

2) with respect to the basis set used in calculating atomic polar and axial tensors, basis sets containing polarisation functions yield superior results to split-valence basis sets and enormously superior results to minimal basis sets;

3) all choices of geometry yield similar results

4) when the force field is derived using basis sets larger than minimal and the atomic polar and axial tensors are calculated using basis sets containing polarisation functions, predicted absorption and VCD spectra are in excellent overall agreement with experiment. The principal disagreement occurs in the region 1100-1170 cm^{-1}. In this region, the results are particularly sensitive to the choices of basis set made in calculating geometry, force field and atomic polar and axial tensors.

Similar conclusions are reached from the study of the C-H stretching region [15]. Theory and experiment are in moderately good agreement. However, here the comparison is complicated by the existence of Fermi resonance involving fundamentals and overtone/combinations.

Discussion

Comparison of VCD spectra predicted via SCF calculations, along the lines described in the previous section, and experimental VCD spectra have by now been carried out for several molecules, including [16]

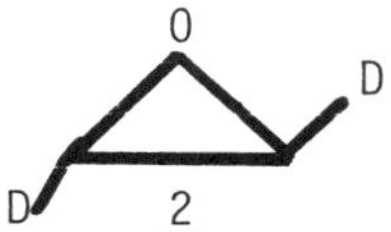

2

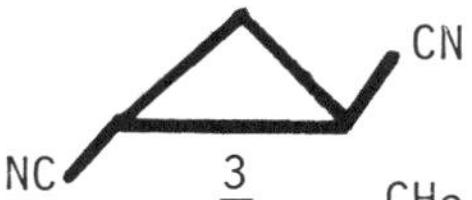

3

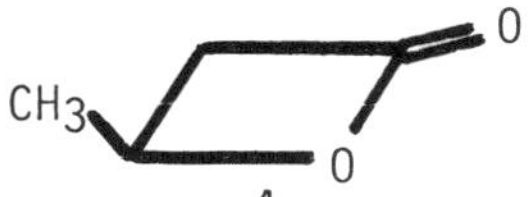

4

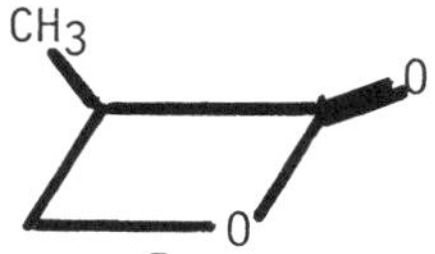

5

With the qualifications that

1) force fields are calculated using split-valence or larger basis sets and scaled, and

2) $P^{\lambda}_{\alpha\beta}$ and $M^{\lambda}_{\alpha\beta}$ tensors are calculated using split-valence-plus-polarisation-functions-on heavy-atoms or larger basis sets,

predicted spectra are <u>overall</u> in excellent agreement with experiment. By this we mean that in the large majority of cases signs (+ or -) are correctly predicted, relative magnitudes of VCD intensities are qualitatively correct, and the absolute magnitudes of rotational strengths are within a factor 2-3. In each case, some parts of the VCD spectrum are better predicted than others. In general, predictions are less effective for vibrations whose normal coordinates are a very sensitive function of the force field. This is most often the case when other modes are close in energy. Thus, predictions are more likely to err, the more congested the region of the vibrational spectrum. In addition, regions of the spectrum subject to significant Fermi resonance are naturally difficult to reproduce using a theory in which anharmonicity is neglected. The C-H stretching region is usually the one most prone to Fermi resonance. Since the C-H stretching region is also typically more congested than any other region, it is generally the region where predictions and experiment differ the most.

On the one hand, the results obtained to date are very encouraging. There is no reason to believe that the same level of success will not occur in other molecules. There is every reason to believe that improvements in theoretical techniques will lead to closer agreement between theory and experiment. On the other hand, the data base currently available is obviously extremely limited.

A substantial expansion of this data base is clearly mandatory. In particular, a wider variety of molecular structures must be studied. Parts of the periodic table not encompassed so far must be included. More varied types of bonding must be examined. Ions, weakly-bound aggregates (dimers, ...) and other different categories of molecules require study. At the present time, we are studying several molecules containing second-row atoms (Cl, S ...) and internal hydrogen bonding, in an effort to broaden the current data base. It is our hope that many more molecules will soon be enlisted in this endeavor.

At the present time, the SCF calculations described above when carried out using the current generation of supercomputers (such as the CRAY-XMP used in our work) are limited to about ~ 150 basis functions. For molecules whose formula is X_nH_{2n}, where X = C,N,0 ..., the number of basis functions when split valence basis sets are used is 13n. For split-valence-plus-polarisation-functions-on-heavy-atom basis sets, the number is 19n. In these two cases, n is thus limited to ~ 12 and ~ 8 respectively. At this time, SCF calculations using basis sets of adequate size are therefore limited to fairly small molecules. This limitation can be expected to be short lived, however. First, the rate of development of both ab initio molecular orbital programming and computer hardware is currently extremely rapid. The frontier of feasibility will undoubtedly be pushed back rapidly. Secondly, it is very likely that approximations can be developed allowing calculations for much larger molecules without serious loss of accuracy. There are two obvious approximations which will probably be the first to be explored. First, semi-empirical methods can be employed in place of the more rigorous ab initio methods. Second, properties of larger molecules can be obtained from calculations on smaller molecular fragments. Consider, for example, atomic polar tensors, $P^{\lambda}_{\alpha\beta}$, defining the change in the molecular electric dipole moment as a result of the displacement of nucleus λ.

Clearly in most molecules $P^{\lambda}_{\alpha\beta}$ will not be sensitive to the nature of atoms in the molecule distant from nucleus λ. Such atoms can then therefore be deleted or replaced by H atoms and a smaller molecule used to calculate $P^{\lambda}_{\alpha\beta}$. In this way, all $P^{\lambda}_{\alpha\beta}$ tensors for a large molecule can be obtained by a series of calculations on fragments of this molecule (no more than N for an N-atom molecule). Of course, these are not the only approaches possible and it would be premature to judge which will be the most successful. It is only sure that this topic will receive intense attention in the near future.

Given the feasibility of predicting the VCD spectrum of molecules of substantial size, a new technique is available for the study of molecular stereochemistry. Of course, vibrational spectroscopy has been used by chemists for many years in studying molecular structure. However this usage has until recently been entirely empirical. Very recently, a priori calculations of vibrational absorption spectra have begun to be used by organic chemists, notably in elucidating the structure of novel, unstable small molecules [17]. A priori calculations of VCD spectra further extends the utility of vibrational spectroscopy. VCD only exists in chiral molecules, of course. However, within this category, the use of VCD substantially enhances the information content of vibrational spectroscopy. Specifically, VCD is uniquely dependent on the absolute configuration of a chiral molecule. In addition, VCD can be expected to be generally more sensitive to molecular conformation than absorption spectra and therefore to be more discriminating in distingushing alternative conformations.

ACKNOWLEDGMENTS

Important contributions to the work described above have been made by R.D. Amos, F. Devlin, K.J. Jalkanen, R. Kawiecki and M.A. Lowe. Research support by NSF, NIH, NATO and the San Diego Supercomputer Center are also gratefully acknowledged.

REFERENCES

1. G. Holzwarth, E.C. Hsu, H.S. Mosher, T.R. Faulkner and A. Moscowitz, J. Am. Chem. Soc. 96, 251 (1974).
2. P.J. Stephens and R. Clark in Optical Activity and Chiral Discrimination, Ed. S.F. Mason, D. Reidel, 1979, p263; T.A. Keiderling, Appl. Spectr. Revs. 17, 189 (1981); L.A. Nafie and D.W. Vidrine, in Fourier Transform Infrared Spectroscopy, Ed. J.R. Ferraro and L.J. Basile, Academic, 3, 83 (1982); P. Malon and T.A. Keiderling, Appl. Spectr. 42, 32 (1988).
3. F. Devlin and P.J. Stephens, Appl. Spectr. 41, 1142 (1987).
4. P.J. Stephens and M.A. Lowe, Ann. Rev. Phys. Chem. 36, 213 (1985).
5. **a)** P.J. Stephens, J. Phys. Chem. 89, 748 (1985).
   b) P.J. Stephens, J. Phys. Chem. 91, 1712 (1987).

6. **a)** M.A. Lowe, G.A. Segal and P.J. Stephens, J. Am. Chem. Soc. 108, 248 (1986).
   b) R.D. Amos, N.C. Handy, K.J. Jalkanen and P.J. Stephens, Chem. Phys. L. 133, 21 (1987).
7. **a)** Reference 6a.
   b) M.A. Lowe, P.J. Stephens and G.A. Segal, Chem. Phys. Lett. 123, 108 (1986).
   c) K.J. Jalkanen, P.J. Stephens, R.D. Amos and N.C. Handy, J. Am. Chem. Soc. 109, 7193 (1987).
   d) K.J. Jalkanen, P.J. Stephens, R.D. Amos and N.C. Handy, Chem. Phys. Lett. 142, 153 (1987).
   e) K.J. Jalkanen, P.J. Stephens, R.D. Amos and N.C. Handy, J. Phys. Chem. 92, 1781 (1988).
   f) K.J. Jalkanen, P.J. Stephens, R.D. Amos and N.C. Handy, J. Am. Chem. Soc. 110, 2012 (1988).
   g) R.W. Kawiecki, F. Devlin, P.J. Stephens, R.D. Amos and N.C. Handy, Chem. Phys. Lett. 145, 411 (1988).
   h) K.J. Jalkanen, R. Kawiecki, P.J. Stephens, and R.D. Amos, J. Phys. Chem. (submitted).
   i) R.W. Kawiecki, F. Devlin, P.J. Stephens and R.D. Amos, J. Phys. Chem. (submitted).
   j) K.J. Jalkanen, P.J. Stephens, A. El-Azhary and T.A. Keiderling, to be submitted.
   k) F. Devlin, K.J. Jalkanen, P.J. Stephens and T. Polonski, to be submitted.
   l) K.J. Jalkanen, P.J. Stephens and R.D. Amos, to be submitted.
8. See reference 7g.
9. W.B. Person and J.H. Newton, J. Chem. Phys. 61, 1040 (1974).
10. See for example: W.J. Hehre, L. Radom, P.R. Schleyer and J.A. Pople, Ab Initio Molecular Orbital Theory, Wiley, 1986.
11. See, for example: R.D. Amos, Adv. Chem. Phys. 67, 99 (1987); E.D. Simandiras et al., J. Am. Chem. Soc. 110, 1388 (1988); J. Chem. Phys. 88, 3187 (1988); J. Phys. Chem. 92, 1739 (1988); R.D. Amos et al., J. Chem. Soc. Faraday Trans. 84, 1247 (1988).
12. See, for example, E.D. Simandiras, R.D. Amos and N.C. Handy, Chem. Phys. 114, 9 (1987).
13. G. Fogarasi and P. Pulay, Ann. Rev. Phys. Chem. 35, 191 (1984); Vibrational Spectra and Structure 14, 125 (1985).
14. R.D. Amos and J.E. Rice, CADPAC, Version 4.0, Cambridge, 1987.
15. Refs. 7b, 7g, 7i, M.A. Lowe, J.S. Alper, R.W. Kawiecki and P.J. Stephens, J. Phys. Chem. 90, 41 (1986) and R.W. Kawiecki, Ph.D. Thesis, U.S.C., 1988.
16. 2 : Refs. 7f and 7l ; 3 : Refs. 7c and 7j; 4 : Ref. 7k; 5 : Ref. 7k.
17. See, for example: B.A. Hess, L.J. Schaad, P. Carsky and R. Zahradnik, Chem. Revs. 86, 709 (1986).

INDEX

RETURN TO → CHEMISTRY LIBRARY
100 Hildebrand Hall 642-3753

B.C.

JUL 18 1990